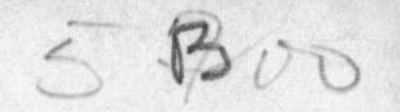

Omotoye Olorode
BOTANY DEPARTMENT, UNIVERSITY OF IFE, ILE-IFE, NIGERIA

Taxonomy of West African Flowering Plants

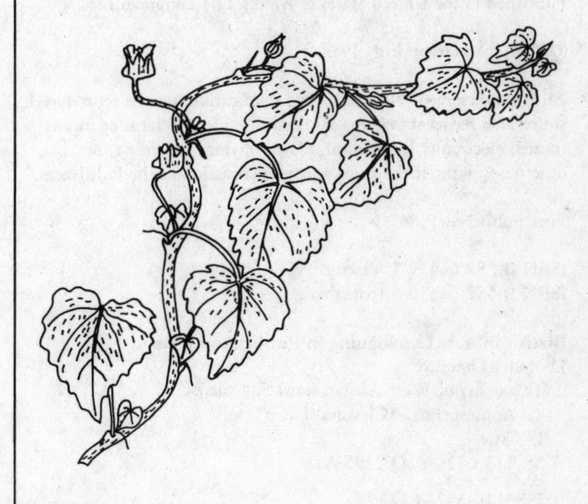

Longman
LONDON AND NEW YORK

Longman Group Limited
Longman House
Burnt Mill, Harlow
Essex CM20 2JE, England and Associated Companies throughout the World.

Published in the United States of America by Longman Inc.

First published 1984

ISBN 0 582 64429 1 (cased)
ISBN 0 582 64430 5 (paperback)

British Library Cataloguing in Publication Data
Olorode, Omotoye
Taxonomy of West African flowering plants.
1. Angiosperms – Classification
I. Title
582.13'012 QK495.A1

ISBN 0–582–64429–1
ISBN 0–582–64430–5 Pbk

Set in 10/11 pt Garamond Roman
Printed in Singapore by MCD Pte. Ltd.

Acknowledgements
The cover photos of *Spathodea campanulata* – African Tulip Tree – were kindly supplied by Ardea Photographics (middle), Biofotos/Heather Angel (top) and Bruce Coleman Limited (bottom).
We are grateful to Hodder & Stoughton Ltd (formerly University of London Press) for permission to reproduce modified extracts from the Appendix to *Introduction to the Flowering Plants of West Africa* by M.S. Nielsen.

Dedication

This book is dedicated to the toiling masses of our people whose conditions make us what we are and whose genuine interests should make us what we ought to be – conscious instruments in their and, indeed, our collective liberation.

Contents

Preface

This book should serve a dual purpose. Firstly, it should assist ordinary plant enthusiasts and non-botanists who need some knowledge of plants to create a firm intimacy with common plants of West Africa. Secondly, the book should provide a broad foundation in taxonomy for students of botany, agriculture and pharmacy at different levels of endeavour.

In the first chapter an attempt is made to survey the historical development of taxonomic thought and to demonstrate the unity of theory and practice as these relate to taxonomic practice and the theory of evolution. Chapter 2 covers general taxonomic practice. The remainder of the book deals with details of the characteristics of selected angiosperm families with as many examples as practicable from diverse vegetational zones. We have included known Nigerian names of the plants in order to encourage our readers to enlarge the community of plant enthusiasts. We hope that this will stimulate learning from the generality of our people, many of whom know much more than we can read from books.

The enthusiasm, patience and collaboration of my past and present students are acknowledged; in particular, Femi Folaranmi, Charity Akunwanne (Angel) and Uche Nwobi helped me with Nigerian names of plants. Messrs Tunde Onifade, T. Alo, L. T. Salami and Mrs C. Labiyi helped with typing various stages of this book. Mr Daniel Edike did all the reproduction at the mimeographing stage of the book; I cannot thank him enough. Messrs G. Ojo, Kunle Adebowale, Samson Adegoke, Isaac Faremi and L. O. Faturoti contributed immensely to the evolution of this book. My colleague, Dr A. O. Olatunji participated in writing portions of the section on the use of descriptions and keys in Chapter 2; this assistance is gratefully acknowledged. I thank all my other colleagues and comrades who have inspired and encouraged this effort.

I would also like to put on record the enthusiasm and encouragement given to me by Rob Francis and Nicola Bennett-Jones (both of Longman Group) and Katherine Novosel. Their suggestions have contributed immensely to whatever merits this book may have. Last, but not least, I must register my appreciation of the patience, solidarity and personal sacrifices of my mother, Bola; my wife, Sike; my children, Gbonju, Akintoye, Oyeyemi and Molara; my sister, Yinyinade and my brother and companion, Iyiade.

The drawings in this book are original drawings by the author; they were made either from live specimens or from herbarium materials.

Ile-Ife, 1983   Omotoye Olorode

Glossary of common botanical terms

Abaxial On the side away from the axis.
Acaulescent Without an evident leafy stem.
Achene Small, dry, one-seeded indehiscent fruit with testa which is free from the seed (as in members of Compositae).
Achlamydeous Without a perianth; naked.
Actinomorphic With radial symmetry, the parts similar in size and shape.
Acuminate Tapering into a long point (Fig. G. 1).
Acute Pointed; forming less than a right angle.
Adaxial On the side next to the axis.
Adnate Referring to dissimilar parts: grown together or united.
Adventitious In an unusual place; often applied to roots or buds.
Aestivation Arrangement of perianth parts in the bud.
Aggregate Referring to fruit: a cluster of fruits produced by a single flower, as in *Sterculia* or *Uvaria*.
Alternate Located singly at a node, as leaves on a stem; situated between other parts, as stamens between petals.
Amphitropous Referring to an ovule: attached near its middle; half-inverted.
Amplexicaul Clasping the stem.
Anatropous Referring to an ovule: inverted, with the micropyle close to the point of attachment.
Androecium The stamens collectively.
Anemophilous Pollinated by wind (see **Entomophilous.**)
Angiosperm A plant producing seeds enclosed by an ovary.
Annual A plant that completes its development in one year or one season and then dies.
Anthesis The time of blooming.
Apetalous Without petals.
Apocarpous (polycarpous) With separate carpels as in *Cola* or *Uvaria*.
Arborescent Becoming almost tree-like in size.
Aril An appendage or outgrowth from the hilum or funiculus of a seed, often spongy or gelatinous, and sometimes enveloping the seed.
Armed Having prickles, spines or thorns.
Article One of the segments of a jointed fruit such as a loment.
Articulate With one or more joints or points of separation.
Attenuate Gradually tapering; drawn out into a narrow portion.
Auricle An ear-like lobe or appendage.
Awn A bristle-like appendage such as occurs on the back or at the tip of glumes and lemmas of grasses.

Axil The angle formed between two organs, as between a leaf and a stem.

Baccate Like a berry.
Barbed With rigid points that are reflexed, or directed backward, as in a fish-hook.
Basifixed Attached by the base.
Bearded With rather long hairs.
Bifid Split about midway into two lobes.
Bilabiate Two-lipped.
Bipinnate Twice-pinnate (Fig. G. 1).
Blade The expanded portion of a leaf, sepal, petal or other part.
Bloom Whitish powdery covering of a surface, easily rubbed off.
Bole Unbranched stem of a tree; main stem of a tree.
Bract A more or less modified leaf subtending a flower or a flower cluster.
Bracteole (bractlet) A secondary bract, often very small.
Bud An undeveloped or dormant branch, leaf or flower, usually enclosed by protective scales.
Bulb A subterranean bud composed of fleshy scales attached to a central or basal stem.
Bulbil A little bulb, a bulb-like body or a small tuber produced above ground on stems or inflorescences serving as a unit of vegetative propagation.

Caducous Falling early; often applied to petals or stipules.
Caespitose Tufted or matted.
Callus A hardened projection; in grasses the hardened base of a lemma at the point of its attachment to the rhachilla.
Calyx (plural: calyces) The outer part of the floral envelope, composed of sepals.
Campanulate Shaped like a bell.
Capitate Like a head; in a dense, more or less rounded cluster.
Capitulum A little head of flowers.
Capsule A simple, dry dehiscent fruit of two or more carpels, and usually many-seeded (Fig. G. 2).
Carpel A megasporophyll; often regarded as a single, modified, seed-bearing leaf.
Caruncle An appendage or protuberance adjacent to the hilum of some seeds as occurs in the castor oil seed.
Caryopsis A dry, single-seeded, indehiscent fruit with fused testa and pericarp; grain (as in grasses).
Catkin (ament) An elongate, deciduous cluster of unisexual, apetalous, and usually bracteate flowers, as in garden *Acalypha*.
Caudicle The thread-like or strap-shaped stalk that connects a mass of pollen to an anther sac in many orchids.
Caulescent With a leafy stem above ground.
Cauline Pertaining to the stem, as cauline leaves.
Cespitose Tufted or matted.
Chaff Small, dry, scale-like bracts.
Choripetalous (polypetalous) With separate petals.
Chorisepalous With separate sepals.
Ciliate With marginal hairs or bristles.
Circumscissile Opening by a transverse circular split so as to release a lid, as in *Celosia* spp.
Clasping Partly surrounding another structure at the base.
Cleft Deeply cut.
Column In orchids: the combined style and stamen structure; in certain grasses: the basal portion of awns; in *Hibiscus*: the stamen tube.
Compound Composed of two or more parts; compound leaves have two or more leaflets; compound pistils have two or more carpels.
Conduplicate Folded in half lengthwise.
Connate United similar parts, as leaves or anthers.
Connective The part of a stamen joining the two anther sacs.
Connivent Aggregating together or touching but not united (as in the anthers of *Solanum* or Compositae)
Contorted Twisted (Fig. G.4).
Convolute Rolled up lengthwise.
Cordate Heart-shaped, with a basal notch (Fig. G. 1).
Coriaceous Leathery.
Corm A modified, usually subterranean, stem that is fleshy and thickened and often bears scale-like leaves.
Corolla The inner set of floral segments, consisting of petals.
Corona A petaloid appendage situated between the corolla and the stamens of some flowers.
Corymb A rounded or flat-topped inflorescence in which the pedicels or branches are attached at intervals on an elongate axis and are of unequal length, the lower ones being longer.
Cotyledon An embryonic leaf of a seedling or a seed.
Crenate With rounded teeth (Fig. G. 1).
Culm The flowering stem of a grass or sedge.
Cuneate Wedge-shaped.
Cuspidate With an abrupt, short, sharp and often rigid point.
Cyathium The small, cup-like, specialised inflorescence of *Euphorbia*.
Cyme An inflorescence in which the central flower of each group is the oldest; in general a loose term for

complex flower clusters that are more or less rounded or flat-topped (Fig. G. 3).

Cypsela An achene with pappus at the stylar end (as in Compositae).

Cystoliths Mineral aggregations in stems or leaves (as in the Acanthaceae).

Deciduous Falling off, as leaves that are shed in the dry season.

Decompound Two or more times compound, usually meaning with many small divisions.

Decumbent With the base prostrate, but the upper parts erect or ascending.

Decurrent Extending downwards, as in leaves having their bases prolonged downward as wings along the stem.

Deflexed Bent downwards.

Dehiscence The method or act of opening or splitting (Fig. G. 2).

Dehiscent Referring to fruit: one that splits open.

Deltoid Broadly triangular, with the base nearly straight and the sides often a little curved toward the apex.

Dentate Toothed, the teeth being acute and directed outward.

Denticulate The diminutive of 'dentate': with small teeth.

Determinate Referring to inflorescence: sometimes applied to those in which the terminal or central flower is the oldest.

Diadelphous Referring to stamens: in two united groups or whorls (Fig. G. 5).

Dichasium An inflorescence having a central older flower and a pair of lateral branches bearing younger flowers (Fig. G. 3).

Dichotomous Forking; branching by pairs.

Didymous Paired or twinned.

Didynamous With four stamens in two pairs, one pair of stamens being longer than the other pair.

Digitate With parts diverging from a common base, as fingers of a hand.

Dimorphic Of two forms.

Dioecious Having unisexual flowers which are produced on separate plants.

Disc An enlargement of the floral axis, often fleshy or glandular; in the Asteraceae; the central part of the head.

Dissected Cut into many fine segments.

Distichous In two rows.

Divaricate Widely spreading or divergent.

Drupe A simple, fleshy fruit with a single seed enclosed in a bony endocrap or pith; a stone fruit.

Echinate Prickly.

Eglandular Without glands.

Ellipitic Like an ellipse; longer than wide and with rounded ends.

Emarginate Notched at the apex (Fig. G. 1).

Endocarp The innermost of the three layers forming the wall of the pericarp of a fruit.

Endosperm Nutritive material or tissue in some seeds outside the embryo.

Ensiform Sword-shaped.

Entire With an unbroken or even margin; without teeth or other indentations.

Entomophilous Pollinated by insects.

Epicalyx A set of bracts adjacent to and resembling a calyx.

Epigynous A flower in which the hypanthium or the perianth is attached to the upper part of the ovary, the ovary then appearing inferior in position.

Epipetalous Referring to stamens: attached to the corolla. (Fig. G. 5).

Epiphyte An independent plant growing upon another plant and not connected to the ground.

Exocarp The outermost of the three layers forming the wall or pericarp of a fruit.

Exserted Projecting beyond, as stamens protruding from the corolla.

Exstipulate Without stipules.

Extrorse Facing outward.

Falcate Curved like a sickle.

Farinose Mealy; covered with a mealy powder as in some grasses.

Fascicle A cluster or bundle.

Filament The stalk of a stamen; any thread-like body.

Filiform Thread-like.

Fimbriate Fringed.

Fistulose Hollow and cylindrical, as the leaves of some onions.

Flabellate Fan-shaped; broadly wedge-shaped.

Flexuous Wavy; curved alternately in opposite directions.

Floccose With tufts of soft hair.

Floret A little flower; in grasses including the lemma.

Floriferous Flower-bearing.

Foliaceous Leaf-like, usually meaning with green coloration.

Foliar Pertaining to leaves or leaf-like parts.

Follicle A dry fruit of one carpel that splits on one side (Fig. G. 2).

Free Not adnate to other parts.

Fruit A ripened ovary, sometimes including other

adherent parts (Fig. G. 2).
Frutescent Becoming shrubby.
Fruticose Shrubby.
Funnelform Shaped like a funnel, with a gradually widening tube.
Fusiform Spindle-shaped, thickened in the middle and tapering to the ends.

Geniculate Bent at a joint; kneed, as in anthers of some Melastomaceae.
Gibbous Swollen or with a protuberance on one side, usually near the base.
Glabrate Becoming glabrous or hairless at maturity.
Glabrescent Same as glabrate.
Glabrous Without hairs; smooth.
Gland A secretory hair or other part that produces nectar or some other liquid.
Glaucous Greyish or bluish in colour because of a coating of minute powdery or waxy particles.
Globose Spherical.
Glomerule A small, compact, more or less rounded cluster.
Glumes A pair of empty scale-like bracts at the base of a grass spikelet.
Grain (caryopsis) The fruit of grasses, seed-like, with a thin pericarp adherent to the seed.
Gynobasic style One that originates between the lobes of a deeply lobed ovary, as in mints (family Labiateae).
Gynoecium The collective term for the female parts of a flower, the pistil or pistils.
Gynophore A stalk or stipe on which an ovary or fruit is elevated above the floral axis.

Habit Gross form or appearance.
Hastate Like an arrowhead but with divergent lobes at the base.
Head A dense inflorescence of sessile or subsessile flowers on a short or broadened axis (Fig. G. 3).
Helicoid In a spiral, like a snail shell.
Herb A plant that dies completely at the end of the growing season, or one that dies to the ground; not woody-stemmed.
Herbaceous Like an herb, not woody; or having a green colour and a leafy texture.
Hilum A scar on a seed indicating point of attachment to the placenta (as in peas).
Hispid With rigid or bristly hairs.
Hirsute With rather stiff or bristly hairs.
Hirtellous Minutely hirsute.
Homogamous With one kind of flower; stamens and stigmas maturing at the same time.
Hood In the Asclepiadaceae, one of the concave segments of the corona.
Hyaline Thin and translucent.
Hypanthium A saucer-shaped, cup-shaped, tubular or sometimes rod-shaped expansion of the floral axis that produces floral organs such as sepals, petals and stamens, from its upper margin.
Hypocotyl The axis of an embryo or seedling below the cotyledons and above the radicle.
Hypogynous Having the flower parts attached near the base of the ovary and free from it; ovary superior.

Imbricate With overlapping edges, as shingles on a roof (Fig. G. 4).
Imparipinnate Once-pinnate with an odd leaflet terminating the leaf (Fig. G. 1).
Incised Cut sharply, irregularly and rather deeply.
Included Not protruding beyond the surrounding structure (see **Exserted**).
Incumbent Referring to cotyledons: placed with their backs to the radicle.
Indehiscent Not splitting open.
Indeterminate Referring to inflorescences: sometimes applied to those in which the terminal or central flower is the last to open.
Indumentum Any covering of a plant surface, especially hairs.
Induplicate Once-pinnate with an odd leaflet terminating folded inwards (in petals and sepals) but not overlapping.
Indurate Hardened.
Inferior Referring to an ovary: situated below the point of attachment of the flower parts.
Inflorescence A flower cluster; a flowering axis.
Inserted Attached to, meaning the point of origin.
Integument The covering of a body, as the coat of an ovule or seed.
Internode The portion of a stem between two adjacent nodes.
Interrupted Having gaps between the parts.
Introrse Referring to an anther: facing inwards.
Involucel A secondary involucre that subtends a part of an inflorescence.
Involucrate Having an involucre.
Involucre A whorl of bracts around the base of an inflorescence.
Involute Rolled lengthwise so as to expose the lower side and conceal the upper side, as in some leaves (see **Revolute**).
Irregular Referring to a flower: having dissimilar parts of the same kind (usually the petals); with bilateral symmetry; zygomorphic.

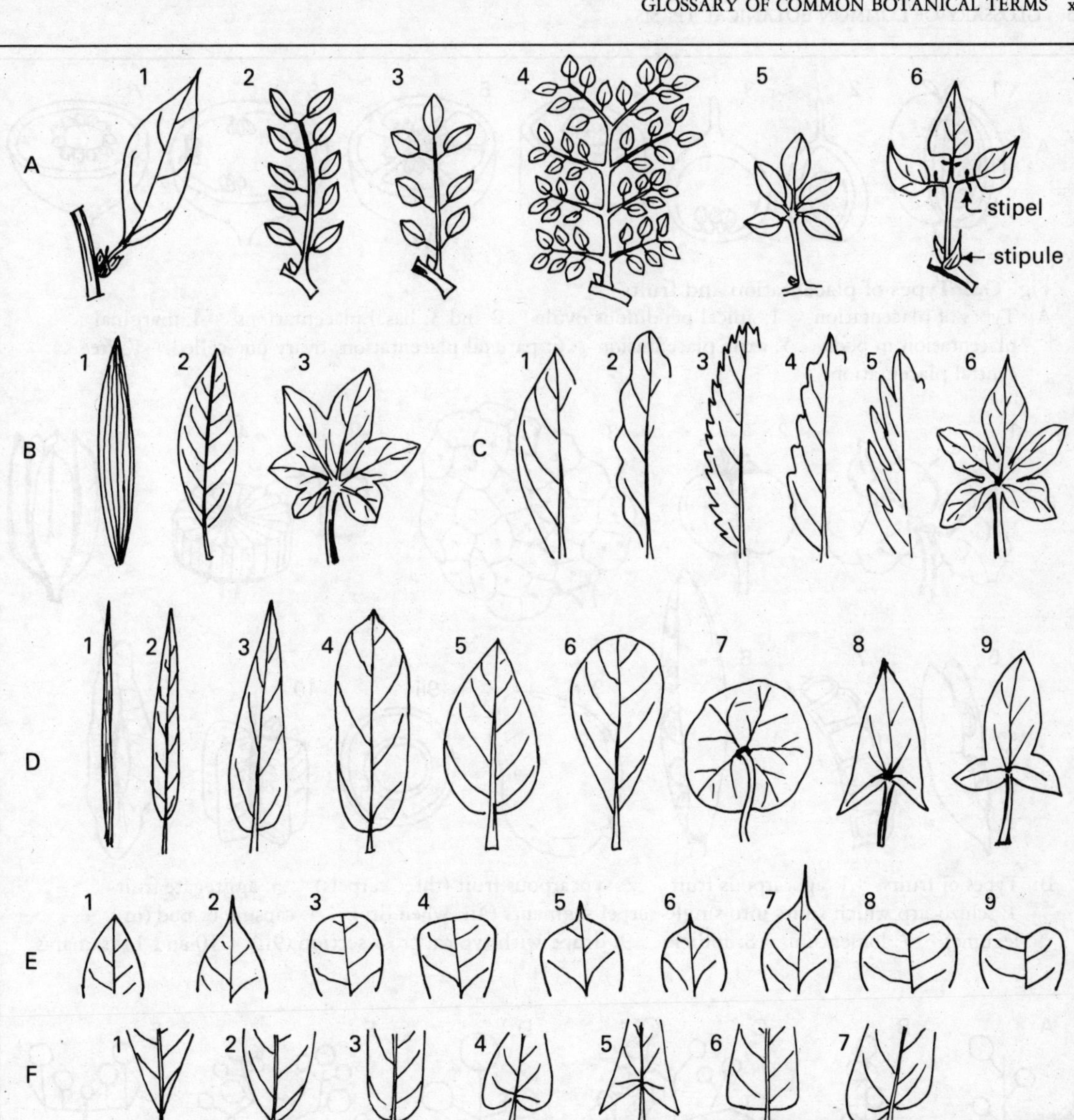

Fig. G.1 Illustrations relating to the description of leaves

A: Simple leaves and compound leaves. 1: simple 2 and 3: pinnately compound (2 is paripinnate while 3 is imparipinnate). 4: bipinnate compound 5: digitately compound 6: trifoliolate compound

B: Leaf venation. 1: parallel-veined 2: pinnately-veined 3: palmately- or digitately-veined

C: Leaf margins. 1: entire 2: undulate 3: serrate 4: lobed 5: parted 6: palmately-lobed

D: Leaf shape. 1: acicular (needle-shaped) 2: linear 3: lanceolate 4: oblong 5: ovate 6: obovate 7: orbicular 8: sagittate 9: hastate

E: Leaf apices. 1: acute 2: acuminate 3: obtuse 4: truncate 5: cuspidate 6: mucronate 7: aristate 8: emarginate 9: retuse

F: Leaf bases. 1: cuneate 2: acute 3: rounded 4: cordate 5: sagittate 6: truncate 7: oblique (inequilateral)

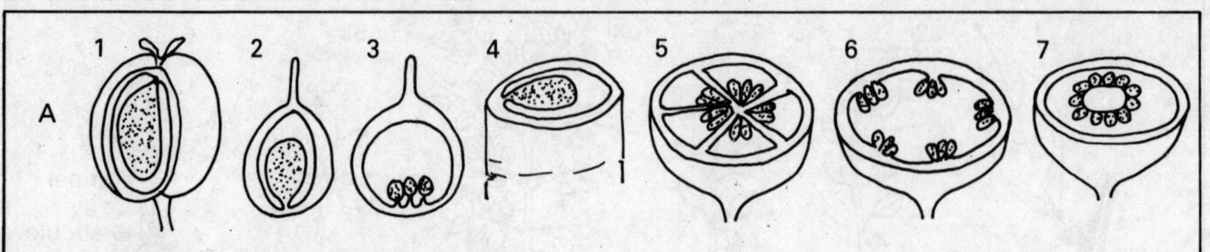

Fig. G.2 Types of placentation and fruits

A: Types of placentation 1: apical pendulous ovule 2 and 3: basal placentations 4: marginal placentation in pods 5: axile placentation 6: parietal placentation, ovary one-celled 7: free central placentation

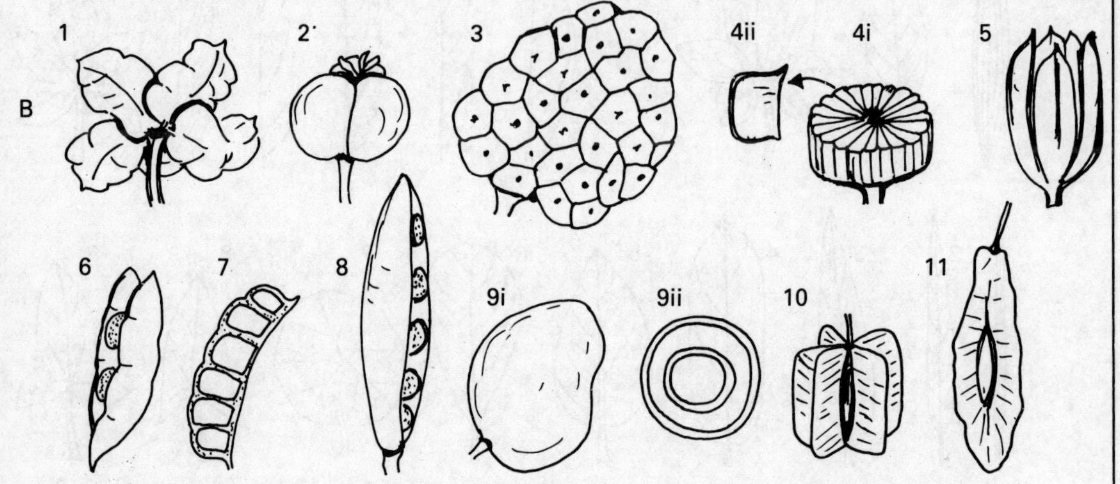

B: Types of fruits 1: apocarpous fruit 2: syncarpous fruit (three carpels) 3: aggregate fruit 4: schizocarp which splits into single-carpel segments (4ii: when dry) 5: capsule 6: pod (in legume) 7: lomentum 8: follicle 9: drupe with typical cross-section (9ii) 10 and 11: samaras

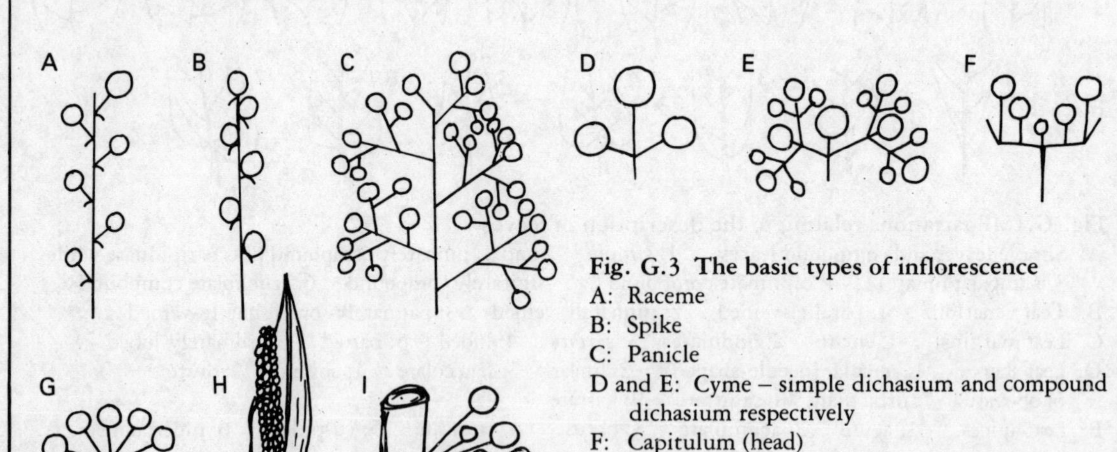

Fig. G.3 The basic types of inflorescence

A: Raceme
B: Spike
C: Panicle
D and E: Cyme – simple dichasium and compound dichasium respectively
F: Capitulum (head)
G: Simple umbel
H: Spadix
I: Fascicle as in cauliflorous inflorescences

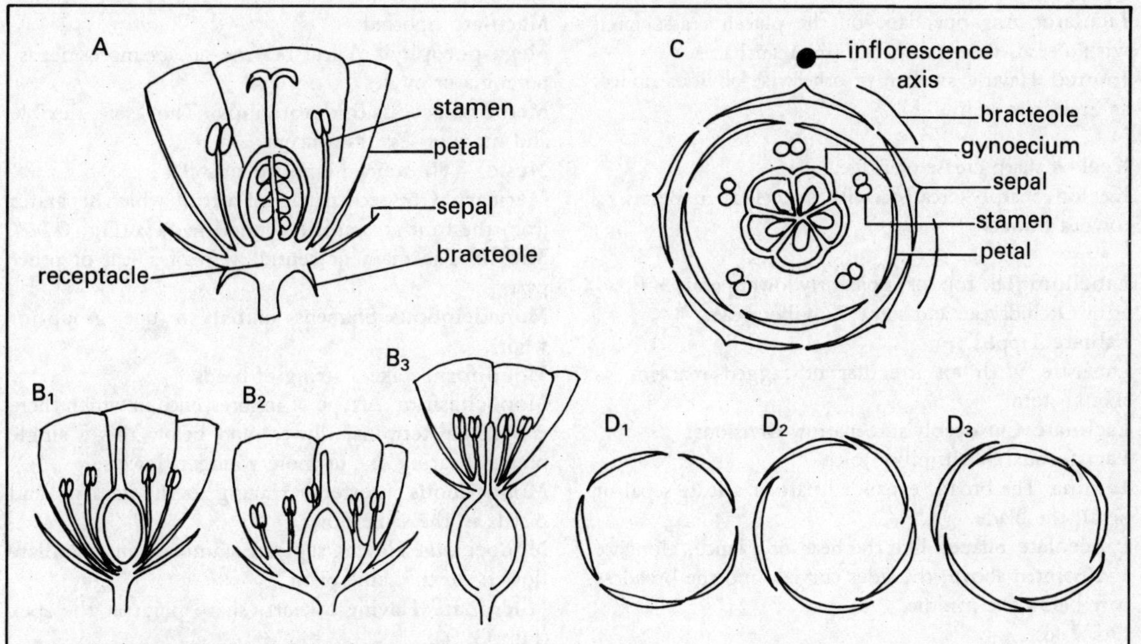

Fig. G.4 Description of the flower

A: Generalised illustration of parts of a flower.

B: Position of the ovary with respect to other parts of the flower. B_1: hypogynous, B_2: perigynous, B_3: epigynous floral parts. Ovary is superior in B_1 and B_2 but inferior in B_3.

C: Typical floral diagram.

D: Aestivation of perianth segments. D_1: valvate D_2: contorted D_3: imbricate

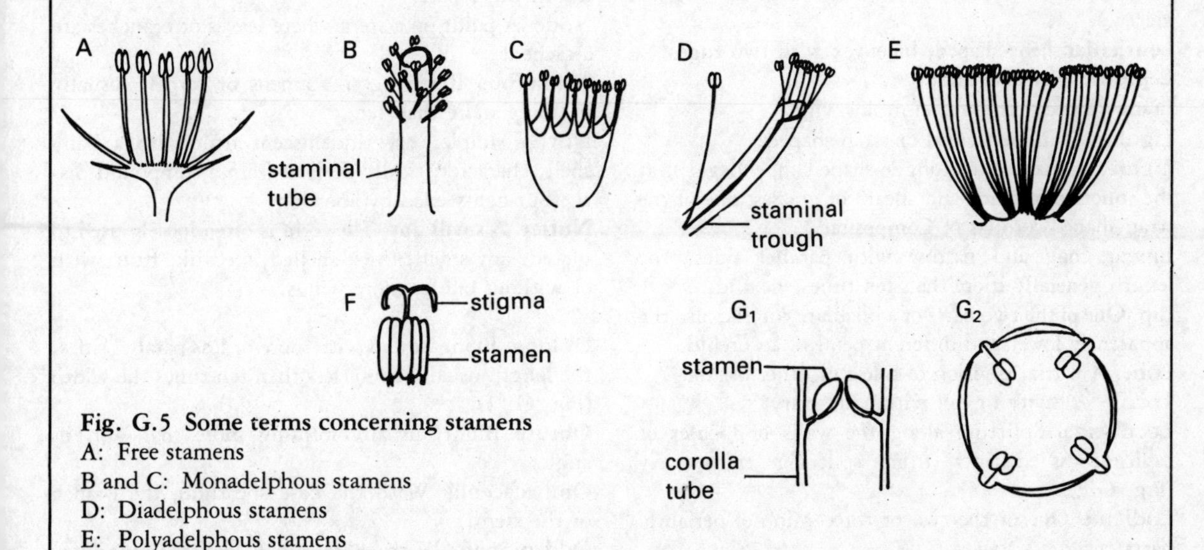

Fig. G.5 Some terms concerning stamens

A: Free stamens

B and C: Monadelphous stamens

D: Diadelphous stamens

E: Polyadelphous stamens

F: Syngenesious anthers

G: Epipetalous stamens shown in vertical section of corolla tube and in floral diagram

Jaculator An appendage on the placenta associated with a seed, to aid dispersal (in Acanthaceae).
Jointed Having swollen or otherwise obvious nodes, as in a grass stem.

Keel A sharp crease or ridge.
Keeled Sharply creased; with a keel as in petals of cowpea flowers.

Labellum The top, or apparently lower petal, of flower of Orchidaceae and some Zingiberaceae.
Labiate Lipped.
Lacerate With an irregular or ragged margin, as though torn.
Laciniate Cut deeply into narrow divisions.
Lactiferous With milky juice.
Lamina The broad, expanded part of a leaf, sepal or petal; the blade.
Lanceolate Shaped like the head of a lance, elongate and pointed above, the sides curved, and the broadest part below the middle.
Lax Loose.
Leaflet One of the divisions of a compound leaf (Fig. G. 1).
Legume A simple, dry dehiscent fruit of one carpel, usually splitting at maturity along two sutures (Fig. G. 2).
Lemma A bract that usually encloses a flower in the spikelet of grasses.
Lenticels Corky spots on the bark (as in the Bignoniaceae).
Lenticular Lens-shaped, biconvex with two edges.
Lepidote Flat and thin.
Liane Woody, tropical climbing vine.
Ligulate Tongue-shaped or strap-shaped.
Ligule A small, often tongue-shaped appendage, as at the junction of blade and sheath of grasses; one of the strap-shaped corollas of Compositae.
Linear Long and narrow with parallel sides, the length generally more than ten times the width.
Lip One of the two parts of a bilabiate corolla; also the apparently lower and different petal of an orchid.
Lobe A partial division of a leaf or other organ.
Locule A cavity or cell within an ovary.
Loculicidal Splitting along the walls of locules or cavities, as distinct from splitting transversely (Fig. G.2).
Lodicule One of the two or three minute 'perianth' parts of a grass flower.
Loment A modified legume having constrictions between the seeds and breaking apart transversely at the constrictions.

Maculate Spotted.
Megasporophyll A leaf bearing one or more megasporangia or ovules.
Membranaceous (membranous) Thin, soft, flexible and more or less translucent.
Mesic Well drained (applied to soil).
Mericarp One seeded part of a fruit which separates from the fruit at maturity, found in *Sida* (Fig. G. 2).
Midrib The main or central vein of a leaf or other part.
Monadelphous Stamens united in one group or whorl.
Moniliform Like a string of beads.
Monochasium A type of inflorescence in which there is a single terminal flower, and below this a single branch bearing one or more younger flowers.
Monoclinous (perfect) Having both stamens and pistils in the same flower.
Monoecious Having separate staminate and pistillate flowers on the same plant.
Mucronate Having a short, sharp point at the apex (Fig. G. 1).
Muricate Having the surface covered with short, sharp projections.

Naked Lacking organs or parts, a naked flower being one that lacks a perianth.
Nectary A gland or glands secreting nectar.
Nerve One of the principal veins of a parallel-veined leaf or other part.
Node A point on a stem where leaves or branches are attached.
Numerous Referring to stamens or carpels: usually meaning more than ten.
Nut A simple, dry, indehiscent fruit with a bony shell, characteristically derived from a compound pistil, but one-seeded by abortion.
Nutlet A small nut. The term is often loosely used to include any small, thick-shelled, seed-like fruit, with or without one or more wings.

Oblong Elongate and with more or less parallel sides, the length usually being less than ten times the width (Fig. G. 1).
Obtuse Blunt, usually forming more than a right angle.
Orchreaceous With the base sheathing the branch or the stem.
Odd-pinnate With a terminal leaflet (imparipinnate).
Oligomerous Having few parts.
Opposite Referring to leaves: in pairs, one on either

side of the node. Referring to stamens: inserted in front of petal lobes and thus opposite them.
Orbicular Circular.
Oval Broadly elliptic, the width more than half the length.
Ovate Egg-shaped, the broadest part being below the middle.

Palea (palet) The inner and usually smaller of two scaly bracts immediately subtending the grass flower in a spikelet.
Palmate (digitate) With parts diverging from a common base, as fingers of a hand.
Panicle An elongate inflorescence with compound branching.
Papilionaceous Descriptive of a flower like that of a pea, having a standard (banner), two wings and two keel petals comprising the corolla; butterfly-like in shape.
Papillose Covered with short, rounded projections.
Pappus The modified and late-maturing calyx of the Compositae arising from the summit of the achene, and consisting of hairs, bristles, scales or awns.
Parietal Produced along the inner side of the ovary wall (Fig. G. 2).
Paripinnate Once-pinnate without an odd leaflet terminating the leaf (Fig. G. 1).
Parted Cut or lobed more than halfway to the middle or base.
Pedicel Stalk of a single flower in an inflorescence.
Peltate Attached by the lower surface, not by the margin, as the leaves of some cocoyams.
Pentamerous Occuring in fives or multiples of five.
Perennial A plant that continues to live year after year.
Perfect (monoclinous) A flower having both male and female reproductive parts (stamens and pistils).
Perfoliate Descriptive of a leaf with the stem apparently passing through it because of a joining of the basal lobes of the blade.
Perianth The calyx and corolla collectively, or either one when only one is present.
Pericarp The ovary wall in the fruiting stage.
Perigynous A type of flower with a hypanthium that arises from the base of the floral axis; ovary superior (Fig. G. 4).
Persistent Remaining attached rather than falling off.
Petal One of the parts of the corolla or inner leaf-like parts of a flower.
Petaloid Resembling a petal in colour or texture, usually delicate and not green.
Petiole The stalk of a leaf.
Phloem Conducting tissue in plants through which products of photosynthesis are distributed throughout the plant body.
Pilose With rather sparse, soft hairs.
Pinnate Like a feather; having the parts arranged in two rows along a common axis.
Pinnatifid Cleft or divided pinnately.
Pistil The female reproductive part of a flower, occupying a central position; in some flowers single and in others several or many.
Placenta A point or line of attachment of ovules within an ovary or of seeds within a fruit.
Plumose Feather-like, having fine, soft hairs along the sides, the hairs divergent from the organ to which they are attached.
Pod A term often loosely applied to any simple, dry, dehiscent fruit.
Pollen Microspores; minute spores produced by the anther of a stamen.
Pollinium A mass of pollen grains adhering to each other and shed as a unit.
Polygamous Producing some perfect and some unisexual flowers on the same plant.
Polypetalous With separate petals.
Polysepalous With separate sepals.
Posterior Next to the axis; in bilabiate corollas the upper lip is posterior and the lower lip is anterior.
Precocious Blooming before the leaves are expanded.
Procumbent Lying on the ground but not rooting at the nodes.
Prostrate Lying flat on the ground.
Puberulent Very finely pubescent.
Pubescent Covered with hairs, especially soft, downy hairs.
Punctate Covered with dots or pits.

Raceme A type of inflorescence with an elongate axis which has simple pedicels along it, the order of blooming of the pedicels being from base to apex (Fig. G. 3).
Racemose Like a raceme; in general any inflorescence capable of indefinite prolongation, having lateral and axillary flowers.
Ray One of the main branches of a compound umbel; a strap-shaped marginal flower in the head of a Compositae when tubular disc flowers are also present.
Receptacle The floral axis to which the various flower parts are attached; the enlarged summit of the peduncle of a head to which the flowers are attached.
Recurved Curved downwards or backwards, as in prickles.

Reflexed Bent downwards or backwards.
Regular (actinomorphic) With radial symmetry; the parts of the same sort being of similar size and shape.
Reniform Kidney-shaped; broader than long, with rounded ends, and with a wide basal sinus.
Repand With a wavy margin.
Reticulate Like a network.
Retrorse Directed downwards.
Retuse With a rounded sinus at the apex (Fig. G. 1).
Revolute Rolled lenghwise so as to expose the upper side and conceal the lower side (see **Involute**).
Rhachilla A little rhachis, particularly the axis of a spike or of a spikelet in grasses.
Rhachis The axis of a spike or of a pinnately compound leaf.
Rhizome A modified underground stem, usually growing horizontally.
Rib One of the main veins of a parallel-veined leaf or other organ.
Rosette A basal cluster of leaves produced on a very short stem.
Rostrate With a beak.
Rotate Referring to a corolla: wheel-shaped; have a short tube and a widely spreading limb, as in a potato flower.
Rugose Wrinkled.
Runner (stolon) A horizontal, above-ground stem that may root at the nodes or apex and develop new plantlets at those places, as in the sweet potato.

Saccate Shaped like a bag.
Sagittate Shaped like an arrowhead.
Salverform A calyx or corolla with a slender tube and widely spreading limb.
Samara A simple, dry, indehiscent fruit, usually one-seeded, and with one or more wings, as in fruits of *Terminalia*.
Saprophyte A plant without green colour that obtains its food from dead organic matter.
Scabrous Rough to the touch because of minute stiff hairs or other projections.
Scandent Ascending; usually applied to semi-erect stems.
Scape A leafless flowering stalk arising from the ground or from a very short stem bearing basal leaves.
Scapose Bearing a scape or produced on a scape.
Schizocarp A fruit that splits apart into one-seeded carpels or parts.
Scorpioid Coiled at the tip.
Secund Turned to one side.
Seed A fertilised ovule, consisting of an embryo, with or without endosperm, and a protective coat.
Sepal One of the parts of the calyx or outer set of floral segments.
Septate Having partitions, as many ovaries and some leaves with obvious, internal cross-thickenings.
Septicidal Referring to a capsule: splitting along the septae or partitions, as in okra.
Septum A partition.
Seriate In series, rows or rings.
Serrate With fine, sharp teeth that are inclined forwards or upwards.
Serrulate Finely serrate.
Sessile Lacking a stalk, as some leaves and flowers.
Seta A bristle.
Sheath A tube-like part surrounding another part, as the lower part of a grass leaf that is wrapped round the stem.
Shrub A plant that is woody and has several main stems, smaller than a tree in height.
Silicule A little silique, usually not much longer than wide.
Silique A simple, dry, dehiscent fruit of two carpels that split apart and leave a thin, persistent partition.
Simple Referring to a fruit: derived from a single flower and a single pistil. Referring to a leaf: having the blade in one piece. Referring to a pistil: consisting of a single carpel.
Sinuate With a wavy margin.
Solitary Single.
Spadix A spike of flowers on a fleshy axis, as in the cocoyam family.
Spathe A single, large, often showy bract enclosing or subtending an inflorescence, as is common in a spadix.
Spatulate Shaped like a spatula, oblong, a little broader towards the upper end, and with a rounded apex.
Spicate Produced in a spike.
Spike An elongated inflorescence of sessile or subsessile flowers.
Spinulose Bearing small spines or thorns.
Spur A hollow, more or less pointed projection, usually from the calyx or corolla, commonly producing nectar in its tip.
Squarrose With parts widely spreading or recurved.
Stamen Pollen-producing organ of a flower, typically consisting of anther and filament; a microsporophyll.
Staminate Bearing stamens and consequently male; usually used in reference to a unisexual flower or plant.
Staminode A sterile stamen.
Standard (banner) The uppermost, large petal of a papilionaceous corolla, as in the cowpea.

Stellate Star-shaped; usually used in reference to hairs, as in some Sterculiaceae.
Stigma The part of a pistil to which pollen adheres and on which it germinates, generally terminal in position, and often enlarged.
Stipe A stalk supporting a single organ, particularly an ovary; also the petiole of a fern leaf.
Stipel A stipule-like appendage at the base of a leaflet of a compound leaf.
Stipitate Having a stipe or stalk.
Stipular Pertaining to the stipules.
Stipules A pair of appendages that may be present at the point of attachment of a leaf to a stem.
Stolon A modified horizontal stem, above-ground, that may root at the nodes, developing new plantlets at those places, as in many lawn grasses.
Stoloniferous Having stolons.
Strigose Having the surface covered with straight, appressed (pressed together) hairs that usually are directed forwards.
Strophiole An appendage of the hilum in some seeds.
Style The stalk-like part of some pistils, connecting the stigma and the ovary.
Subtend Arise directly below.
Succulent With a fleshy or juicy texture or composition that is usually resistant to drying.
Suffrutescent Woody or shrubby at the base but not throughout.
Sulcate Longitudinally grooved.
Superior Referring to an ovary: having a position above the point of attachment of the other flower parts, as in hypogynous and perigynous flowers.
Suture A seam or line along which splitting may take place.
Syconium A hollow, multiple fruit.
Symmetrical Referring to a flower: having the same number of each kind of part.
Sympetalous Having the petals partly or completely united to each other.
Synandrium United stamens in the 'cocoyam' family.
Syncarpous With united carpels (Fig. G. 2).
Syngenesious Referring to stamens or anthers: united by the anthers in a ring.

Tendril A part of a stem or leaf modified into a slender, twining, holdfast structure.
Terete Circular in cross-section, cylindrical, rod-shaped.
Terrestrial Growing on the land.
Testa The outer seed coat.
Tetrad A group of four, particularly the four pollen grains from one pollen mother cell.
Tetradynamous With six stamens, four of them longer and two of them shorter.
Tetramerous Occuring fours or multiples of four.
Thalloid Like a thallus, undifferentiated into stems and leaves.
Thallus A plant body that is not differentiated into stems and leaves.
Throat The place in a calyx or corolla of united parts where the tube and limb come together.
Thyrse A term loosely used to describe a compact panicle; more accurately, a complex group of dichasia resembling a panicle.
Tomentose Woolly, covered with curly, matted hairs.
Torose Thickened, elongate and having more or less regular constrictions.
Torus The receptacle or floral axis.
Trigonous Three-angled.
Trimerous Occuring in threes or multiples of three.
Truncate Having the base or apex flattened as though cut off.
Tubercle A small, swollen structure, usually different in colour from the part to which it is attached, and often with a hardened texture.
Tubular Shaped like a tube; also descriptive of corollas that have a well-developed tubular portion but little or no limb portion.
Tufted Forming clumps. (See **Cespitose**).
Tunicate Referring to a bulb: having the leaves arranged in circles when viewed in cross-section, as an onion.
Turgid Swollen and firm.

Umbel A flat-topped or rounded inflorescence having flowers on pedicels of nearly equal length and attached to the summit of the peduncle, the characteristic order of blooming being from the outside toward the centre.
Unarmed Hooked at the end, as some spines or bristles.
Undershrub A perennial plant having stems that are woody only in the basal part, the upper part dying back.
Undulate Having a wavy margin (Fig. G. 1).
Uniseriate In one series or whorl.
Unisexual (diclinous) Of one sex only, either male or female, staminate or pistillate.
Utricle A one-seeded fruit with a thin wall, often dehiscent by a lid.

Valvate With the edges coming together but not overlapping (see **Imbricate**).
Valve A portion of the wall of a fruit or other part

that separates from the remaining part or parts at maturity.
Velutinous Velvety.
Vein A bundle of externally visible transporting tissue in a leaf or other organ.
Venation The system or pattern of veins in an organ.
Ventral The lower side of a flat organ, or the adaxial side of carpel. The ventral suture of a carpel bears the seeds along its inner edge.
Ventricose Enlarged on one side, as some bilabiate corollas.
Vernation The arrangement of parts in a bud.
Verrucose Warty.
Versatile Attached by the middle and free to swing, as some anthers.
Verticillate In a whorl.
Villose (villous) Covered with fine, long hairs that are not tangled.
Virgate Wand-like; descriptive of a slender, erect, straight, leafy stem bearing only short or no branches.
Viscid Sticky, causing foreign particles to adhere to it.

Whorl A group of three or more parts at a node.
Wing A flat, usually thin appendage on a seed or fruit; also each of the two lateral petals of a papilionaceous corolla, as in the cowpea.

Xerophyte A plant adapted to very dry situations.

Zygomorphic Referring to a flower or corolla: having the parts of the same kind of different sizes or shapes so as to be bilaterally, but not radially, symmetrical.

SECTION 1

The Science of Taxonomy

CHAPTER 1

Introduction

The angiosperms

Before giving the definitions and objectives of angiosperm taxonomy, we will explain what the angiosperms are and what position they occupy in the plant kingdom. We cannot go into the various criteria underpinning all the systems used in classifying the plant kingdom. We may, for our purpose, adopt the classification that places the 'flowering' plants in the division Spermatophyta (see Fig. 1.1) – the seed plants: Other divisions under this scheme include Lycophyta (the club moss division) and the Pteridophyta (the fern division). The Spermatophyta includes the Gymnospermae (the seed plants with 'naked' seeds) and the Angiospermae (the seed plants with seeds inside 'bags').

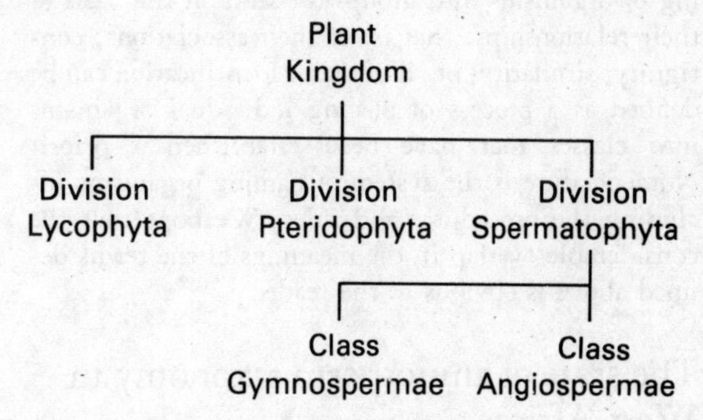

Fig. 1.1 Classification of the angiosperms

The angiosperms include most of the important elements of tropical vegetation; they comprise all the plants designated as monocotyledonous or dicotyledonous. From this, the importance of the angiosperms is very obvious. They include most tropical timber trees, most root crops and vegetables, all pulses and cereals, all fibre-producing plants (cotton, sisal etc.), most edible fruits, all pasture species, most weeds and a large number of medicinal plants.

Definitions and objectives of angiosperm taxonomy

Human beings in all cultures have tended to systematise their knowledge of their environment. One of the important components of this environment is different kinds of animals and plants. Variation is implicit in the word 'kinds' used in the last sentence. This variation is not haphazard, because certain characters are associated with certain other characters. Although certain character associations are *possible*, they are never observed: this creates morphological discontinuities which make certain organisms look alike and make others morphologically dissimilar. The purpose of angiosperm taxonomy, then, is to develop a system of classifying the angiospermous plants in a way that all their differences and similarities are set out in an ordered manner.

Taxonomic practice includes *identification, classification* and *nomenclature*, while *systematics* is a very important weapon for taxonomic theory and practice. We adopt the same definitions of some of these concepts as those adopted (from Simpson, 1961) by Sokal and Sneath (1965). Systematics is the scientific study of the kinds and diversity of organisms and of any and all relationships among them. Classification is the ordering of organisms into groups (or sets) on the basis of their relationships, that is, of their associations, contiguity, similarity, or all of these. Identification can be defined as a process of placing individual organisms into classes that have been established a priori. Nomenclature is the system of naming organisms including the procedures and rules. We hope that the considerable overlap in the meanings of the terms defined above is obvious to the reader.

The state of angiosperm taxonomy in West Africa

The importance of angiosperms in the Tropics has been mentioned on page 1. The role of plant taxonomy in the utilisation and management of vegetable resources, in weed control and in pest management (since many of the pests depend directly or indirectly on plants) is easy to appreciate.

Although plant taxonomists in West Africa (like other scientists or scholars) do not live in isolation from the scientific tradition of other parts of the world, they must recognise that they are operating in a peculiar historical context. This historical context is the one in which the temptation to engage in 'fashionable' scientific pursuits (as 'they' are doing in the major scientific centre of the world) is very strong. The danger of succumbing to this temptation is that the foundation on which 'fashionable' scientific pursuit can be pivoted is neglected. The state of plant and animal taxonomy (especially the latter) is an eloquent example of this neglect – a deficiency which must be rectified without delay. It is not that nothing has been achieved in the taxonomy of West African plants. Indeed, considerable work has been done and is in progress. However the impetus, facilities and institutional framework for the necessary rapid advance are inadequate.

The development of taxonomic thought

The practice of plant taxonomy has two interdependent aspects – the aspect that has to do with local usage (folk taxonomy) and the aspect that has to do with the formalised, or 'scientific', study of plant groups. Each aspect has its amateurs and its professionals; however, it is on formalised taxonomic ideas that extensive written literature exists.

Folk taxonomy

In all cultures, there is some system of ordering plant and animal life, so that there is a system of reference to particular organisms. Some of the systems of naming plants are based on the experience of the community with respect to utilisation of the plants. Other systems of reference rest on the perceived relationships among the plants.

Among the Yoruba of Nigeria, we find examples of purely utilitarian and artificial (see page 3) plant classification. Most common vegetables among the Yoruba can be referred to by the common name 'Efo' which will be appropriately a supra-generic category: thus all species of *Lactuca* (Compositae) are 'Efo yanrin' and all edible species of *Amaranthus* are 'Efo tete'.

It is at the generic and infra-generic levels that the powers of abstraction, observation as well as recognition of taxonomic relationships are demonstrated in folk taxonomy. This is true to such an extent that formal taxonomy depends heavily on folk taxonomy and there is a very close correspondence between the generic delimitation and the specific delimitation in folk taxonomy and formal taxonomy. Let us examine some examples from common taxonomic practice among some Nigerian language groups. Among the Hausa, the generic name 'farun' corresponds to the genus *Lannea* (Anacardiaceae – the mango family) which has a few savanna species; the plant *Lannea kerstingii* has a corresponding Hausa name – 'farun biri'; 'biri' is thus a specific epithet. Among the Igbo, *Cola nitida* (Sterculiaceae) and *Cola acuminata* have corresponding Igbo names – 'Oji Hausa' and 'Oji Igbo' (Hausa kola and Igbo kola respectively); 'Hausa' and 'Igbo' are thus specific epithets for the genus name 'Oji'. Among the Yoruba, the corresponding generic name for *Terminalia* (Combretaceae) is 'Idi'; thus there are 'Idi igbo' and 'Idi odan' (*Terminalia ivorensis* and *Terminalia glaucescens* respectively); 'igbo' and 'odan' are specific, ecologically derived epithets meaning forest and savanna respectively.

From Aristotle to new systematics

Rigorous study of plant classification is believed to have had its roots among the Greeks. Specifically, Aristotle (384–322 BC) and some of his pupils such as Alexander the Great and Theophrastus (*c.* 370–287 BC) were known to have taken a considerable interest in plant life. From the time of Aristotle to the present, three main lines of thought are discernible in plant classification: *essentialism, empiricism* and *evolutionism*. Essentialism and empiricism are basically pre-Darwinian concepts (i.e. occuring before the publication of *The Origin of Species* by Charles Darwin in 1859) while evolutionism is post-Darwinian. As we will see below, these movements of thought were not in water-tight compartments: some evolutionist opinions were expressed before Darwin and there is still considerable amount of essentialist and empiricist influence in taxonomy today.

Essentialism

Essentialism in taxonomic practice and thought derived from Aristotelian logic which is based on the belief that an object or thing has certain attributes (its essence) that make it what it is; this kind of logic is very important in mathematics. Thus a rectangle or triangle can be precisely defined in terms of its essence. Essentialism strives to discover the essence or true nature of things. By implication certain attributes of a thing in the essentialist system have greater importance or weight attached to them a priori than other attributes and, applied to plant taxonomy, this approach gives greater weight to certain morphological features than others. Theophrastus, Caesalpinus, Linnaeus and de Candolle were well known essentialists.

Theophrastus (*c.* 370–287 BC) classified plants primarily according to their habit (trees, shrubs and herbs) and according to whether they were cultivated or wild plants. He recognised certain diagnostic criteria such as adnation or separation of parts, the position of the ovary relative to other floral parts and types of inflorescences. Caesalpinus (1519–1603) also adopted the criterion of habit in classification. He saw the importance of seeds and fruits in classification and recognised groups that correspond to families such as Cruciferae, Compositae and Leguminosae.

Karl Linnaeus (1707–1778) published the *Genera Plantarum* in 1737 and he published the *Species Plantarum* in 1753. In these he put the binomial system (which was already in use before him) on a sound footing. He became famous for the *sexual system* of classification. The sexual system like other essentialist systems is considered artificial (as opposed to natural) because it is usually based on a few characters which are considered by the classifier to be important a priori. Linnaeus divided seed-plants into 23 classes on the basis of the number and arrangement of their stamens and the classes were divided into orders on the basis of the number of their styles.

A. P. de Candolle (1778–1841) is best known for his gigantic work, *Prodromus*, on plant classification which runs into seventeen volumes. Apart from being basically an essentialist, de Candolle emphasised the usefulness of constant characters in taxonomic practice.

The empiricists

The system of classification espoused by the empiricists is called the 'natural system'. This system uses as many characters as possible in classification without assuming, a priori, importance of certain characters. The best known empiricists before the Darwinian era include John Ray and Michel Adanson. The natural system of classification is, in fact, often referred to as 'Adansonian'.

John Ray (1628–1705) established the concept that all parts of a plant are useful in classification. He recognised the distinction between dicotyledon and monocotyledons and the use of seeds and fruits in taxonomy. Apart from making use of the division of plants according to habit, he used leaf types in classification. More importantly, Ray recognised that classifications must be dynamic rather than static – this is a basic tenet of empricism.

Michel Adanson (1727–1806) made a significant contribution to the principle of not only using as many characters as possible in plant classification, but also giving weight to any of them a priori) which by implication ensured that those genera that share the fact that the taxonomic systems used in Europe up till then were not suitable for dealing with the complexity of tropical flora. He therefore devised a natural system (one that uses as many characters as possible without giving weight to any of them a priori) which by implication ensured that those genera that share the greatest number of characters in common, i.e. that show the greatest amount of similarity, were grouped together.

Although Adanson's system was not adopted by most of his contemporaries, it influenced subsequent taxonomic thought considerably. The numerical taxonomy of today which makes use of all sorts of mathematically defined similarity coefficients and statistical methods for measuring affinity is a kind of neo-empiricism or neo-Adansonianism. The use of computers has made it much easier to deal with a really large number of characters in the estimation of affinity (see page 6).

Darwinism and post-Darwinian taxonomy

In 1859, Charles Darwin published *The Origin of Species*. Darwin's observations led him to conclude among other things that:

a) organisms have the potential for increasing in number, but that the limit of this potential is never realised because a number of factors are often limiting; organisms therefore have to compete for these limiting factors;

b) since organisms vary in their capability to compete for the limited resources of the environment, only the most suited will survive the competition; survival in the face of limiting factors is, therefore, not random;

c) the constant struggle for survival leads to evolution, i.e. gradual and cumulative adaptive changes.

Along with Darwin's synthesis of the theory of evolution, palaeontological and palaeobotanical studies demonstrated that some organisms that were abundant at one time in earth's history became extinct (as a result of competition or unfavourable response to other environmental factors). It became clear to Darwin's contemporaries and those who succeeded him that members of a species are not uniform in their visible form and their responses to environmental pressures; thus species are populations with differing degrees of variability and are not unvarying entities. It should be obvious that it is the actual or potential variability of the members of a species that makes responses to environmental changes possible.

The influence of Darwin's theory of evolution on plant classification was enhanced by the rediscovery of Gregor Mendel's laws of heredity (genetics) at the turn of this century. This rediscovery of Mendel's work created a basis for understanding the mechanisms that produced the phenomena that Darwin observed. Organisms were seen as being related and there was speculation about the evolutionary closeness of organisms and the pattern of descent among morphologically close organisms. The general intellectual atmosphere created by these studies and speculations led many eminent post-Darwinian taxonomists to arrange the families of plants according to their conceptions of the order of descent or phylogeny. Thus some families were considered primitive while other families were considered advanced, or derived; by implication certain characters were considered primitive, or unspecialised, while others were considered advanced, specialised or derived.

Before we discuss the use to which post-Darwinian taxonomists put Darwin's theory (page 6) we should elaborate briefly on how new species arise (the process of speciation). This should give us an insight into how higher categories arise. When we talk of higher categories (genus, family, order etc.) in this context we do not intend to imply certain knowledge on our part as to how these categories are to be defined. The terms are used, rather, to denote levels of difference or identity among plants or groups of plants. At the same time we are not necessarily endorsing the nominalist position that taxonomic categories are artificial and arbitrary constructs that have no relationship to real life.

Variation and speciation

We have discussed the importance of genetical variation among organisms from the point of view of capability to respond to changes in the environment. We illustrate this phenomenon in Fig. 1.2. Suppose we have a population of plants containing individuals that have different tolerances to light, such that some individuals will perform well in complete shade, some in partial shade and some under full light (curve A in Fig. 1.2). If the habitat is tampered with such that only the full light condition becomes available, the members of the population that possess hereditary material that makes them tolerate full light are likely to move into the area and increase in number, whereas the other populations are likely to decrease in number (curve C in Fig. 1.2). Two distinct populations may result if the partial-shade habitat were completely removed so that only the extreme habitats – complete shade and full light – became available (curves D and E).

We could ask how it is possible for two populations that derive from a base population to become so different as to be recognised as distinct varieties or species or even higher categories. In other words, how do new species and varieties arise?

The major prerequisite for cumulative genetic (and therefore morphological) differences to arise among populations of common origin, is that such populations should be isolated from each other. This prevents the exchange of hereditary materials between the populations. Thus the daughter populations should become reproductively isolated from one another.

The mechanisms that ensure or enhance reproductive isolation among populations are called 'isolating

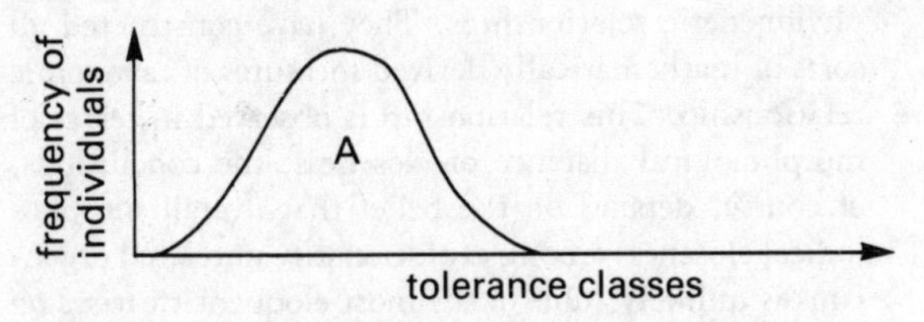

Environment where complete shade, partial shade and full light are available

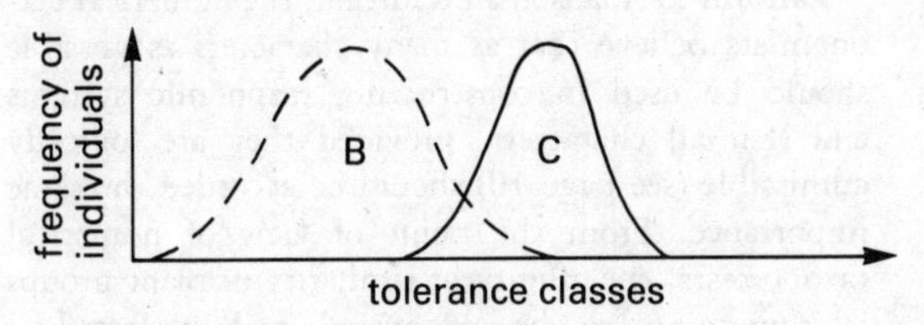

Environment where full light condition only is available

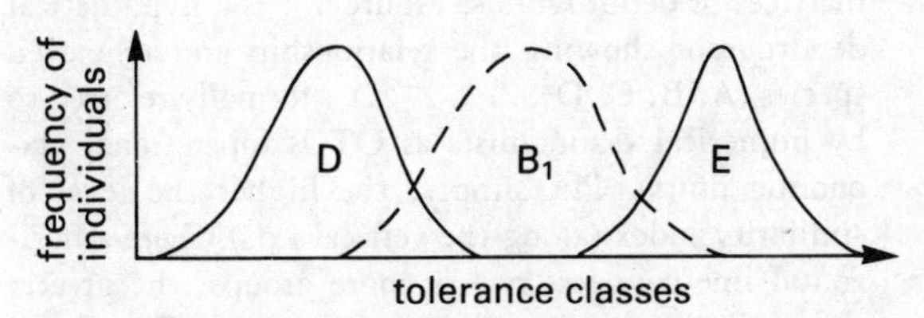

Environment where complete shade and full light are available

Fig. 1.2 Changes in population structure mediated by changes in environment

mechanisms'. Two major groups of such mechanisms are known. These are the mechanisms that prevent pollination and/or fertilisation (pre-zygotic mechanisms) and those mechanisms that act after fertilisation (post-zygotic mechanisms). Spatial or geographical isolation prevents pollination. Related plants which are normally capable of genetic exchange, may even grow together but still not exchange hereditary material because different pollen vectors (insects, birds etc.) visit particular flowers preferentially. These are the so-called pre-zygotic mechanisms. Even when pollination and fertilisation succeed, the offpsring may be weak or sterile. Once populations which have a common origin become isolated, the possibility of independent evolution of the derived populations is greater. Independent adaptive changes are restricted to the population in which the changes first occur and, since such independent adaptive changes are cumulative, the members of one population become recognisably different from the members of the other populations. It is believed that reproductive isolation, as described above, along with cumulative, heritable, adaptive changes act together in the formation of a species, i.e. in the process of speciation.

The phylogenetic systems

In phylogenetic systems plant families are arranged according to presumed relationships by descent. Even though they are post-Darwinian, the phylogenetic systems rely very heavily on pre-Darwinian taxonomic data, at least in the sense that the systems remain essentially artificial. This is demonstrated by the fact that the disagreements between the two major schools of thought on phylogenetic systems arose almost entirely because of their concepts of the nature of the presumed primitive angiospermous flower. These two schools were the Englerian school and the ranalian school.

The leaders of the Englerian school were Adolf Engler (after whom the system was named) and Karl Prantl; the system itself was a modification of the work of Eichler (1839–1887) who worked in Germany. According to this system, primitive flowers are unisexual, wind-pollinated and monochlamydeous (i.e. with perianth not differentiated into petals and sepals). Families with such attributes will include Urticaceae, Casuarinaceae, and Piperaceae. The Englerians also considered that the monocots and dicots have independent gymnospermous origins and that the dicots consist of many different lines of descent (polyphyletic) from various gymnosperms.

The ranalian school of thought was originated by Bessey, who based his work on that of Bentham and Hooker, and whose work was in turn adopted and modified extensively by John Hutchinson. As the name of the school of thought implied, the ranalians believe that the Ranales are the most primitive angiosperms; the ancestral angiosperms are supposed to be entomophilous, hermaphroditic, with numerous stamens and carpels and all derived from a single line of descent (monophyletic) from an ancestral gymnosperm.

It should be emphasised that there is considerable agreement on certain issues between the Englerians and the ranalians. Both systems agree for example that an inferior ovary is more advanced than a superior ovary and that gamopetaly (joined petals) is more advanced than polypetaly (separate petals).

The families discussed in the second and third parts of this book are arranged according to Hutchinson's system. The basic reason for this is that Hutchinson, in collaboration with other workers, has done monumental work on the plants of West Africa, culminating in the volumes of *Flora of West Tropical Africa*, which are crucial literature to taxonomists in this part of the world.

New systematics and the new empiricists

Even a casual glance at the phylogenetic systems shows that they concern plant groups at suprageneric (above the level of genus) levels; they deal with presumed phylogenetic relationships among families and/or orders of plants. However, with new systematics the focus was to shift to the infragenetic levels. The explosion of knowledge in plant geography, plant variation (at infrageneric levels), genetics, cytology, plant chemistry etc. after the first few decades of the twentieth century led to active interest among plant biologists in plant evolution. What Julian Huxley (in 1940) called 'New Systematics' was born.

The detailed evolutionary relationships among certain species were worked out, population structures were studied, breeding patterns were studied, hybridisation studies were carried out. Geographical variations in certain species were investigated and transplantation experiments were carried out to find the influence of latitude, soil etc. on certain morphological attributes. Specifically, recently in West Africa detailed systematic studies have been done by J. K. Morton on the genus *Physalis* in Sierra Leone. H. G. Baker and Y. Ewusie in Ghana have worked on a number of weed species, Edwards worked on the genus *Eupatorium* (Compositae) in Nigeria and Oyewole on Liliaceae (also in Nigeria). Considerable genetical and biosystematic (as the new systematics is also called) work has also been done in our laboratories at Ife on many weed species of Nigeria.

It should be obvious from the foregoing that new systematics throws much light on the structure of infrageneric categories but not on the relationships of higher categories. Part of this short-coming of new systematics derives from its implicit intent to be phylogenetic from an experimental point of view, in the sense that there is almost always a search for an *observable or demonstrable evolutionary process*.

The new empiricists or neo-Adansonians, as they are variously refered to, have attempted to study taxonomic relationships without necessarily implicating phylogenetic relationships. They have constructed all sorts of mathematically derived measures of taxonomic relationship. This relationship is observed in terms of morphological distance or closeness; the conclusions, of course, depend on the belief that overall morphological closeness among evolutionarily unrelated organisms is unlikely. One of the most eloquent treatises on this subject is *Numerical Taxonomy: The Principles and Practice of Numerical Taxonomy*, written by P. H. A. Sneath and R. R. Sokal in 1973.

Faithful to Adansonian tradition, the numerical taxonomists believe that as many characters as possible should be used in constructing taxonomic systems and that all characters, provided they are logically admissible (see page 10) should be accorded the same importance. From the point of view of numerical taxonomists, the affinity or similarity of plant groups is something to be discovered and perfected – hence the empirical nature of the theory and practice.

The various indices of similarity used by numerical taxonomists yield summary diagrams called similarity matrices or dendrograms. Figure 1.3 is a hypothetical dendrogram showing the relationship among eleven species (A, B, C, D .... K) – formally referred to by numerical taxonomists as OTUs (operational taxonomic units). Put simply, the higher the level of similarity index (along the vertical axis) where a horizontal line connects two or more groups, the greater the similarity among groups so connected. Thus B and C are more similar (they show greater affinity) than E and F or B and F; similarly, A and F show greater affinity than A and K while H and I show as much similarity as B and C.

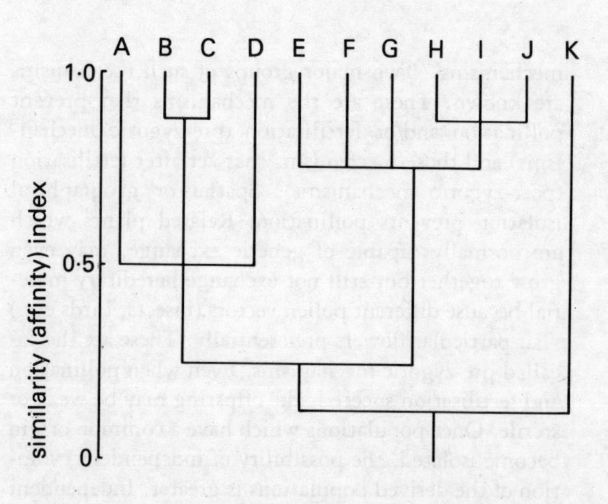

Fig. 1.3 A typical dendrogram

Selected references

Baker, H. G. and Stebbins, G. L. (eds). 1965. *The Genetics of Colonising Species*. Academic Press: New York, USA.

Benson, L. 1957. *Plant Classification*. D. C. Heath and Co.: Boston, USA.

Cronquist, A. 1968. *The Evolution and Classification of Flowering Plants*. Nelson: London, UK.

Davis, P. H. and Heywood, V. H. 1963. *Principles of Angiosperm Taxonomy*. D. Van Nostrand Co. Inc.: Princeton, New Jersey, USA.

Hutchinson, J. and Dalziel J. M. 1954–1972. *Flora of West Tropical Africa*, Vols I, II, III. Crown Agents for Overseas Governments and Administrations: London, UK.

Simpson, G. G. 1961. *Principles of Animal Taxonomy*. Columbia University Press: New York, USA.

Sneath, P. H. A. and Sokal, R. R. 1973. *Numerical Taxonomy: The Principles and Practice of Numerical Classification*, W. H. Freeman & Co.: San Francisco, USA.

Sokal, R. R. and Sneath P. H. A. 1965. *Principles of Numerical Taxonomy*. W. H. Freeman and Co., San Francisco, USA.

CHAPTER 2

Principles and practice of taxonomy

Hierarchic categories: the case of angiosperms

One of the consequences of human attempts to impose order on a seemingly unordered organismal world is the delimitation of categories under which smaller categories are often subsumed. As one would expect, the delimitation of the categories is not haphazard; the greater the affinity among a collection of organisms the lower the category to which they all belong.

Obviously, in spite of the likelihood of agreement among workers, much subjectivity arises in the delimitation of hierarchic categories. It takes international agreements to itemise and standardise taxonomic hierarchies. The commonly adopted taxonomic categories of angiosperms used today derive from the International Code of Botanical Nomenclature. Let us take a common West African plant, *Aspilia africana*, and see how it fits into the hierarchic categories modified from the proposals in the code mentioned above. Obviously, *Aspilia africana* belongs in the plant kingdom: for all practical purposes, any organism belongs to either the plant kingdom or the animal kingdom. The major categories are as follows:

Kingdom ....Plant Kingdom
Division ....Spermatophyta (seed plants)
Class...............Angiospermae
Subclass...........Dicotyledoneae
Order..............Asterales
Family ....Compositae (or Asteraceae)
Subfamily ....
Tribe...............Heliantheae
Subtribe............
Genus............ *Aspilia*
Section ....
Series................
Species ....*africana*
Subspecies ....
Variety ....*africana*

In the presentation above there are a number of categories that are omitted from the proposals of the Code: these include categories like subgenus, subsection, subseries, subvariety and certain sub-subvarietal categories like forma and subforma. The reader's attention is also called to the customary endings of each category: thus orders end in -ales, families end in -aceae, and subfamilies end in -oideae, tribes end in -eae, while subtribes end in -inae. There are certain unusual endings that have been conserved through convention and usage, as in the family Compositae or the grass family Gramineae.

Although the taxonomic hierarchy of groups of plants may be changed (say from family to order), the hierarchic relationships or sequences presented above *may not be altered*. Thus, what constitutes a category depends on the worker, but the categories that are available does not. For example some workers have divided the plant kingdom into three subkingdoms (Prokaryota, Mycota and Chlorota) with the Anthophyta, which corresponds to Angiospermae, as one of the 13 divisions of Chlorota. Other workers give monocots and dicots class categories (Liliatae and Magnoliatae respectively). Similarly, the Dicotyledoneae have been subdivided into 'subgroups' on the basis of the relative prominence of certain floral parts (thus we have Corolliflorae – prominent corolla, Ovariflorae – prominent ovary etc.). This apparent chaos among students of plant taxonomy with respect to the delimitation of categories tends to reinforce the nominalist claim of arbitrariness of categories. While this claim may have some truth at the generic and suprageneric levels, the reality of the species category, especially from the point of view of the biological concept of species, is generally accepted.

Nomenclature

Nomenclature, simply defined, means a system of naming. Botanical nomenclature is a system of naming plants. As we are aware, botanists require a system of naming plants which enables them to standardise, not only the terms that denote taxonomic ranks, but also the scientific names which refer to taxonomic groups among plants.

The guidelines for plant classification and nomenclature are provided in the form of a code (International Code of Botanical Nomenclature) which is revised and adopted at International Botanic Congresses, each of which is named after the town or city in which the congress was held. Thus we have the Montreal Code, the Edinburgh Code, the Seattle Code. The main purpose of the code, according to the Seattle Code published in 1972, is 'the provision of stable methods of naming taxonomic groups, avoiding and rejecting the use of names which may cause error or ambiguity or throw science into confusion. Next in importance is the avoidance of useless creation of names'.

The code consists of 'Principles' and 'Rules and Recommendations'. The Rules and Recommendations set out the details of the Principles with examples and illustrations. The Principles given in the Seattle Code, are enumerated below to give us a feeling of the issues raised.

Principle 1 Botanical nomenclature is independent of zoological nomenclature. The code applies equally to names of taxonomic groups treated as plants whether or not these groups were originally assigned to the plant kingdom.

Principle 2 The application of names of taxonomic groups is determined by means of nomenclatural types (see page 10).

Principle 3 The nomenclature of a taxonomic group is based upon priority of publication.

Principle 4 Each taxonomic group with a particular circumscription, position and rank can bear only one correct name, the earliest that is in accordance with the Rules, except in specified cases.

Principle 5 Scientific names of taxonomic groups are treated as Latin regardless of their derivation.

Principle 6 The Rules of nomenclature are retroactive unless expressly limited.

For obvious reasons, the detailed rules of botanical nomenclature are beyond the scope of this book. However, we would like to point out the use of authority citations. It is often appropriate to give them at the first mention of a species in a scientific work. Care must be taken to ensure they are correctly given. The authorities for all the plants mentioned in this book are given in Appendix II.

Typology and the type concept

One of the consequences of essentialist thought in plant classification is the tacit acceptance of the 'typical' representative of a taxonomic group – the member of the group which 'best presents the essential characteristics of that group'. Even though the fact of organic variation has been evident to naturalists through the ages, the strong temptation towards reductionism has affected human thought. It is precisely the same temptation that forged concepts such as 'average', 'usual form', 'representative specimen' and so on. Thus for each family we have the 'type genus': the type genus for the family Annonaceae is *Annona* and the type genus for the family Combretaceae is *Combretum* while the type family for the order Rosales is Rosaceae.

Typology as a concept or typological thinking, refers to the tendency to contemplate a taxonomic group as a monolith which does not vary. When variation is noticed it is considered an aberration because (especially according to the views of early botanists) true species are considered to be fixed and unvarying.

When a new species or variety is described, the description should always refer to a type specimen which is often deposited in a designated, 'recognised', herbarium. We therefore have the 'type method' (typification) in routine taxonomic practice. Different terms are used to describe such type specimens depending on their status. A *holotype* is the specimen or element used or designated by the author at the time of valid publication as the *nomenclatural* type. As long as the holotype exists, it automatically fixes the application of the name concerned. When the holotype is missing or was not designated by the author at the time of valid publication, a specimen or element selected from the original material to serve as a nomenclatural type is called a *lectotype*. An *isotype* is a duplicate of the holotype (and part of the same collection as the holotype). A *syntype* is any one of two or more specimens cited by the author simultaneously as the nomenclatural type or cited by the author when no holotype was designated. When, and as long as, all the material on which the name of a taxon is based are missing, a specimen or other element selected to serve as a nomenclatural type is selected and is called a *neotype* (see Chapter II, Section 2 of the Seattle International Code of Botanical Nomenclature).

Taxonomic characters

A taxonomic character is an attribute of an organism that can be considered to vary independently of any other attribute. The independence referred to here is logical rather than functional; thus while we can consider hairiness of a stem and hairiness of leaves as independent logically, the attributes are not necessarily functionally independent.

There is usually also the problem of compartmentalising characters. What is a single character? A single character is one that cannot be reasonably subdivided except when changes in the technique of observation make such a subdivision desirable or possible. For example, before the study of biochemical pathways showed that certain pigmentations result from a series of enzymatic processes, such pigments were considered single characters. But since lack of pigmentation may have resulted from distinct enzymatic blocks along the pathway, absence of pigment may be subdivided on the basis of which enzyme was absent.

While some characters provide information for obtaining estimates of affinity among organisms, certain characters are not admissible. Such characters may be meaningless, they may be invariant or two or more of them may show logical, or empirical correlation. A character like the number of secondary roots or the number of leaves on a branch of a plant is meaningless; so also are responses of a plant which are not heritable or constant. Similarly, using the number of carpels as a taxonomic character in a collection of plants that are known to have only two carpels is a useless exercise. Nor can we reasonably consider two characters as independent when from our experience such characters always occur together – this is what is meant by empirical correlation. Similarly, logically correlated attributes such as the radius and the diameter of a round leaf can not be treated as different attributes.

The concepts of homology and analogy

When we say two organs are homologous, we mean they are the same organ in two different organisms; thus we consider that the leaf of *Andropogon tectorum* and the leaf in *Sorghum bicolor* are homologous. On the other hand, when we say two organs are analogous, we mean they are two different organs serving the same function in different organisms: an example is the 'leaf' in mosses and the true leaf in clubmosses. The clarification of these terms is important because most of the inference we can make about evolutionary pathways are dependent on such clarifications or conceptual limitations of the clarifications. Thus we may question whether the leaves in yams and grasses are actually homologous organs or we may question the usefulness of contemplating whether fins in fish and legs in humans are analogous!

Since our observation of the organic world compels us to resolve certain evolutionary questions or at least contemplate them, we have to make use of the implications of homology and analogy to deal with the following cases:

a) Obviously 'unrelated' or distantly related (in a strictly relative sense) organisms that show a striking morphological similarity (convergence) in a certain respect.
b) Obviously related organisms that show certain striking (adaptive) morphological differences (divergence).
c) Related lines of descent that have independently acquired similar attributes.

Differentiating between divergence on one hand, and convergence and parallel evolution on the other, also requires an understanding of the concepts of *monophyly* and *polyphyly*. Organisms that arise from a single line of descent are said to be monophyletic; those that have different (many) lines of descent are said to be polyphyletic. Thus convergence and parallel evolution among a group of organisms may imply polyphyly or divergence, in the remote past, of the constituents. On the other hand, divergence invariably implies monophyletic origin of the diverging groups (Fig. 2.1).

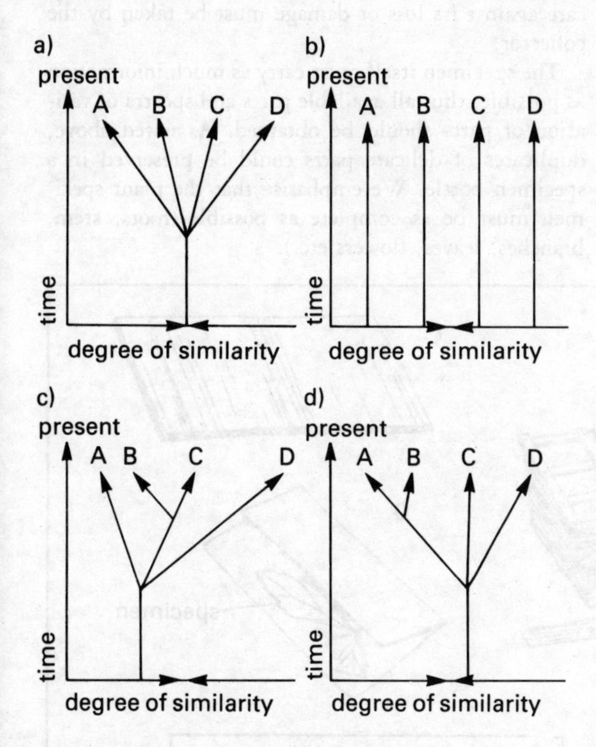

Fig. 2.1 Origin and relatedness of four hypothetical taxa: A, B, C and D
a) monophyletic origin of A, B, C and D
b) polyphyletic origin of A, B, C and D
c) similarity of A and B due to convergence of B on A
d) same similarity of A and B as in (c) above, but in this case due to recent divergence of A and B

Presumed evolutionary trends of angiospermous characters

This section is very important because of the synthetic framework which it provides for understanding the angiosperm world. These trends were proposed by John Hutchinson and Charles Bessey (Porter, 1967). The following are some of the main presumptions:

a) trees and shrubs are more primitive than herbs and climbers;
b) hermaphroditism is more primitive than unisexuality, while dioecism is more advanced than monoecism;
c) a spiral arrangement of floral leaves (perianth) is more primitive than a whorled (circular) arrangement;
d) the inflorescence is more advanced than solitary flowers;
e) a reduction in the number of segments of floral parts (oligomery) is more advanced than the presence of numerous segments (polymery);
f) free petals, free stamens etc. are more primitive than connate petals, stamens etc.
g) apocarpy is more primitive than syncarpy;
h) actinomorphy is more primitive than zygomorphy;
i) hypogyny is more primitive than perigyny and epigyny;
j) heteromorphism of floral and vegetative parts is more advanced than homomorphism.

Field and herbarium techniques

Field work is very important to anyone who is interested in plant classification and identification. Since two individuals of a species are rarely identical, field work continually opens new vistas of variation to the plant scientist. Because of this variation it is always necessary to compare new collections with known collections and for this a systematic specimen bank (herbarium) is required.

In the following subsections, we summarise some of the methods and necessary steps in plant collection, preparation for storage and storage itself.

Hints on plant collection

The following are some of the basic items of equipment that are needed for plant collection and field work:

a) a pen knife or secateurs for cutting branches and tough specimens;
b) a cutlass (matchet) for cutting woody specimens, for removing bark from trees and for digging up roots or underground stems of specimens;
c) plastic bags of various sizes into which collected plants may be put to keep them fresh;
d) specimen bottles of various sizes containing formalin or 70% ethanol (alcohol) for preserving delicate plant structures such as flowers;
e) labels of various sizes with string to attach them;
f) some device (see Fig. 2.2A) for obtaining specimens from tall trees or climbers;
g) hard-covered collection notebooks, a ball point pen and a pencil;
h) a plant press for extended field trips.

The collection notebook is a crucial record book for the field worker. As we noted above, it should have hard cover and possibly be protected by a waterproof wrapper. Each plant collected should have a collection number (accession number) recorded in the field notebook and written on a label which should be securely attached to the specimen. The notebook record for each collected specimen should include date of collection; name of the plant (if known); location of the collection; soil conditions and biotic community of the location; abundance of the specimen at the location; notes on ephemeral characteristics of the specimen (e.g. flower colour). The collection numbers used should follow each other in strict sequence and once a collector has started a numbering sequence, it should be maintained; numbers should never be repeated. Because the collector's record is invaluable, meticulous care against its loss or damage must be taken by the collector.

The specimen itself must carry as much information as possible; thus all available parts and spectra of variation of parts should be obtained. As noted above, duplicates of delicate parts could be preserved in a specimen bottle. We emphasise that the plant specimen must be as complete as possible (roots, stem, branches, leaves, flowers etc.).

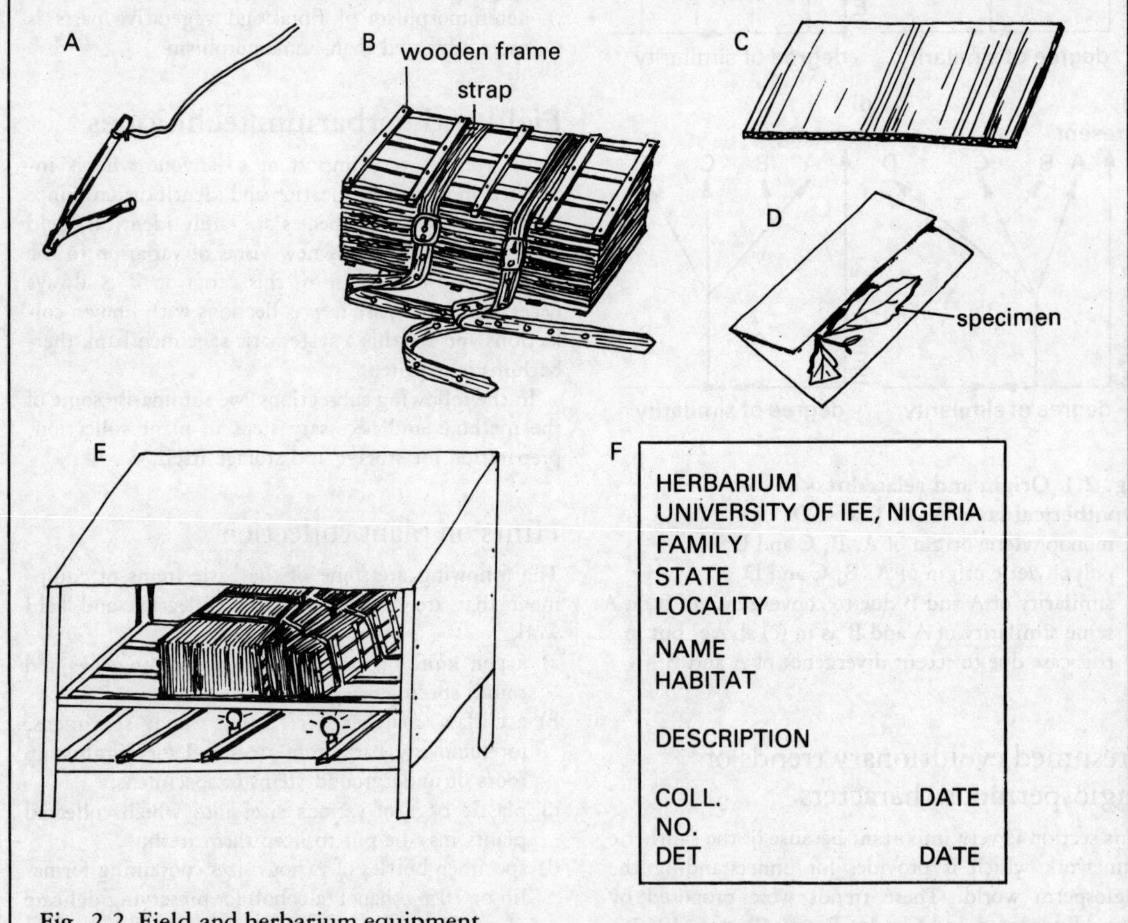

Fig. 2.2 Field and herbarium equipment

A device for obtaining specimens from tall trees or lianes. It consists of a forked piece of wood which is attached to a length of string. The collector holds the end of the string and throws the piece of wood in order to catch the specimen he wants.

B loaded plant press

C sheet of cardboard with corrugated lining

D absorbent folder with plant specimen inside

E drying cabinet with presses inside

F specimen label

Pressing, drying and preservation of specimens

The plant press (Fig. 2.2B) is an important piece of equipment for a plant collector. It consists of two wooden frames (each about 45 cm × 30 cm) between which specimens are kept in absorbent paper folders (Fig. 2.2D) for partial drying and flattening out. The folders themselves (which may be double sheets of newspaper) are often placed between cardboard sheets (about 42 cm × 28 cm) that have a corrugated internal lining (Fig. 2.2C). These boards allow a free flow of air which helps the specimens to dry.

The specimens should be pressed as soon as possible after collection. It is therefore necessary to take presses along on extended field trips. A plant to be pressed is placed inside the folded paper sheet in such a way that unnecessary folding of its parts is avoided, although specimens which are larger than the absorbent folder may be carefully bent so as to be accommodated. A board with corrugated lining is placed on the lower wooden frame and the folder containing the specimen is placed on it. The press is loaded so that absorbent folders alternate with corrugated board until a convenient height is attained. The upper wooden frame is put in place and the press is fastened with strong ropes or straps.

The press is dried in the sun or over a source of low heat (Fig. 2.2E). Where electricity is not available, the drier may be so designed that hurricane lamps can be put under it to supply adequate heat. For succulent plants or succulent parts of plants, the absorbent folder may need to be changed daily to hasten drying and to avoid the growth of mould on the specimens. It is useful to treat the specimens with boiling water before pressing.

Even after drying, plant specimens may still be attacked by insects and moulds. They should, therefore, be treated with preservatives – this is called poisoning. The following preservatives may be applied by spraying or with a brush:

a) 2% solution of mercury (II) chloride in ethanol
b) saturated solution of 1,4-dichlorobenzene.

Other poisons are available. Utmost care is needed in handling them all.

After the specimen has been protected against insect or mould attack and when it is quite dry, it is mounted on a special mounting board (42 cm × 28 cm) with thread or by strips of gummed paper. Sellotape is not recommended because it does not last. The mounting board should carry a label with adequate information about the specimen (Fig. 2.2F). A small hand-made envelope containing materials such as seeds may be gummed on the mounting board.

The herbarium

The herbarium is the place where preserved plant materials are deposited. The herbarium is useful for research, for identification of materials and for comparison of materials with previous collections. A herbarium may be a general one in which all families of plants are deposited: this is the type found in most universities, schools, colleges or homes of individual plant enthusiasts. There are specialised herbaria which handle particular plant families (e.g. the grass herbarium at Ahmadu Bello University Institute of Agricultural Research, Nigeria) or plants from particular regions. Even in herbaria that handle general materials, there may be specialised sections (such as the weed herbarium at the University of Ife, Nigeria).

Researchers usually deposit plant materials in recognised herbaria for future reference. All recognised herbaria have code letters by which references are made to them; thus the forestry herbarium in Ibadan, Nigeria is referred to as FHI, the University of Ife herbarium is referred to as IFE and University of Kansas herbarium, USA is referred to as KANU. Different herbaria all over the world cooperate in research and in the exchange or loan of plant materials.

No herbarium was built in a day. Well-known herbaria attained fame as a result of painstaking effort. Schools and colleges are therefore encouraged to have their own herbaria and strive, in conjunction with larger herbaria, to build them up.

Dried and mounted specimens are stored in cabinets with tight-fitting doors. The storage cabinets are provided with partitions in which the families may be arranged in alphabetical order or (more usually) in some presumed phylogenetic order. Under each family, the species (each in a folder) are arranged in alphabetical order in a genus folder; the genus folder being slightly larger than the species folders so that the latter may fit conveniently in the former. Air-conditioning of the herbarium is important in the Tropics and a dehumidifier keeps humidity low and so curbs fungal growth. Insect repellant such as naphthalene and 1,4-dichlorobenzene crystals can be placed in the cupboards.

Students starting a course in plant taxonomy should attempt to collect, identify and prepare, for inclusion into the herbarium, at least twenty-five different plants each representing a different family or tribe.

Plant identification

Plant identification is the process of determining if an

'unknown' plant is identical with or similar to another known plant (or group of plants). From this definition it follows that the organisation of the knowledge about the plants in a certain locality is a prerequisite for meaningful work on the identification of the plants of the area.

Through the painstaking work of professional and amateur botanists many of the plants existing in many parts of the world, including West Africa, have been listed and/or described. Such listings and descriptions are available in books, manuals and floras. A flora contains a list and record (place of collection, history of identification, herbaria where specimens are located etc.) of plants known to exist in a particular geographic area. Such written records do not exhaust the facilities for a worker: considerable unwritten knowledge of plants exists in most localities.

Those wishing to attempt the identification of plants either as a hobby, or as part of their research and studies, must have some knowledge of plant morphology. Technical terms are used to describe the various features present on any particular plant. In order that such descriptions can be used to relate the unknown plant to its appropriate group or identity, an identifier must be conversant with the terminology employed. Many floras, manuals and taxonomic publications include a glossary explaining the terms or words used in the description of plants they contain. For example, the glossary on pages 17–28 of Volume I of *Flora of West Tropical Africa* may be consulted whenever this flora is used to identify West African plants.

In taxonomic practice it is customary to combine the identification of a plant with the determination of the correct name. The two processes go hand in hand.

Plant identification or confirmation of the identity of a specimen is generally accomplished by one or more of the methods given below. The selection of the method depends on the experience of the worker and the individual circumstances. The methods used are:

a) comparison with already identified plants (e.g. herbarium specimens);
b) comparison with adequate photographs, drawings and illustrations;
c) descriptions and keys in manuals, floras etc.

Comparison with already identified plants

Use of herbarium specimens

In the herbarium the identification of an unknown plant may be attempted and/or confirmed by direct comparison with already preserved and identified specimens in the same category. This method presumes that a plant that may look like the unknown one has probably been collected and identified and is available in the herbarium concerned. If it is impossible to identify the specimen in a local herbarium, other herbaria do offer identification services and these should be consulted.

It is advisable for this type of identification to be done with the assistance of an experienced herbarium worker.

Use of botanical and horticultural gardens

Where they exist, botanical and horticultural gardens offer considerable facilities for identification. To this end, schools, colleges and universities are encouraged to establish such gardens. Failing this, nature reserves could be designated in the locality of the schools. In such gardens or reserves, the names and families of the plants may be carefully displayed on the plants. Like herbaria, nature reserves and gardens cannot be built in one day; it takes sustained effort.

Use of illustrations

An unknown plant may be identified by comparing it with appropriate photographs, drawings and/or paintings. In well-established herbaria, museums and libraries several illustrations of plants are available with which comparisons can be made. Many cultivated plants which are exotic to West Africa can be identified by referring to illustrations in dictionaries and encyclopaedias of plants.

The identification of plants using illustrations should be attempted with considerable caution and with the assistance of an experienced worker.

Use of descriptions and keys

When attempting the identification of an unknown plant, the usual thing is first to refer to a flora, manual or any other available taxonomic publication relating to the plants of that particular region or area. For instance, for the identification of any plant from any West African country, one should first consult *Flora of West Tropical Africa*. Other manuals or books dealing with plants from more specific areas may be available, for example, *The Flora of Nigeria* dealing with grasses (Stanfield, 1970) and sedges (Lowe and Stanfield 1974). Also, we have *Nigerian Trees* (Keay *et al*. 1960 and 1964).

These floras, manuals and books often contain descriptions and keys. Descriptions are detailed listings of all the avaialble and known characteristics of any plant from a particular area. By comparing the available descriptions of plants in a particular category with those of an unknown plant the identification of the latter may be achieved.

Keys are devices in which a few characteristics of the plants are so arranged that the features of a known and an unknown plant can be compared in a systematic manner.

Types of keys

There are two types of keys: natural and artificial (see Chapter 1). In a natural key many characteristics of the plant are involved and many of these may not be easily observable on the plant. For example, chromosome number, chemical constituents etc. may be given in the key. In contrast few, but easily observable, features of the plant are employed in the construction of an artificial key. Since relatively few characteristics are used there is an intuitive elimination of characters that are considered not convenient to use or that are considered not diagnostic. Artificial keys are more commonly used than natural keys and are found in most floras and manuals.

There are two forms of artificial key: dichotomous and numerical keys.

Dichotomous keys

In using a dichotomous key one works through successive stages in which one compares the unidentified plant with features described in the key. At each stage there is a pair of contrasting characteristics (or propositions) called 'leads'. The features of the unknown plant are compared with the two sets of contrasting characteristics and one accepts the lead that more or less tallies with the observed features on the unknown plant and rejects the other lead. One then moves on to the pair of leads which follow on from the lead just accepted. One thus works through the key until one gets to the plant whose features are comparable with those of the unknown. Then we say the unknown plant is identified or matched with a plant which is already known.

Dichotomous keys are set out in two ways; they may be indented or bracketed.

Indented keys

In the indented key, each subsequent pair of leads is written as an indented pair under the preceding lead. For example:

1a Plants either trees or shrubs
  2a Leaves compound .... species A
  2b Leaves simple .... species B
1b Plants herbaceous
  3a Leaves net-veined.................... species C
  3b Leaves parallel-veined .... species D

Bracketed keys

In a bracketed key, the alternative to each lead is written immediately after it. Reference is made to the next sublead by numbers written at the end of the preceding lead or sublead. In both indented and bracketed keys, *a* and *b* can be used to distinguish each alternative of each pair of leads. The following is an example of a bracketed key:

1a Plants trees or shrubs 2
1b Plants herbaceous 3
2a Leaves compound .... species A
2b Leaves simple.............................. species B
3a Leaves net-veined .... species C
3b Leaves parallel-veined .... species D

Numerical keys

As the name implies, the characteristic of the plants are represented (coded) in numerals. This kind of key is used in *Flora of Nigeria* (D.P. Stanfield (1970) and D.P. Stanfield and J. Lowe (1974)). For example, we could code certain diagnostic characteristics of the leaves of cassia, cowpea, pawpaw, cashew, mango and guinea corn as follows:

Compound 1
Simple... 2
Pinnate 3
Obovate 4
Palmately-lobed 5
Parallel-veined 6
Net-veined 7
Linear 8

These numbers indicate the position occupied by each characteristic in a series of 0s and 1s. If a character is present in a plant a 1 is put in the relevant position, but if it is absent a 0 is put in that position. This is done for all the characteristics listed and one produces a code composed of 1s and 0s. For example:

10100010 .... Cassia
01001010 .... Pawpaw
01000101 .... Guinea corn

There are variations of this method. In one variation the leaf characters in the last paragraph are coded as follows:

Leaves compound 1
Leaves pinnate 2
Leaves obovate 3
Leaves linear 4
Leaves palmately-lobed............................ 5
Leaves parallel-veined 6

The code 0 is used for absence of a character and the numeral corresponding to the character for presence of that character. Thus the pawpaw leaf will have the code: 000050, while the cassia leaf will have the code: 120000.

Letters rather than numbers may be used in some codes, increasing the number of characters that can be used. With the above codes, any of the set of leaves provided could be easily identified.

The use of alternative choices in identification as in regular dichotomous keys can easily be applied to numerical keys. However, in most floras and manuals, artificial dichotomous keys are used. For this reason, it is important that we learn how to use this conventional type of key.

Construction of keys

The construction of a key or keys should usually be done by an experienced worker. But students starting a course in taxonomy can grasp some of the procedures involved in making a key and subsequently try their hands on the construction of simple keys.

Table 2.1 Arranging data for the construction of a key

Species \ Characters	A	B	C	D	E
Leaf compound	–	–	+	+	+
Leaf simple	+	+	–	–	–
Blade lobed	+	–	–	–	–
Leaf pinnate	–	–	+	+	+
Plant hairy	+	–	–	+	+
Plant annual	+	–	–	–	+
Plant a tree	–	+	+	+	–

The different plant species for which a key is to be made must first of all be examined very critically and carefully. All the available and observable features must be described and recorded for each species. One of the most satisfactory methods of assembling and extracting data for the construction of a key is by entering the features in a tabular form as illustrated in Table 2.1.

To begin the construction of a key, check the entry of features and select one important and reasonable feature which separates the species into two groups. The size of the group in terms of the number of elements in each does not matter.

Take for instance 'compound leaf'. The first two leads will be:

1a Leaf simple (species A and B)
1b Leaf not simple, i.e. compound (species C, D and E)

Take the lead 1a, and determine which feature(s) can separate A from B. These are: 'blade lobed', 'hairy plant' and 'plant annual.' Therefore, the next sublead under 1a will read like this:

1a Leaf simple
  2a Blade lobed; plant hairy; plant annual ....A
  2b Blade not lobed; plant glabrous; plant a tree ....B

Similarly C, D and E can be separated into 2 other groups by using distinguishing features. For instance the next lead under 1b, will read:

3a Plant glabrous C
3b Plant hairy .... D and E

We have to separate D and E also. Under the lead 3b we will have:

4a Plant a tree; perennial .... D
4b Plant annual .... E

The final rearrangement of the key will therefore look like this:

1a Leaf simple
  2a Blade lobed; plant hairy; plant annual ....A
  2b Blade not lobed; plant glabrous; plant a tree ....B
1b Leaf compound (i.e. not simple)
  3a Plant glabrous C
  3b Plant hairy
    4a Plant a tree; perennial .... D
    4b Plant an annual .... E

There are some rules guiding the construction of keys. Some of these are stated below:

1 Select characters that are in opposition to one another so that the two leads of each set will be contrasting propositions, one of which will fit the situation and the other not apply.
2 The initial word of each lead in a pair of leads must be identical. That is if the first lead of a pair starts with the word 'flowers', the second lead of the same pair should begin with 'flowers'.
3 Use clearly defined variation and avoid overlapping limits. For example, it is bad to use overlapping measurements such as:

 1a Leaves 4 cm – 10 cm long
 1b Leaves 7 cm – 13 cm long

Identification practice

An artificial key to the families of the dicotyledons of West Africa is set out in Volume II of *Flora of West Tropical Africa* and to the monocotyledons in Volume III. Under the different families there are keys to the genera, and under each genus there are keys to the species within that genus; a modified form of the keys used in Neilsen (1965) is appended at the end of this book.

With the appropriate flora or manual and the unidentified plant at hand, and possibly with the aid of a hand lens (×10), the first step is to examine the parts of the plant properly. Examine the stem, leaves, flowers, fruits and, if available, the roots and underground parts. The aim of this important examination is to acquaint the identifier with the characteristics of the plant. The features of the specimen may then be compared with those set out in the key.

Let us practise using the simple procedure set out below.

Suppose you are provided with a branch containing the flower of a plant in the family Malvaceae whose generic identity you do not know. You may find it useful to examine the specimen and answer the following questions.

1 Is the flower regular or irregular (i.e. actinomorphic or zygomorphic)?
2 Is there an epicalyx of bracts outside the calyx or is it absent?
3 Are there sepals and petals, or only sepals, or only petals; or are the sepals and petals absent?
4 How many sepals are there? Are they free from one another right to the base or are they at least partly joined together?
5 How many petals are there? Are they free from one another right to the base or are they at least partly joined together?
6 How many stamens are there? Are the filaments free from each other or are they joined together, partly or completely? Are they joined to the petals, or not?
7 Do the petals, sepals and stamens arise below the ovary, or on a level with it (ovary superior), or do they arise above it (ovary inferior)?
8 How many partitions of the ovary or young fruit are observed in cross-section?
9 Is there one ovule in each section of the ovary, (one seed in each section of the fruit) or are there many? If there are many ovules or seeds, are they attached to the central part of the ovary or fruit (placentation axile), or are they attached to the wall (placentation parietal), or are they attached to the base of the ovary (erect) or the top of the ovary (pendulous)?
10 What type of fruit is found on the specimen?

When you have answered those questions, then you can use the key to the genera of West African Malvaceae provided below.

Key to the genera of Malvaceae in West Africa

Source: Flora of West Tropical Africa (1963)

1a Epicalyx absent
 2a Carpels transversely divided, contracted in the middle, the upper half spreading stellately in fruit, two-seeded, the lower seed tomentose; leaves entire, cordate; flowers small in panicles .... *Wissadula*
 2b Carpels not transversely divided
  3a Staminal column provided with anthers up to the top or nearly so
   4a Carpels – ten or more, with more than one ovule in each cell; leaves always cordate .... *Abutilon*
   4b Carpels – up to ten, with one ovule in each cell; leaves usually not cordate .... *Sida*
  3b Staminal column destitute of anthers at the top; some of the leaves trifoliolate .... *Hibiscus*
1b Epicalyx present
 5a Styles – twice as many as the carpels; staminal column destitute of anthers at the apex
  6a Flowers not in heads or at any rate not surrounded by an involucre

7a Bracteoles of epicalyx connate at the base and adnate to the calyx; carpels with hooked spines .... *Urena*

7b Bracteoles of epicalyx free to the base or nearly so; carpels without hooked spines but sometimes with barbate bristles .... *Pavonia*

6b Flowers in heads surrounded by an involucre of bracts .... *Malachra*

5b Styles – the same number as the carpels or style more or less undivided

8a Style undivided or nearly so

9a Calyx truncate

10a Leaves with minute scales, otherwise glabrous; epicalyx of three to five bracteoles; fruit indehiscent; seeds pubescent .... *Thespesia*

10b Leaves with abundant stellate hairs; epicalyx of (five to) nine to fifteen bracteoles; fruit dehiscent; seeds glabrescent .... *Azanza*

9b Calyx five-lobed; seeds cottony/hairy; dwarf shrub with glandular leaves .... *Cienfuegosia*

8b Style divided into separate stigmas

11a Ovary with three or more ovules in each cell; carpels – usually five, dehiscent

12a Epicalyx of five or more free or partially connate usually narrow bracteoles; seeds various .... *Hibiscus*

12b Epicalyx of three broad foliaceous, more or less cordate, usually deeply toothed or laciniate bracteoles; ovules numerous; seeds covered with long cottony hairs .... *Gossypium*

11b Ovary with one ovule in each cell

13a Epicalyx of seven to ten small or obsolete linear bracteoles; ovary five-celled; fruits depressed, prominently five-angled; carpels dehiscent .... *Kosteletzkya*

13b Epicalyx of one to three narrow free bracteoles; carpels – eight to twelve subindehiscent, separating from the central axis .... *Malvastrum*

Look first at the two statements labelled 1a and 1b. Note that these are alternatives because either there is an epicalyx present outside the sepals or there is no epicalyx.

If your flower has several bracts outside the calyx, it means that the second of the two alternative leads (i.e. 1b) applies to your plant. You can now ignore the first statement and all the parts contained underneath it, i.e. 2a to 3b inclusive.

Now look under 1b, you will find another couple of leads spaced (indented) one step from the 1b entry. Those are 5a, 'Styles – twice as many as the carpels ....', and 5b, 'Styles – the same number as the carpels ....'

Let us assume that the style of your flower is divided into five branches. To check the number of carpels cut a transverse section of the ovary and count the number of spaces (loculi) inside it. Let us assume that in this case there are five.

In your specimen there are five carpels and five stigmas (style branches), so the number of style branches is equal to the number of carpels. Therefore lead 5b applies to your specimen. Ignore 5a and all under it.

You will notice again that immediately below lead 5b there is another lead indented one step from 5b. There are two comparable leads, 'Style undivided or nearly so', and 8b, 'Style divided into separate stigmas'.

Suppose lead 8b applied to your specimen, then ignore all under lead 8a. Move to 8b. Again under 8b you will find two contrasting leads. Assuming that 11a agrees more with the features of your specimen than does the other lead then ignore everything under lead 11b. Now observe the epicalyx and count the number of bracteoles. Suppose that your specimen has five bracteoles and thus agrees with lead 12a, then your specimen has keyed out as belonging to the genus *Hibiscus*.

Any of the keys to families, genera or species can be used in a similar way. At any one time, always compare the features set out in the alternatives very carefully and keenly with those observed on your specimen. Select one of the alternatives each time, and go on to tackle the subsequent leads until you arrive at the identity of the unknown specimen.

Supposing you have used the key to the families of the dicotyledons and you have arrived at the name of the family; the next step is to turn to the section in the flora or manual where the family is treated. There, by means of the key to the genera, as we have done, you determine the generic name of the specimen. After this, and by the use of the key to the species, the specific identity (species) of the unknown plant is ascertained by repeating the same procedure.

These procedures are used to find not only the identity of the plant but also the double name (binomial) attached to the plant and the authority who originally described the plant. For example, in *Hibiscus cannabinus* L. the first two words are the generic and specific names respectively, and L. (an abbrevation for Linnaeus) is the name of the authority who first described and published an account of this plant.

For many reasons the identity and name of a plant obtained solely by use of a manual may be incorrect, incomplete or both. When identifying any unknown plant with the aid of keys in a flora or manual, always check the description of each group or level arrived at (e.g. family, genus, species) to ensure that there is no mistaken identity. If you have any doubts after reading the description, repeat the process of trying to identify, paying attention to all details in the leads.

A reasonable agreement is expected between the characters observed on the unknown plant and those provided in the description of the plant it is presumed to be. Whenever there is a marked difference between these two (plant and description) a misunderstanding probably exists, and it may be necessary to repeat the procedure to clear the doubt. For successful identification, the worker must be able to determine close agreement between the features mentioned in the keys and descriptions and those on the specimens. If the facilities of a herbarium are available, the identified specimen should be compared with correctly identified specimens in the herbarium.

Selected references to Chapter 2

Benson, L. 1957. *Plant Classification*. D. C. Heath & Co.: Boston, USA.

Davis, P. H. and Heywood, V. H. 1963. *Principles of Angiosperm Taxonomy*, D. van Nostrand Co. Inc.: Princeton, New Jersey, USA.

Hutchinson, J. and Dalziel, J. M. 1954–1972. *Flora of West Tropical Africa*. Vols I, II, III. Crown Agents for Oversea Governments and Administrations: London, UK.

Keay, R. W. J., Onochie, C. F. A. and Stanfield, D. P. 1960. *Nigerian Trees*, Vol. I. Federal Government Printer: Lagos, Nigeria.

Keay, R. W. J., Onochie, C. F. A. and Stanfield, D. P. 1964. *Nigerian Trees*, Vol. II. Department of Forest Research: Ibadan, Nigeria.

Lowe, J. and Stanfield, D. P. 1974. *Flora of Nigeria: Sedges*. Ibadan University Press: Ibadan, Nigeria.

Porter, C. L. 1967. *Taxonomy of flowering plants*, (2nd ed.). W. H. Freeman & Co.: San Francisco, USA.

Sokal, R. R. and Sneath, P. H. A. 1963. *Principles of Numerical Taxonomy*. W. H. Freeman & Co.: San Francisco, USA.

Stafleu, F. A. *et al*. (eds). 1972. *International Code of Botanical Nomenclature*. International Association for Plant Taxonomy: Utrecht, Netherlands.

Stanfield, D. P., 1970. *Flora of Nigeria: Grasses*. Ibadan University Press: Ibadan, Nigeria.

SECTION II

Description of Selected Angiosperm Families

Introduction

This section is divided into five chapters. The first four chapters deal with the important families of dicotyledonous plants in West Africa while the last chapter deals with the major families of monocotyledonous plants.

It should be obvious to the reader that the partitioning of the dicots into four, even though convenient, is artificial; it is, in fact, one of many possible ways of delimiting the constituent families of dicotyledonous plants. There was considerable temptation to similarly group the monocotyledonous families. Such a grouping would have resulted in very many chapters of very different lengths. All the monocot families are therefore discussed in Chapter 7.

Within each of the chapters the families are grouped into orders. The orders are as adopted by G.L. Stebbins in his book. *Flowering Plants – Evolution above the Species Level* (1974).

The readers should pay close attention to the trends and patterns of variation among related families and lower taxa. Special attention must be paid to reductions in numbers of floral parts (floral leaves, stamens, carpels), transitions in position of the ovary, transitions from actinomorphy to zygomorphy, degrees of adnation of floral parts, variation in habit (i.e. herbaceous or woody, erect or otherwise) etc. The reader is advised to refer again to the presumed evolutionary trends discussed in Chapter 2; it is, at least, a very important aid for synthesis of observations across taxa.

The synopsis of the representative families in each group (i.e. in each chapter) should also be given close attention. The synopses should serve two critical purposes. Firstly, they should prepare the reader's mind for points to note in the collection of families. Secondly, they should serve as an aid in revision, particularly as an aid to synthesis. The families in each group are numbered a, b, c etc. and these letters are used to identify them in the synopses; this is for economy of effort.

The illustrations are conceived in such a way that the spectrum of variation in each family is spanned as much as possible, taking into account the need to cover species from various geographical and vegetational regions. Indeed, we expect the illustrations to aid plant recognition and identification even by users of the book who are not interested in the depth which professional botanical efforts demand.

Apart from half flowers of 'typical' species, floral diagrams are given; floral formulae which are designed to elucidate the nature of the perianth segments especially, are given in the legends. For example, the floral formula $K_{(5)}\ C_5\ A_{10}\ G_{\underline{(2)}}$ means the flower has five sepals (K_5), five petals (C_5), ten stamens (A_{10}) and two carpels (G_2). K_5 means the sepals are free while $K_{(5)}$ means they are united; brackets around the number of parts is interpreted the same way. A line under the number of carpels, e.g. $G_{\underline{1}}$, means that the ovary is superior while a line above the number of carpels, e.g. $G_{(\overline{2})}$, indicates an inferior ovary. Although bracts are shown in the floral diagram, they are not included in the corresponding floral formula. A line connecting the tops of two parts, e.g. $K_5\ \overline{C_5\ A_5}\ G_2$, or two parts enclosed in square brackets e.g. $K_5\ [C_5\ A_{(5)}]$, indicates adnation of the parts so connected. In plants such as kola, where the floral leaves are not differentiated into petals and sepals, the former are referred to as perianth segments; thus $P_5\ A_{10}\ G_{\underline{10}}$ means five perianth segments, ten stamens and ten carpels.

CHAPTER 3

Dicotyledonous plants with corolla free or absent, ovary superior

Representative families

3·1 MAGNOLIALES
 a Annonaceae
3·2 PIPERALES
 b Piperaceae
3·3 RANUNCULALES
 c Ranunculaceae
3·4 CARYOPHYLLALES
 d Amaranthaceae
3·5 MALVALES
 e Sterculiaceae
 f Bombacaceae
 g Tiliaceae
 h Malvaceae
3·6 URTICALES
 i Ulmaceae
 j Moraceae
 k Urticaceae
3·7 ROSALES
 l Rosaceae
3·8 FABALES
 m Mimosaceae
 n Caesalpiniaceae
 o Papilioniaceae
3·9 EUPHORBIALES
 p. Euphorbiaceae
3·10 SAPINDALES
 q Meliaceae
 r Sapindaceae
 s Anacardiaceae

Synopsis

The ovary is superior in all the families.

Petals are always present in a, f, g, h, m, n, o, q and s; petals are always absent in b, d, i, j and k; petals may be present or absent in c, e, p and r. Note that the absence of petals (apetalous condition) is believed to be more advanced than the petalous condition.

Carpels and stamens are numerous, free and spirally arranged in a and c. Stamens are numerous but free in c, l and m and some members of p; numerous and joined in f and h; few and joined in d, e and q. There are either one or two stamens in b and j thus constituting an important departure in this group of families.

Flowers are usually actinomorphic but zygomorphic in all of o and some members of l and n. Floral segments are free (not joined). Flowers are solitary in a and some members of h, but they are organised in definite inflorescences in most members of the remaining families. Readers should recall that solitary flowers are considered more primitive than flowers in inflorescence. Unisexual flowers and/or hermaphroditic (i.e. simultaneously male and female) flowers are encountered in b, d, e, i, k, l, m, n, o, p, r and s in the same inflorescence. Unisexual flowers are found in separate inflorescences in some members of j while some members of j and p are dioecious.

Leaves are usually compound in m, n and o, simple in the other families (except in a few members of f, j, l, q, r and s) and spirally or alternately arranged (except in very few members of p).

3·1 Magnoliales

3·1a Annonaceae

This family is composed of small trees, shrubs and climbers which range in distribution from forest environments to savanna. The leaves are entire and alternately arranged; they are exstipulate usually with prominent nerves and the upper surface is a deeper green than under surface.

The flowers are bisexual, radially symmetrical and with superior ovaries. The perianth is trimerous (sepals 3, petals 3 + 3) and free. Stamens and carpels are numerous and they are spirally arranged on a prominent receptacle. Carpels are joined or free.

The fruits may be composed of free carpels (one- to many-seeded follicles or berries); in addition fruits may either arise from many joined carpels forming a mass with the receptacle or the carpels may be joined together to form a berry.

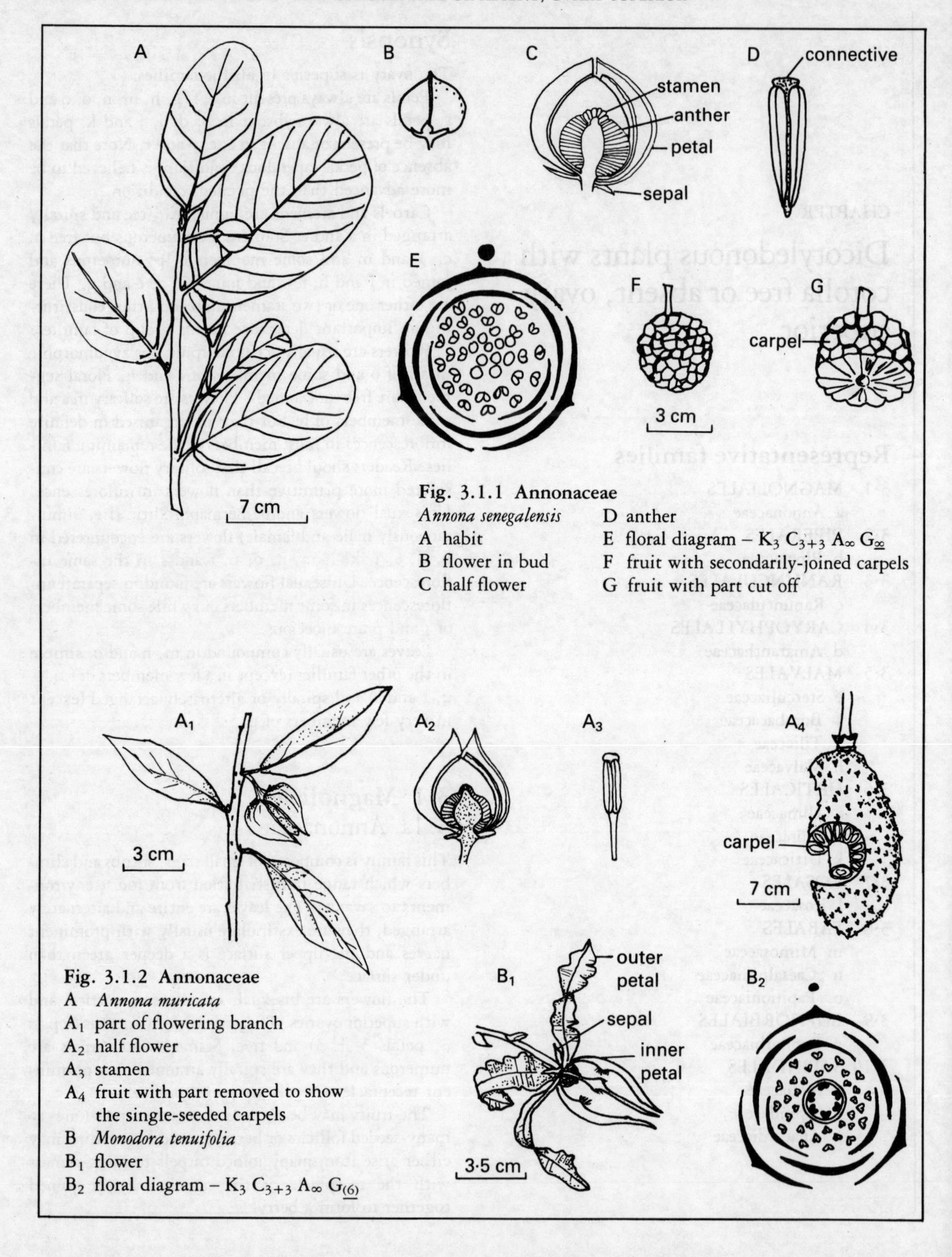

Fig. 3.1.1 Annonaceae
Annona senegalensis
A habit
B flower in bud
C half flower
D anther
E floral diagram – $K_3\ C_{3+3}\ A_\infty\ G_{\underline{\infty}}$
F fruit with secondarily-joined carpels
G fruit with part cut off

Fig. 3.1.2 Annonaceae
A *Annona muricata*
A_1 part of flowering branch
A_2 half flower
A_3 stamen
A_4 fruit with part removed to show the single-seeded carpels
B *Monodora tenuifolia*
B_1 flower
B_2 floral diagram – $K_3\ C_{3+3}\ A_\infty\ G_{\underline{(6)}}$

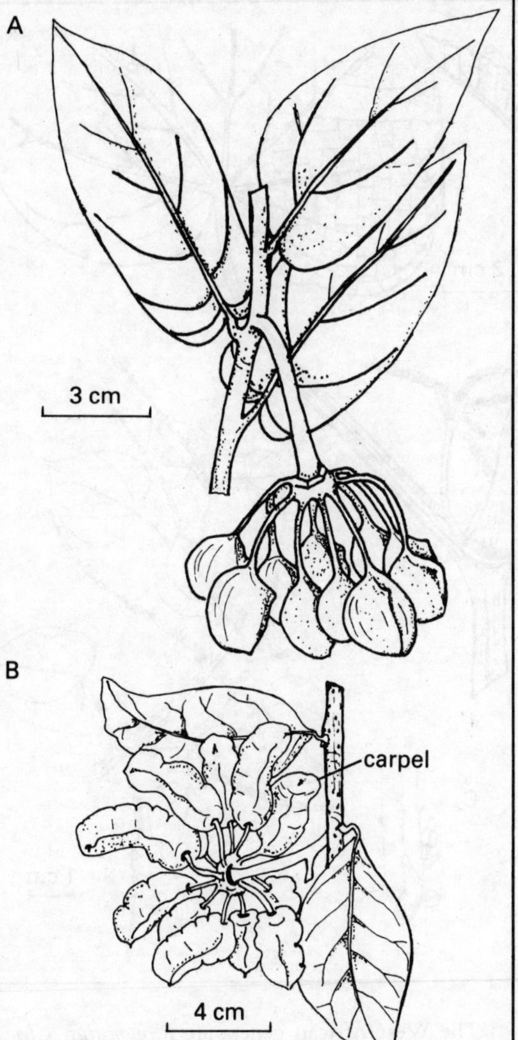

Fig. 3.1.3 Annonaceae
A *Uvaria afzelii* – fruiting branch
B *Uvaria chamae* – fruiting branch

Members of the family include many species of importance in indigenous medicine. *Annona senegalensis* produces yellow edible fruits and *A. muricata* is the well-known Soursop. *Dennettia tripetala* is another edible species of forest zones.

Common species include *Annona senegalensis, A. muricata, Monodora tenuifolia* and *Uvaria chamae*. *Annona senegalensis* (Fig. 3.1.1) is a common savanna shrub with alternate leaves, characterised by prominent veins; its flowers are typical with thick sepals and petals. *A. muricata* (Fig. 3.1.2A) is a common cultivated tree of medium size; its fruits and flowers are typical. The cottony pulp of the fruit is used as a vegetable.

Monodora tenuifolia (Fig. 3.1.2B) may also grow into a medium-sized forest tree with very attractive flowers that are produced when the plant sheds its leaves. The flower stalk has bracts while the flower has three sepals. The outer petals are distinctly different from the internal petals and the fruit is a berry formed from several joined carpels.

The genus *Uvaria* is exemplified by *U. chamae* and *U. afzelii* (Fig. 3.1.3), both of which are regrowth shrubs in forest regions; the general form of the fruit is the same in both except that the carpels of *U. chamae* are many-seeded.

3·2 Piperales

3·2b Piperaceae

Members of this family are herbs or shrubs which have erect, scandent or climbing stems. They are sometimes epiphytic and are found mostly in forest environments. The leaves are alternate, rarely opposite. Stipules are absent or adnate to the distinct petioles.

Piper guineense is used for medicinal purposes. Other species of *Piper* are also cultivated for medicinal and other purposes.

The flowers are produced in terminal or axillary spikes which may be umbellate. Perianth segments are reduced or absent in the flowers which are unisexual or hermaphroditic. There are two to six stamens in each flower; the stamens are free. The ovary is superior, unilocular and one-seeded. The stigmata are short and there may be one to five of them.

The fruit is a berry with a succulent or dry fruit coat; the seeds are small.

Piper guineense is a common plant in dry and wet forest environments. The clasping roots produced at the nodes give support to the weak stem. The pattern of the veins is distinct (Fig. 3.2.1A). *P. umbellatum* is a scandent, herbaceous or (rarely) woody plant of wet forest environments with characteristic stipulate leaves, a ridged stem and an umbellate inflorescence (Fig. 3.2.1B).

Peperomia pellucida is a common plant of wet and shaded environments especially in the forested zones. The leaf-opposed spicate inflorescences, the filmy stems and leaves and the branching patterns of the stem are diagnostic (Fig. 3.2.1C).

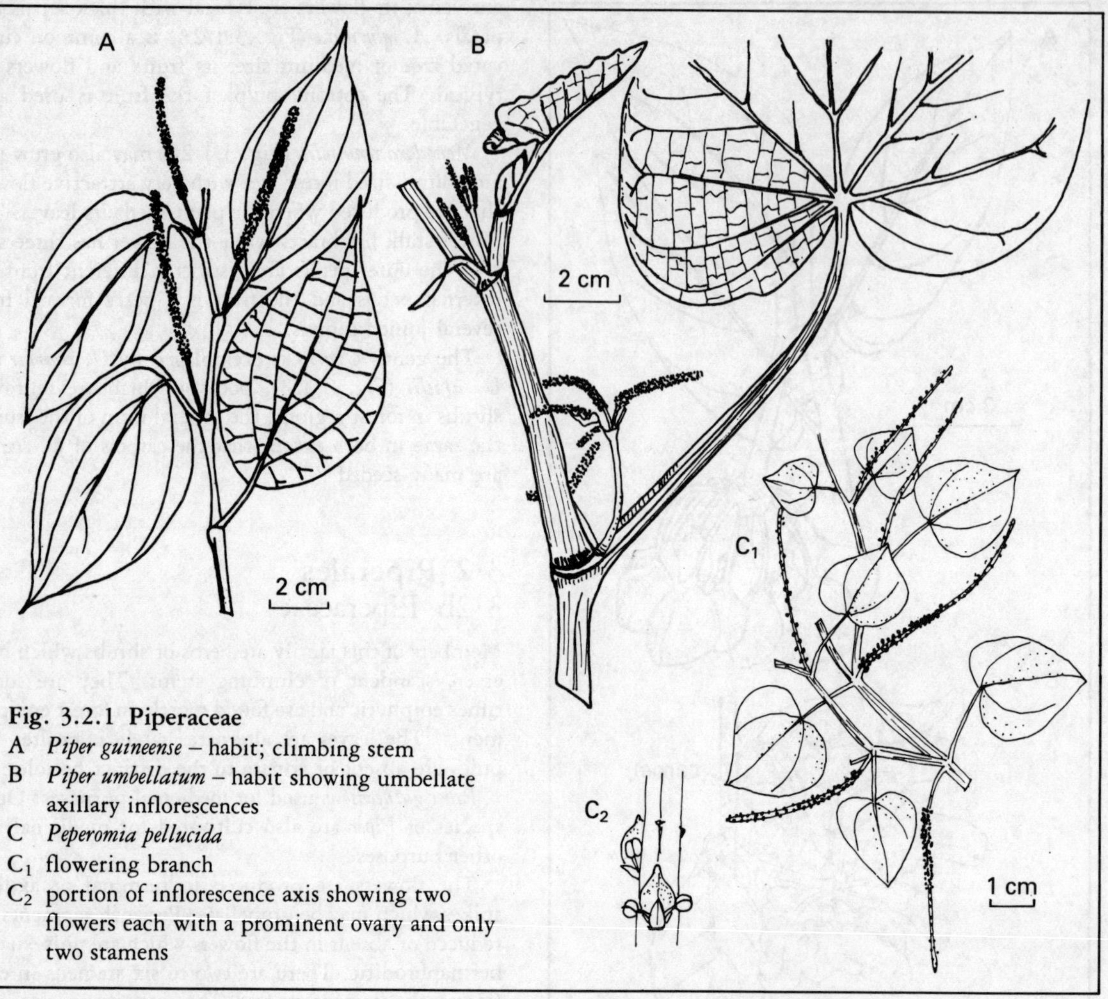

Fig. 3.2.1 Piperaceae
A *Piper guineense* – habit; climbing stem
B *Piper umbellatum* – habit showing umbellate axillary inflorescences
C *Peperomia pellucida*
C_1 flowering branch
C_2 portion of inflorescence axis showing two flowers each with a prominent ovary and only two stamens

3·3 Ranunculales

3·3c Ranunculaceae

These are mostly perennial herbs with alternate, exstipulate leaves; or (rarely) shrubs or climbers with opposite exstipulate leaves. Representatives of the family are distributed widely in environments ranging from forest regions through derived savanna to dry savanna. Stem and leaves have short, simple hairs.

Flowers are bisexual, and radially symmetrical with superior ovaries. Petals are present or absent. The perianth has four or five segments. Stamens and carpels are numerous; the carpels are free or (rarely) partially connate with persistent long styles having conspicuous hairs.

The fruit is a group of follicles or of dry achenes with elongate styles which may be plumose.

The West African genera are *Ranunculus, Clematis, Thalictrum, Clematopsis* and *Delphinium.* Species of *Ranunculus* and *Delphinium* are mostly restricted to cool highland habitats. Common species include *Clematis grandiflora* (a forest climber), *C. hirsuta* (a widely distributed climber in the derived savanna, Fig. 3.3.1) and *Clematopsis scabiosifolia* (a dry savanna species with rootstock producing shoots after fires, Fig. 3.3.2).

3·4 Caryophyllales

3·4d Amaranthaceae

Members of this family are annual or perennial, erect or prostrate herbs; seldom climbers or shrubs. They have opposite or alternate, exstipulate leaves. Many agricultural and other weeds are represented in this family. However, some members of this family, in

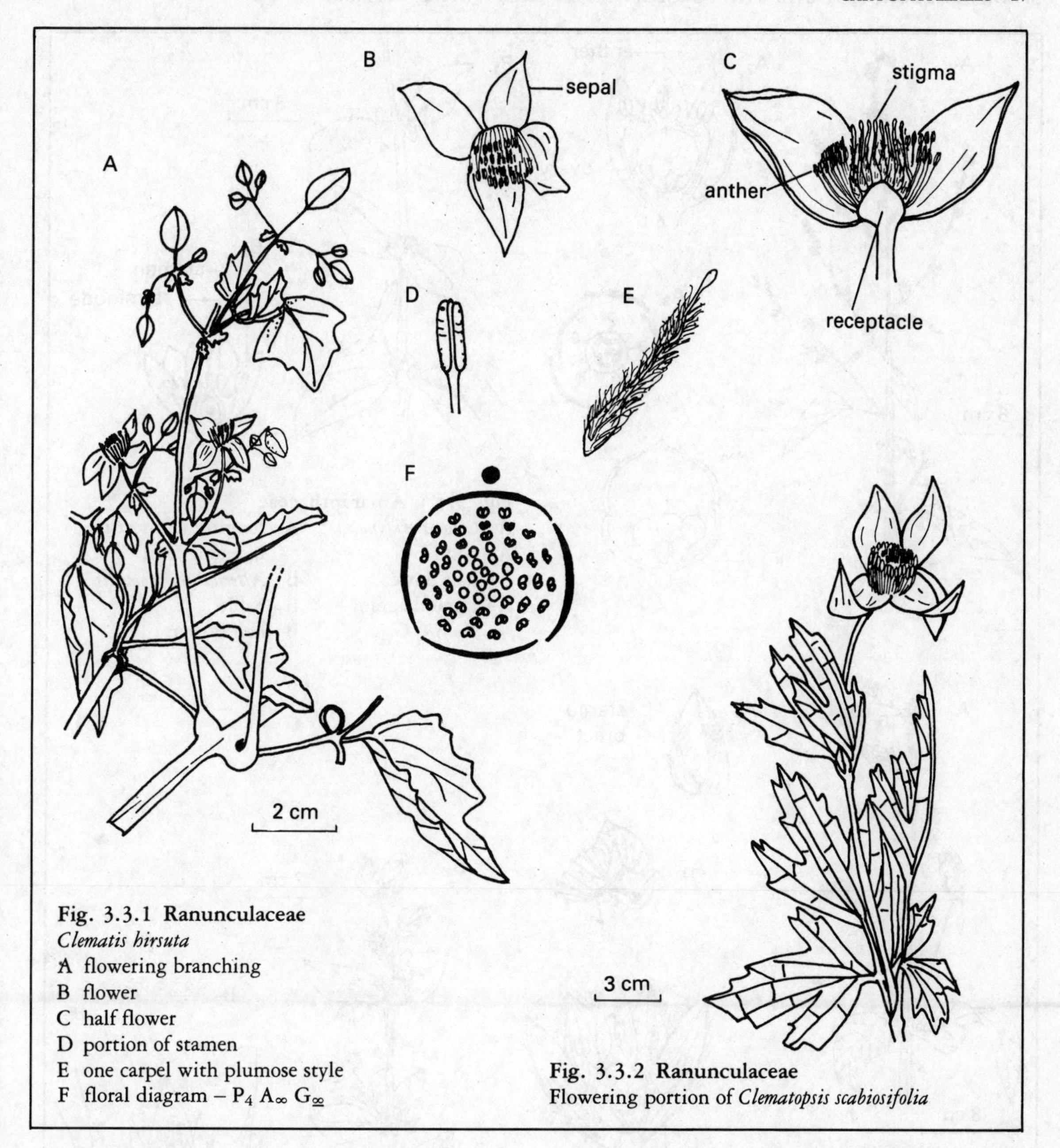

Fig. 3.3.1 Ranunculaceae
Clematis hirsuta
A flowering branching
B flower
C half flower
D portion of stamen
E one carpel with plumose style
F floral diagram – $P_4\ A_\infty\ G_{\underline{\infty}}$

Fig. 3.3.2 Ranunculaceae
Flowering portion of *Clematopsis scabiosifolia*

particular *Celosia* and *Amaranthus*, are important vegetable species in Nigeria. Some species of *Celosia* are also cultivated as ornamentals.

The flowers are actinomorphic and usually hermaphroditic; they are small and apetalous in spikes, globose heads or racemes. The flowers are often borne in the axils of bracts or bracteoles which may be brightly coloured and/or hooked. There are three to five sepals which are imbricate, often dry and membranous. There are usually five stamens opposite the sepals; they are often joined basally into a short tube; staminodes may be present between the fertile stamens. The ovary is superior; it has a short style with a capitate or divided stigma. The ovary is one-celled with a solitary ovule or several ovules on basal funicles (Fig. 3.4.1).

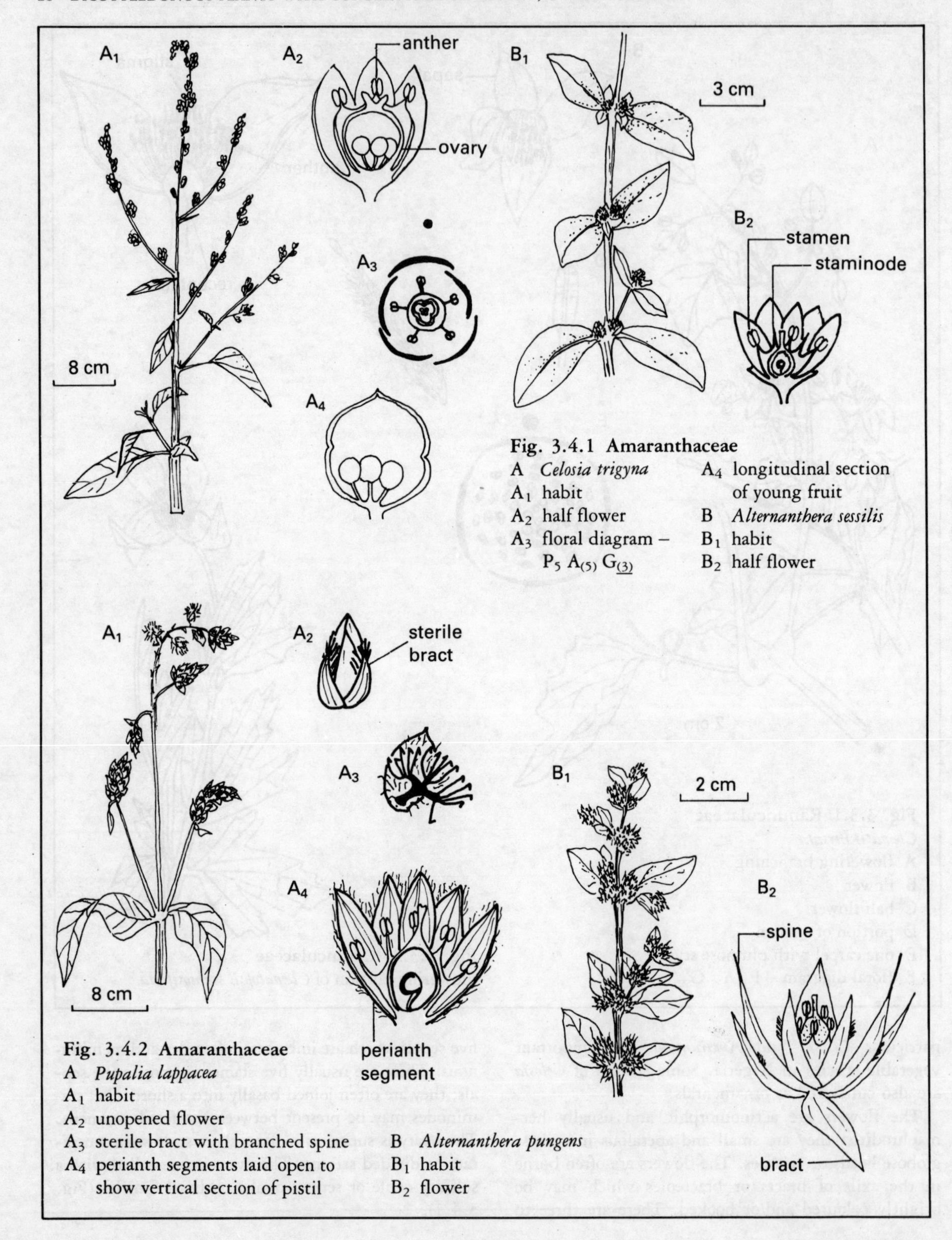

Fig. 3.4.1 Amaranthaceae

A *Celosia trigyna*
A_1 habit
A_2 half flower
A_3 floral diagram – P_5 $A_{(5)}$ $G_{\underline{(3)}}$
A_4 longitudinal section of young fruit
B *Alternanthera sessilis*
B_1 habit
B_2 half flower

Fig. 3.4.2 Amaranthaceae

A *Pupalia lappacea*
A_1 habit
A_2 unopened flower
A_3 sterile bract with branched spine
A_4 perianth segments laid open to show vertical section of pistil
B *Alternanthera pungens*
B_1 habit
B_2 flower

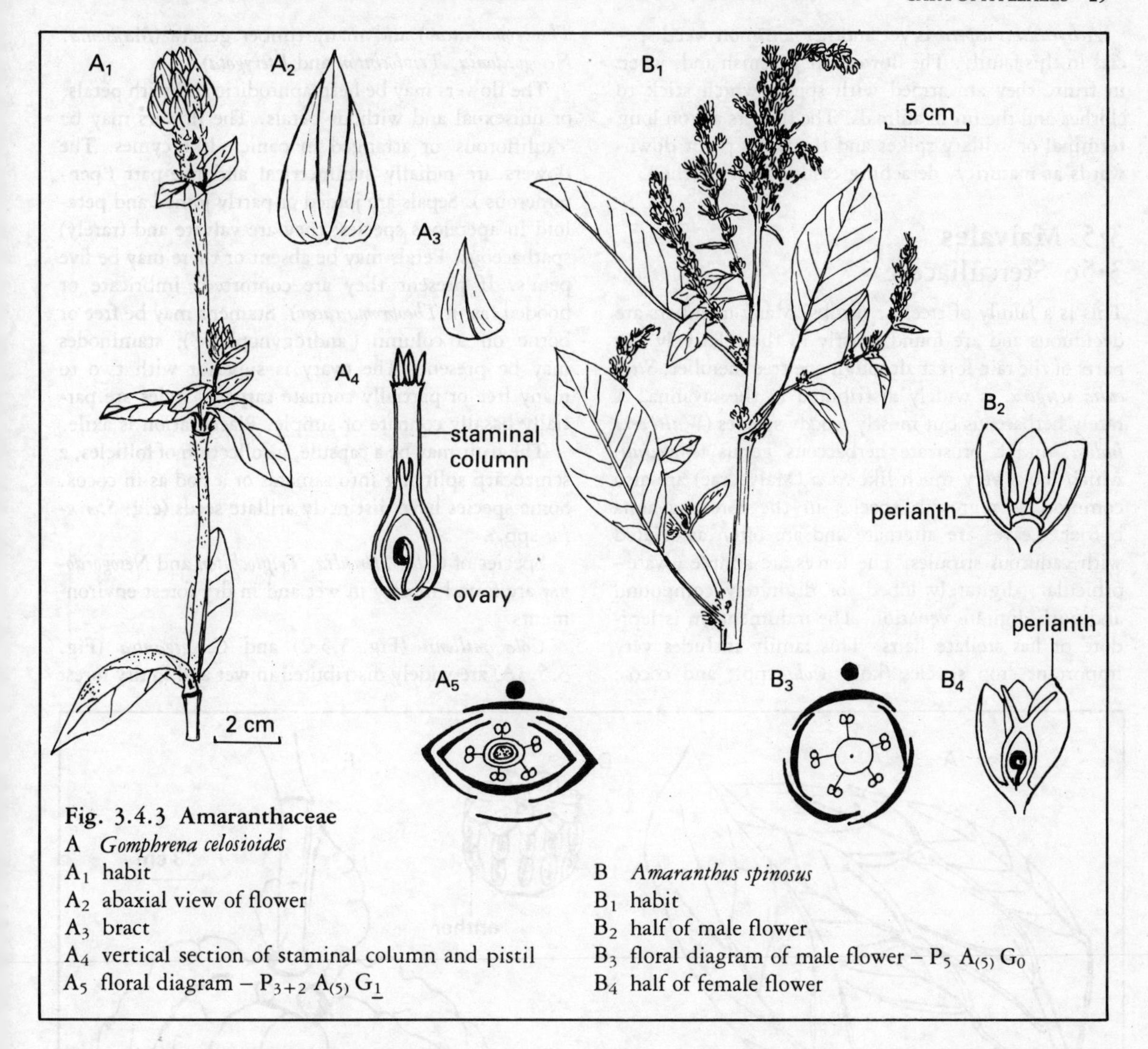

Fig. 3.4.3 Amaranthaceae

A *Gomphrena celosioides*
A_1 habit
A_2 abaxial view of flower
A_3 bract
A_4 vertical section of staminal column and pistil
A_5 floral diagram – $P_{3+2}\ A_{(5)}\ G_{\underline{1}}$

B *Amaranthus spinosus*
B_1 habit
B_2 half of male flower
B_3 floral diagram of male flower – $P_5\ A_{(5)}\ G_0$
B_4 half of female flower

The fruit is dehiscent (splitting by a lid) or indehiscent. The seeds are ellipsoid, globose or compressed and smooth; the embryo is semi-circular and endospermous.

Celosia trigyna is a common weed in open forest and savanna environments. Its habit varies from erect to straggling or decumbent. The flowers are silvery white with pink stamens. The plant is usually harvested for use as a green vegetable. The details of floral structure of *C. trigyna* are shown in Fig. 3.4.1A.

The genus *Alternanthera* is represented by four species in West Africa. It is a herbaceous genus which occurs in waste areas. It is usually prostrate with silvery or golden axillary clusters of flowers which have pointed or spiny bracts. Two species, *A. sessilis* and *A. pungens*, are shown in Fig. 3.4.1B. and Fig. 3.4.2B.

Pupalia lappacea is also a weed of waste places. It is herbaceous with flowers in terminal spikes. One of the sterile bracts of a flower develops a characteristic branched spine (Fig. 3.4.2A).

Gomphrena celosioides is a common weed on lawns and open places. The leaves have a conspicuous, grey indumentum. The inflorescence is a globose head with silvery white flowers, each of the latter being subtended by a persistent abaxial bract (Fig. 3.4.3A).

Amaranthus is an important genus of weeds and edible green vegetables. *A. spinosus* is a common weed widespread in forest and savanna. It is armed with spines and the flowers are unisexual (Fig. 3.4.3B). *A. hybridus* is a commonly cultivated vegetable species.

Achyranthes aspera is yet another common weed species in this family. The flowers are greenish and, when in fruit, they are armed with spines which stick to clothes and the fur of animals. The flowers are on long terminal or axillary spikes and the fruits point downwards at maturity, detaching easily when touched.

3·5 Malvales

3·5e Sterculiaceae

This is a family of trees or shrubs. Many members are deciduous and are found mostly in the relatively dry parts of the rain forest although one tree member, *Sterculia setigera*, is widely distributed in the savanna. A rarely herbaceous but mostly woody species (*Waltheria indica*) and a prostrate herbaceous genus (*Melochia*) which looks very much like *Sida* (Malvaceae) are also common as regrowth species in the forest-savanna border. Leaves are alternate and are often associated with caducous stipules. The leaves are simple (ovate–orbicular, digitately lobed) or digitately compound and with digitate venation. The indumentum is lepidote or has stellate hairs. This family includes very important crop species (kola: *Cola* spp.; and cocoa: *Theobroma cacao*) and many timber genera (*Mansonia, Nesogordonia, Triplochiton* and *Pterygota*).

The flowers may be hermaphroditic and with petals, or unisexual and without petals. The flowers may be 'cauliflorous' or arranged in panicle-like cymes. The flowers are radially symmetrical and five-part ('pentamerous'). Sepals are joined or partly joined and petaloid in apetalous species; they are valvate and (rarely) spathaceous. Petals may be absent or there may be five petals. If present they are contorted, imbricate or hooded (as in *Theobroma cacao*). Stamens may be free or borne on a column ('androgynophore'); staminodes may be present. The ovary is superior with two to many free or partially connate carpels. Styles are partially basally connate or simple. Placentation is axile.

The fruit may be a capsule, a collection of follicles, a schizocarp splitting into samaras or a pod as in cocoa. Some species have distinctly arillate seeds (e.g. *Sterculia* spp.).

Species of *Cola, Sterculia, Triplochiton* and *Nesogordonia* are found mostly in wet and in dry forest environments.

Cola millenii (Fig. 3.5.2) and *C. gigantea* (Fig. 3.5.3A) are widely distributed in wet and in dry forest

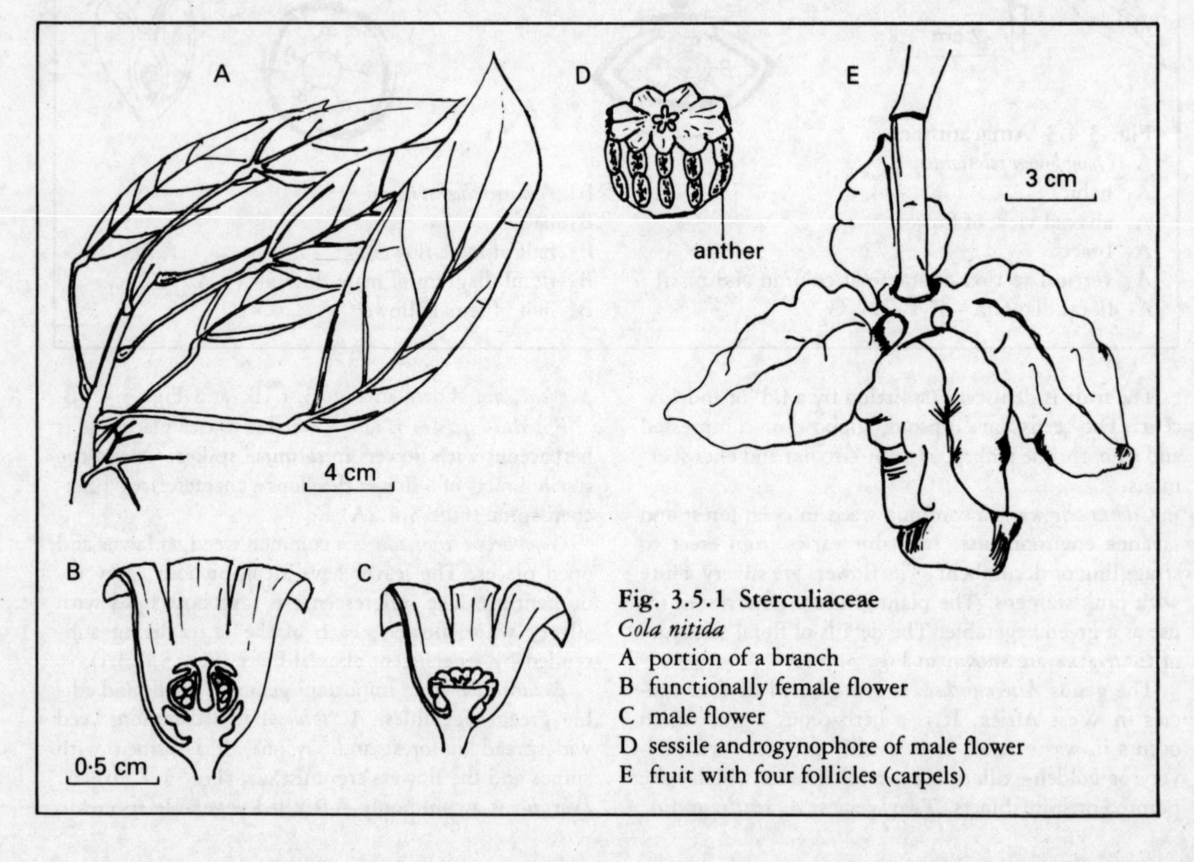

Fig. 3.5.1 Sterculiaceae
Cola nitida
A portion of a branch
B functionally female flower
C male flower
D sessile androgynophore of male flower
E fruit with four follicles (carpels)

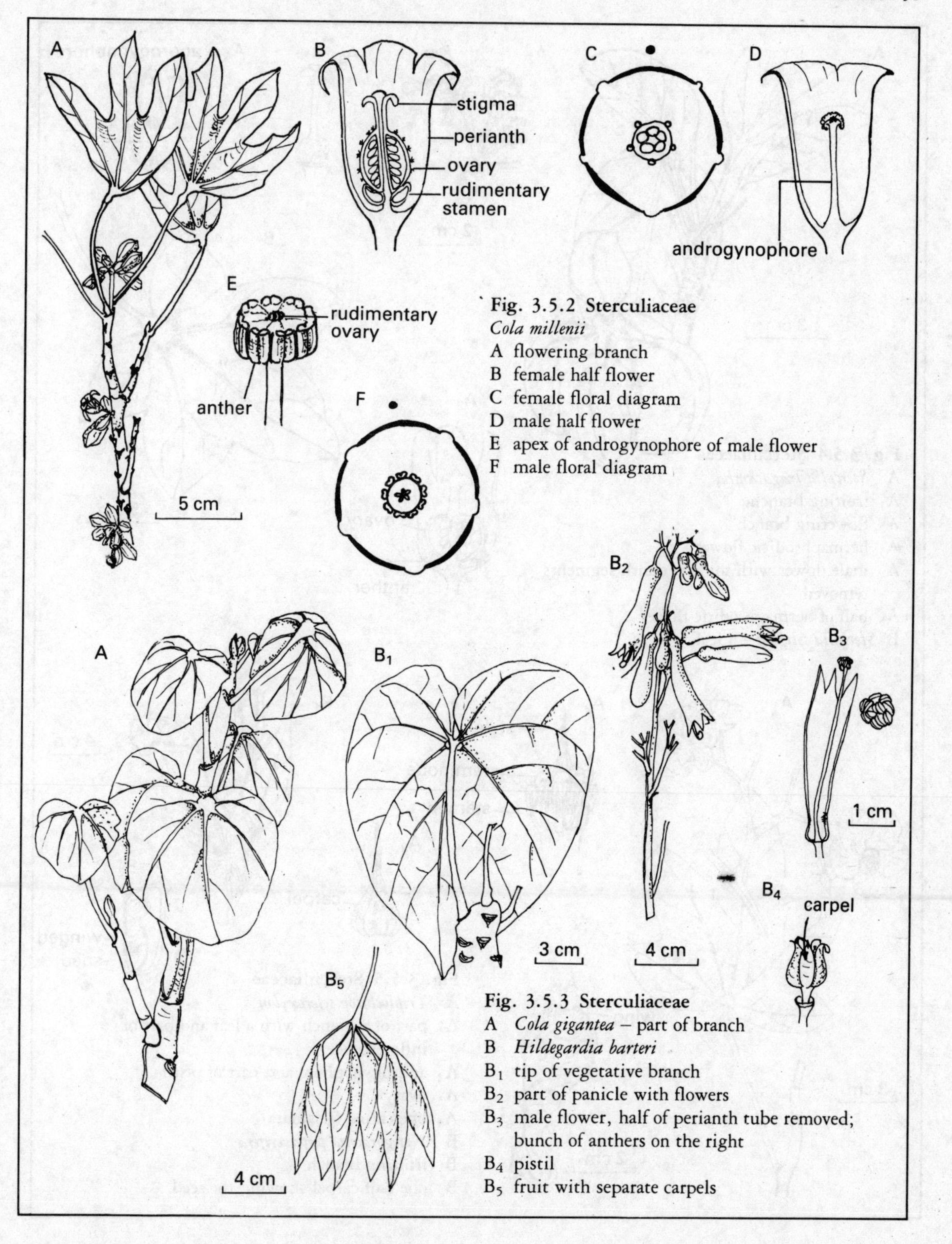

Fig. 3.5.2 Sterculiaceae
Cola millenii
A flowering branch
B female half flower
C female floral diagram
D male half flower
E apex of androgynophore of male flower
F male floral diagram

Fig. 3.5.3 Sterculiaceae
A *Cola gigantea* – part of branch
B *Hildegardia barteri*
B_1 tip of vegetative branch
B_2 part of panicle with flowers
B_3 male flower, half of perianth tube removed; bunch of anthers on the right
B_4 pistil
B_5 fruit with separate carpels

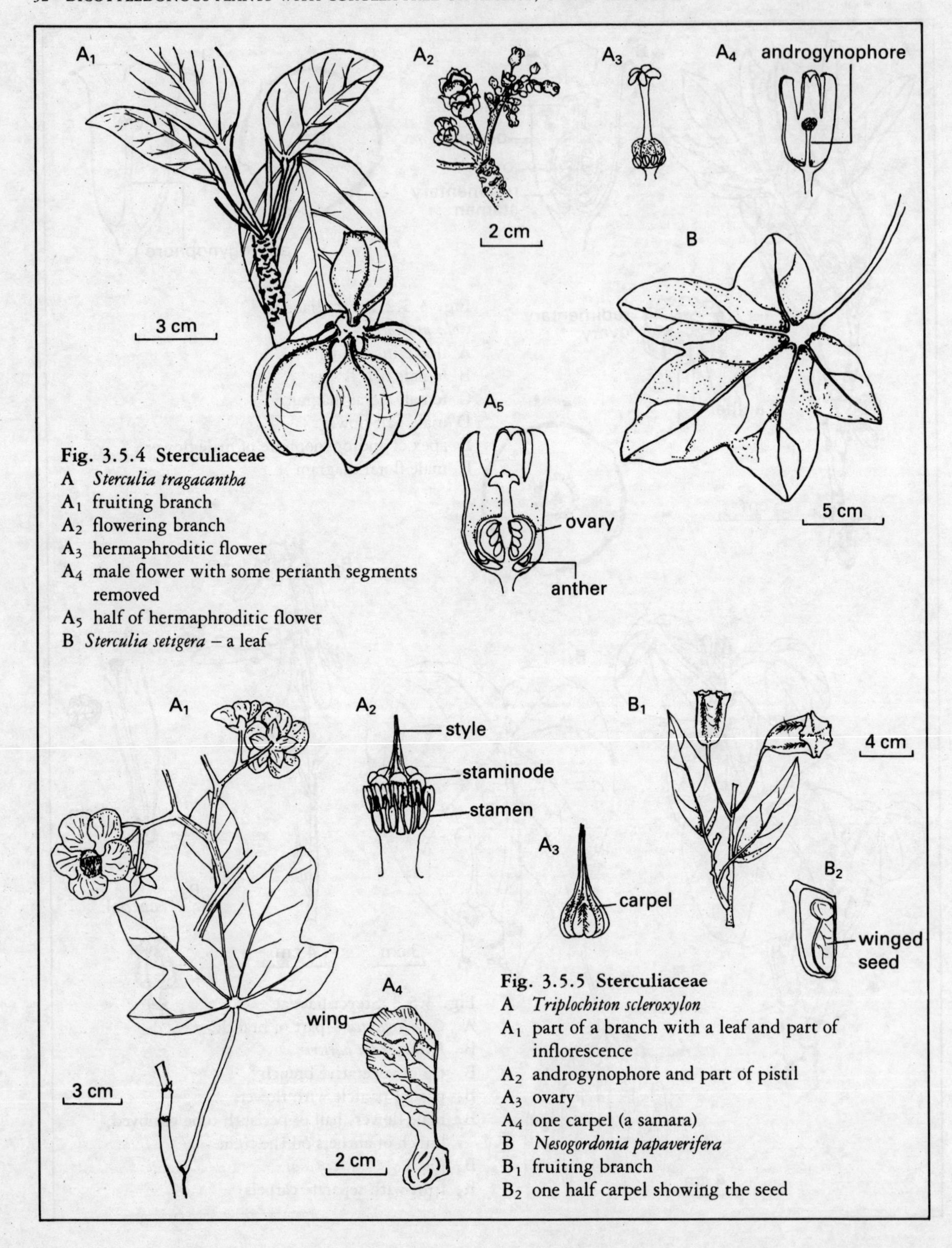

Fig. 3.5.4 Sterculiaceae

A *Sterculia tragacantha*

A_1 fruiting branch

A_2 flowering branch

A_3 hermaphroditic flower

A_4 male flower with some perianth segments removed

A_5 half of hermaphroditic flower

B *Sterculia setigera* – a leaf

Fig. 3.5.5 Sterculiaceae

A *Triplochiton scleroxylon*

A_1 part of a branch with a leaf and part of inflorescence

A_2 androgynophore and part of pistil

A_3 ovary

A_4 one carpel (a samara)

B *Nesogordonia papaverifera*

B_1 fruiting branch

B_2 one half carpel showing the seed

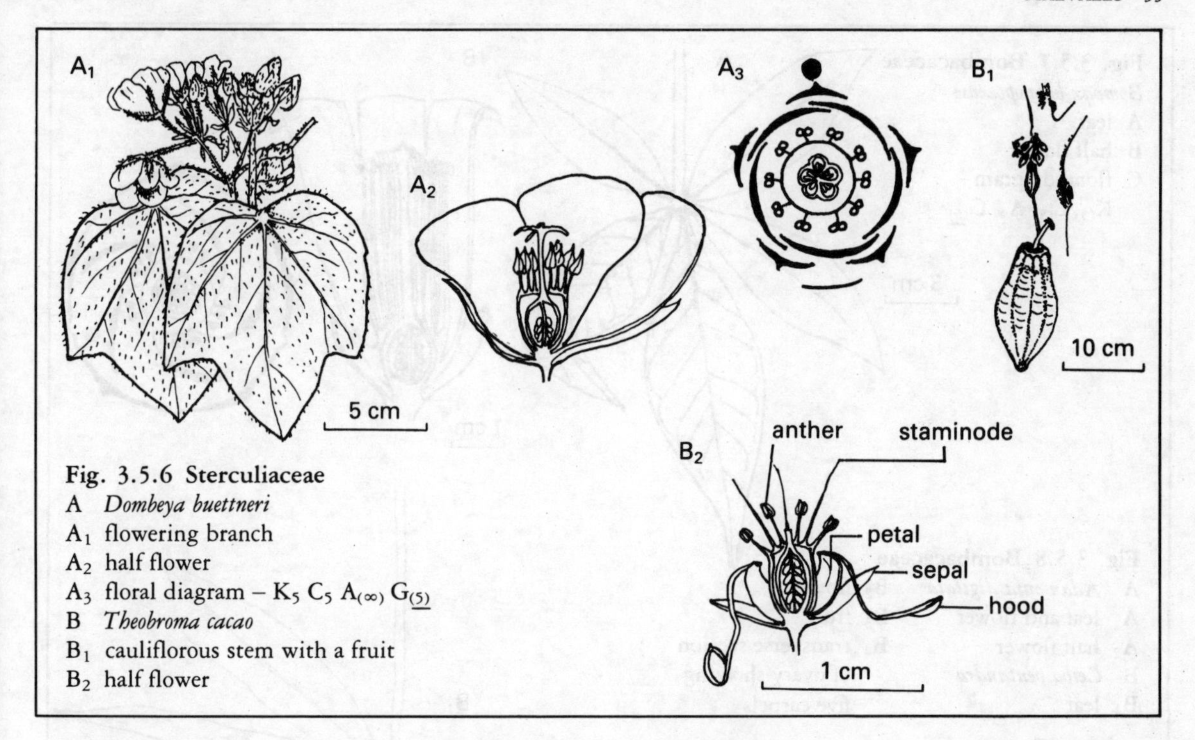

Fig. 3.5.6 Sterculiaceae
A *Dombeya buettneri*
A_1 flowering branch
A_2 half flower
A_3 floral diagram – $K_5\ C_5\ A_{(\infty)}\ G_{\underline{(5)}}$
B *Theobroma cacao*
B_1 cauliflorous stem with a fruit
B_2 half flower

environments. Mature fruits of *C. gigantea* are rust brown while those of *C. millenii* are reddish orange.

Sterculia tragacantha is a medium-sized three of dry forest and forest-savanna mosaic with broadly ovate leaves. Its velvety kola-like fruits are olive green when immature and red when ripe (Fig. 3.5.4A). *S. setigera* (Fig. 3.5.4B) is a savanna tree with palmately-lobed leaves and kola-like fruits.

Triplochiton scleroxylon is a deciduous timber-producing tree found in habitats ranging from wet forest to the forest–savanna border (Fig. 3.5.5A). The leaves are digitately lobed and the fruit splits into samaras. Its flowers are hermaphroditic and petals are present. The fruit of *Nesogordonia papaverifera* is a distinctive capsule splitting to release samaras (Fig. 3.5.5B).

Hildegardia barteri is a medium-sized tree of drier forest regions especially forest regions of higher elevation. The bark is green and soft-looking; the red unisexual flowers (Fig. 3.5.3B) are produced in the dry season before the leaves.

Dombeya buettneri is a regrowth shrub of dry forest and forest–savanna mosaic with very hairy digitately-lobed leaves and creamy hermaphroditic flowers (Fig. 3.5.6A). The bark is tough and is often used as rope. Two other species of *Dombeya* occur in the savanna.

As noted above, *Melochia* is a herbaceous genus of Sterculiaceae and it is often a prostrate herb of waste places.

Waltheria indica is a herb or small shrub which is an important regrowth species of forest–savanna mosaic; its axillary flowers have one-celled ovaries each with two seeds.

3·5f Bombacaceae

Bombacaceae is a small family of usually large trees represented in forest and savanna. They have alternate, digitately-divided, simple or compound leaves (Fig. 3.5.7) associated with deciduous stipules. One species *Ceiba pentandra* produces kapok (the white cottony material used for stuffing mattresses). In the savanna its young leaves are dried and powdered for food.

The flowers are usually large, showy, hermaphroditic and actinomorphic. The calyx is of joined sepals and is usually valvate in bud. Petals are large, usually elongate and joined basally to each other and to the base of the stamens. The corolla is made up of five imbricate segments. Stamens are numerous and free or basally united with the petals. The anthers are reniform, linear and with smooth pollen. The ovary is superior and with two to five united carpels. The style may be simple, lobed or capitate. Each ovary cell usually has many ovules in axile placentation.

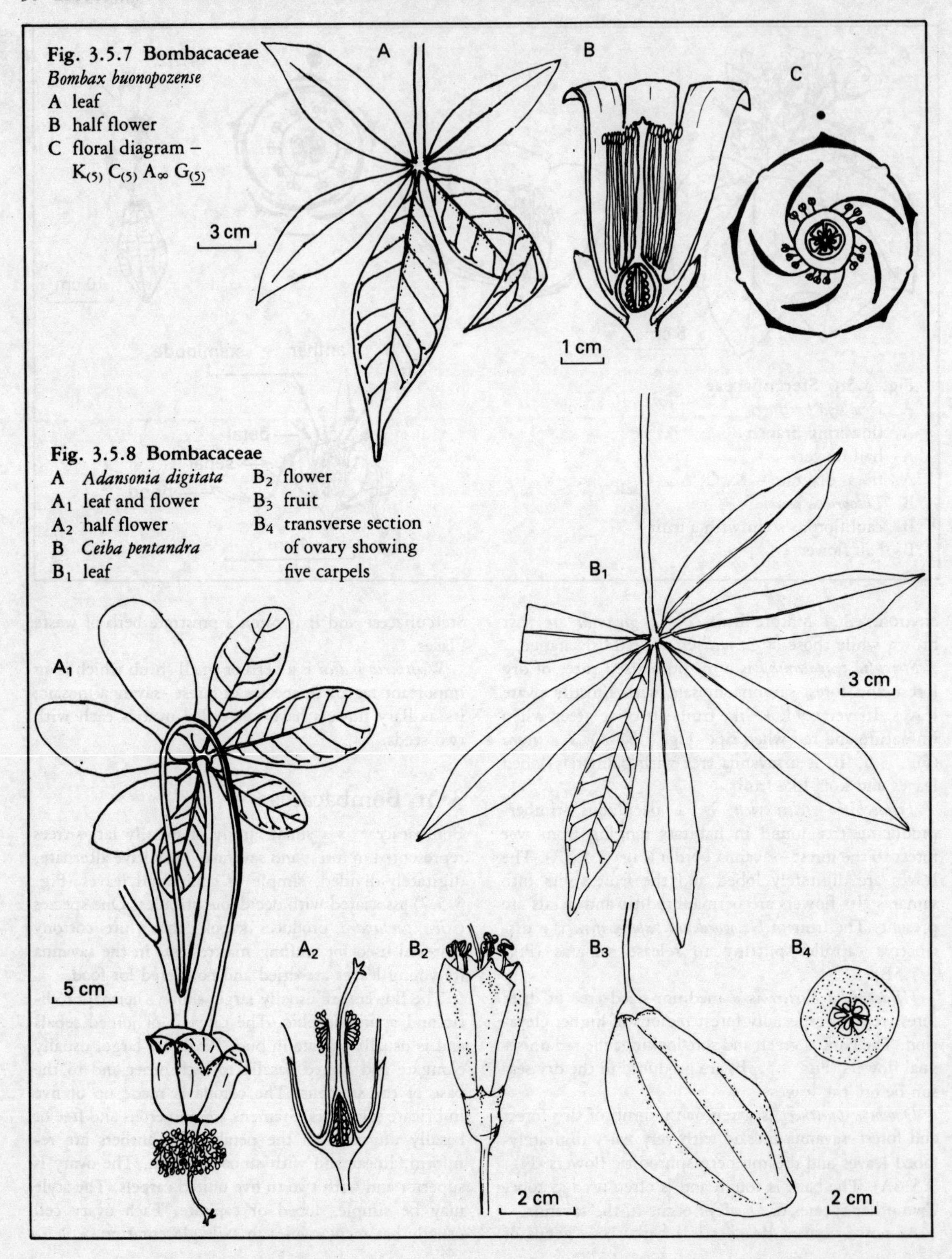

Fig. 3.5.7 Bombacaceae
Bombax buonopozense
A leaf
B half flower
C floral diagram – $K_{(5)}\ C_{(5)}\ A_{\infty}\ G_{\underline{(5)}}$

Fig. 3.5.8 Bombacaceae
A *Adansonia digitata*
A_1 leaf and flower
A_2 half flower
B *Ceiba pentandra*
B_1 leaf
B_2 flower
B_3 fruit
B_4 transverse section of ovary showing five carpels

Fruits are dehiscent (loculicidal) or indehiscent, with the seeds usually embedded in cottony hairs originating from the walls of the fruit.

Three species of *Bombax* are known in West Africa. *B. brevicuspe* is a tall deciduous rain forest species with white flowers. *B. buonopozense* flowers in the dry season after shedding its leaves; it is found from the drier forest areas to the northern boundaries of Guinea savanna. The flowers are bright red (Fig. 3.5.7). *B. costatum* is a savanna species with smaller, usually brick-red, flowers also produced after the leaves have been shed.

Adansonia digitata (baobab) is widespread in the savanna. It has characteristic swollen trunks which store water. The white flowers are borne solitarily on pendulous stalks (Fig. 3.5.8A) and the fruits are large velvety structures (up to about 30 cm long) which hang down from long stalks. Local drinks are made from the pulp of the fruits in some parts of Nigeria.

Ceiba pentandra is the white silk cotton tree found from forest to forest–savanna mosaic environments. The tree is deciduous producing flowers after leaves are shed. The flowers are about 4 cm long with brown velvety petals (Fig. 3.5.8B). They have 15 stamens which are joined into five bunches of three stamens each. Fruits of *C. pentandra* mature and split to release black seeds with white or brown floss.

3·5g Tiliaceae

This is a family composed of trees, shrubs and herbs with alternate, simple leaves. The leaves usually have three digitate nerves arising from the base of the blade while pinnate nerves arise from the midrib. Stipules may be present or absent. The family is represented by weeds and regrowth species both in forest and savanna environments. *Corchorus olitorius* is a very important vegetable species in southern Nigeria.

The flowers are yellow or white, radially symmetrical, hermaphroditic, tetramerous or pentamerous and have numerous stamens which are free or connate at the base. The inflorescence is a cyme. The joined carpels are two to many in number and may be pivoted on a column. The ovary is superior and with two to many ovules per cell in axile placentation.

The fruit may be a capsule, it may be baccate or drupe-like and it may be beaked.

Grewia is a large genus represented in the forest and savanna areas by small trees or sometimes straggling shrubs. The savanna species include *G. mollis* and *G. villosa* with yellow and reddish brown flowers respectively; the fruits of *G. villosa* are warty and densely hairy and the fruits of *G. mollis* are black when ripe. *G. pubescens* and *G. carpinifolia* are straggling regrowth shrubs in forest and derived savanna environments; the

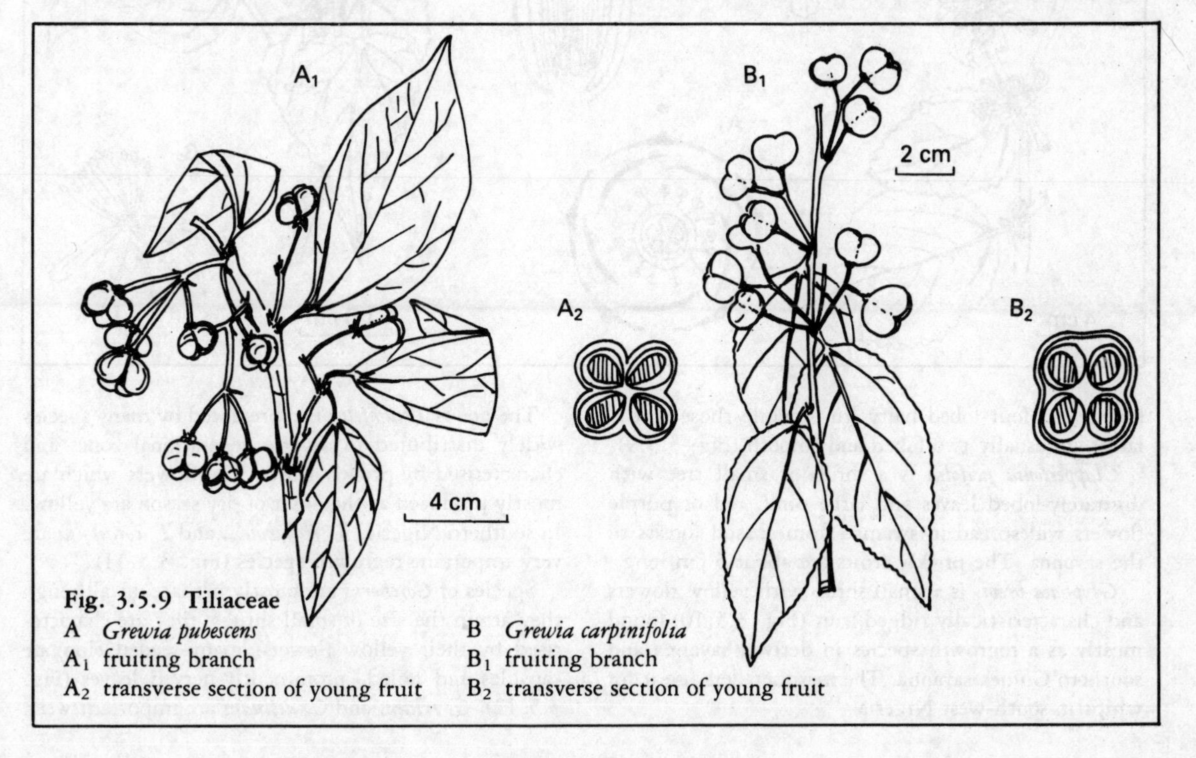

Fig. 3.5.9 Tiliaceae
A *Grewia pubescens*
A_1 fruiting branch
A_2 transverse section of young fruit
B *Grewia carpinifolia*
B_1 fruiting branch
B_2 transverse section of young fruit

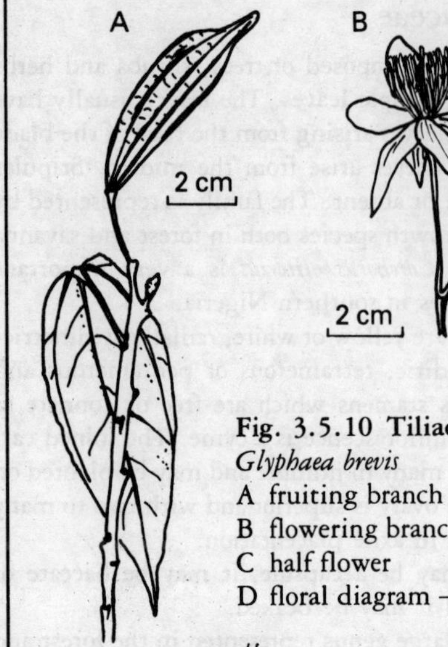

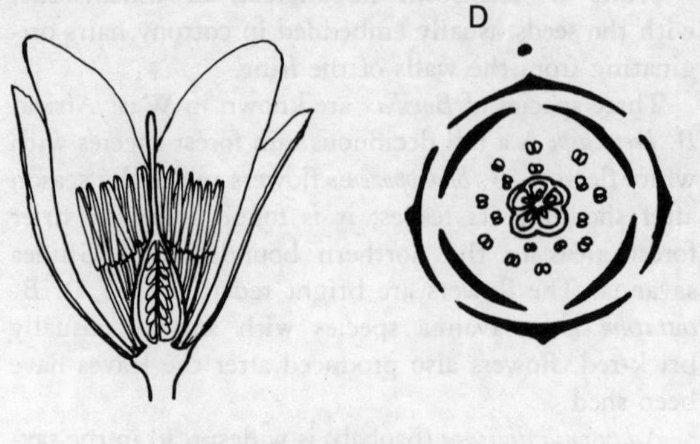

Fig. 3.5.10 Tiliaceae
Glyphaea brevis
A fruiting branch
B flowering branch
C half flower
D floral diagram – $K_4\ C_4\ A_\infty\ \underline{G_{(5)}}$

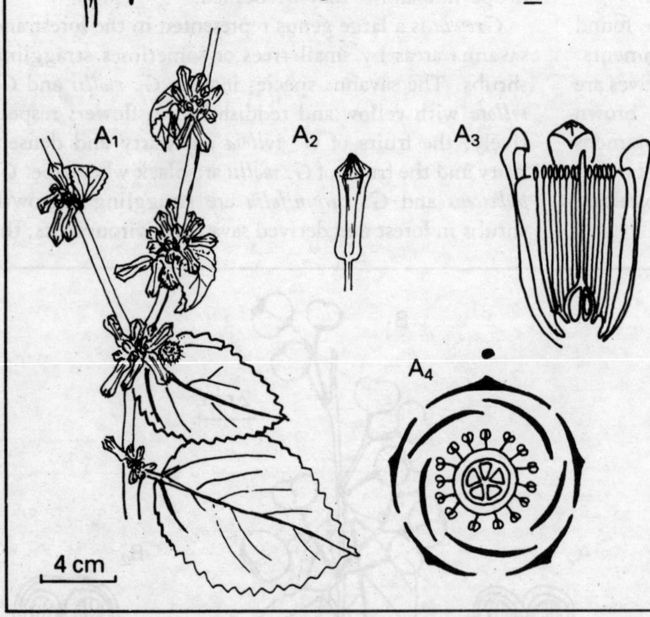

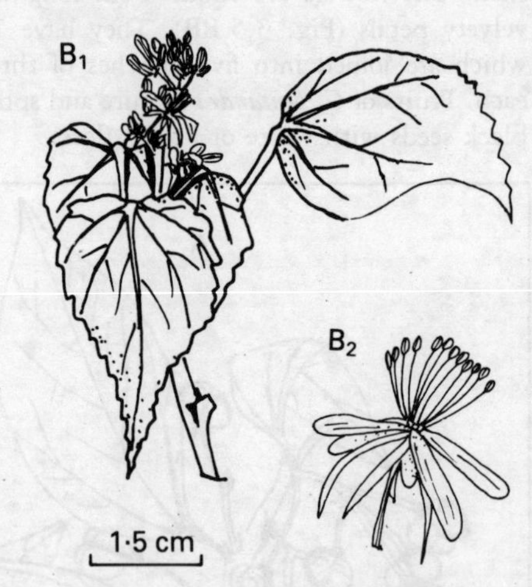

Fig. 3.5.11 Tiliaceae
A *Triumfetta rhomboidea*
A_1 flowering branch
A_2 flower in bud
A_3 half flower
A_4 floral diagram – $K_5\ C_5\ A_{(\infty)}\ \underline{G_{(5)}}$
B *Triumfetta cordifolia*
B_1 flowering branch
B_2 flower

former has four-lobed hairy fruits while those of the latter are usually two-lobed and smooth (Fig. 3.5.9).

Clappertonia ficifolia is a shrub or small tree with digitately-lobed leaves and large pink, red or purple flowers widespread in swamps from coastal forests to the savanna. The prickly fruits are about 5 cm long.

Glyphaea brevis is a small shrub with yellow flowers and characteristically ridged fruit (Fig. 3.5.10) found mostly as a regrowth species in derived savanna and southern Guinea savanna. The masqueraders use it for whips in south-west Nigeria.

The genus *Triumfetta* is represented by many species widely distributed in diverse vegetational zones and characterised by prickly fruits. The flowers which are mostly produced at the onset of dry season are yellow. In southern Nigeria. *T. rhomboidea* and *T. cordifolia* are very important regrowth species (Fig. 3.5.11).

Species of *Corchorus* are mostly herbaceous although they attain the size of small shrubs; they are characterised by their yellow flowers, many-seeded elongate capsules and tailed, prominently-nerved leaves (Fig. 3.5.12). *C. tridens* and *C. aestuans* are important weed

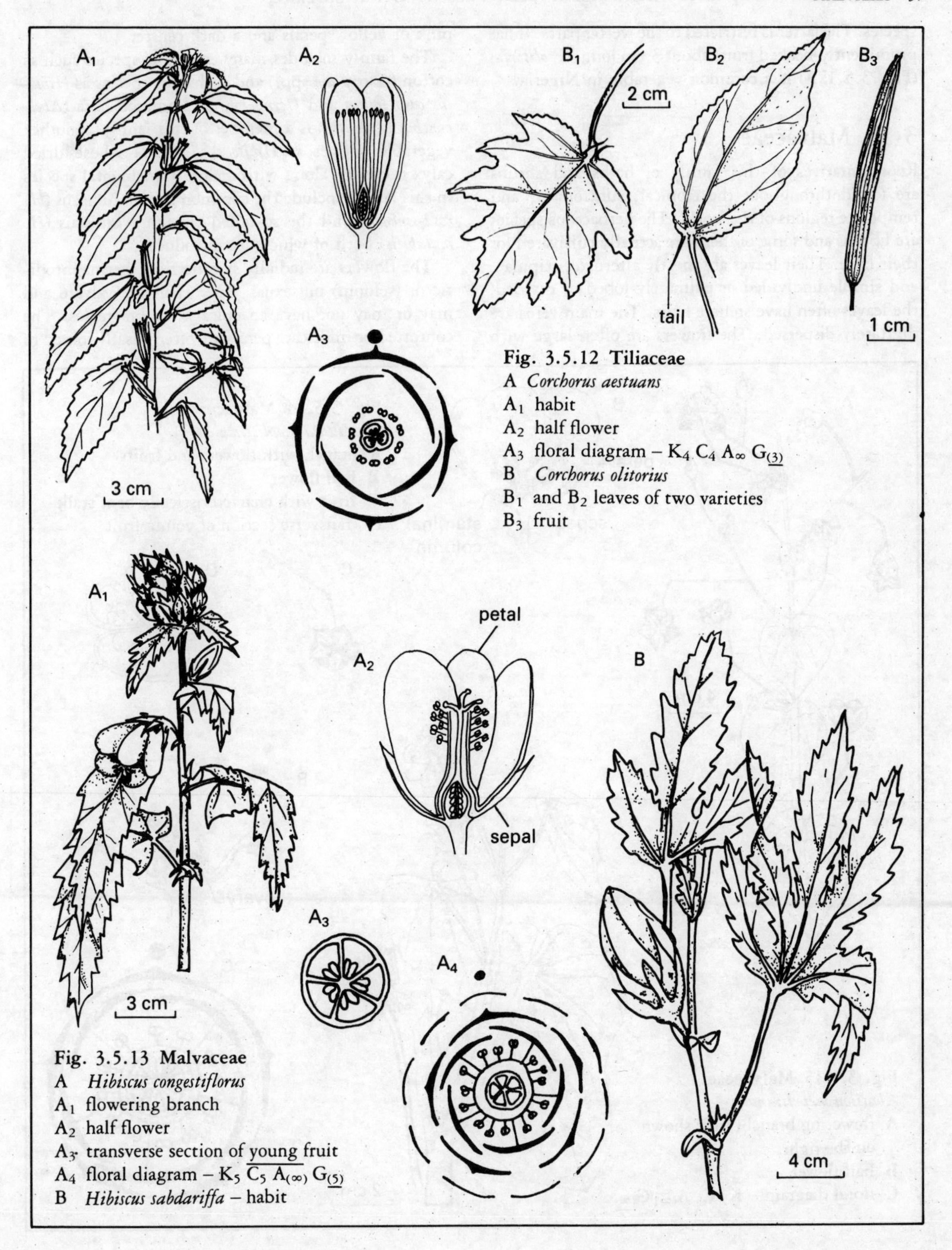

Fig. 3.5.12 Tiliaceae
A *Corchorus aestuans*
A_1 habit
A_2 half flower
A_3 floral diagram – $K_4\ C_4\ A_\infty\ G_{\underline{(3)}}$
B *Corchorus olitorius*
B_1 and B_2 leaves of two varieties
B_3 fruit

Fig. 3.5.13 Malvaceae
A *Hibiscus congestiflorus*
A_1 flowering branch
A_2 half flower
A_3 transverse section of young fruit
A_4 floral diagram – $K_5\ \overline{C_5\ A_{(\infty)}}\ G_{\underline{(5)}}$
B *Hibiscus sabdariffa* – habit

species. The latter is restricted to the wetter parts. It has prominently-beaked fruits about 3 cm long. *C. olitorius* (Fig. 3.5.12A) is a common vegetable in Nigeria.

3·5h Malvaceae

Representatives of this family of herbs and shrubs are found throughout the tropical, subtropical, and temperate regions of the world. The herbaceous species are fibrous and some of them are actually cultivated for their fibre. Their leaves are mostly alternate, stipulate and simple undivided or palmately-lobed or divided; the leaves often have stellate hairs. The main veins are digitately dispersed. The flowers are often large with pink or yellow petals and a dark centre.

The family includes many economic species such as cotton (*Gossypium* spp.) and such fibre species as *Hibiscus cannabinus* and *Urena lobata*. Common okra (*Abelmoschus esculentus*) is a member of this family. Another vegetable species is *Hibiscus sabdariffa* whose dried calyx is cooked along with melon. Ornamental species in this family include the common garden hibiscus (*H. rosa-sinensis*) and the so-called changing hibiscus (*H. mutabilis*) both of which are introduced.

The flowers are radially symmetrical, hermaphroditic or (seldom) unisexual. The sepals are valvate and may or may not have an epicalyx or bracteoles. The contorted or imbricate petals are often basally joined to

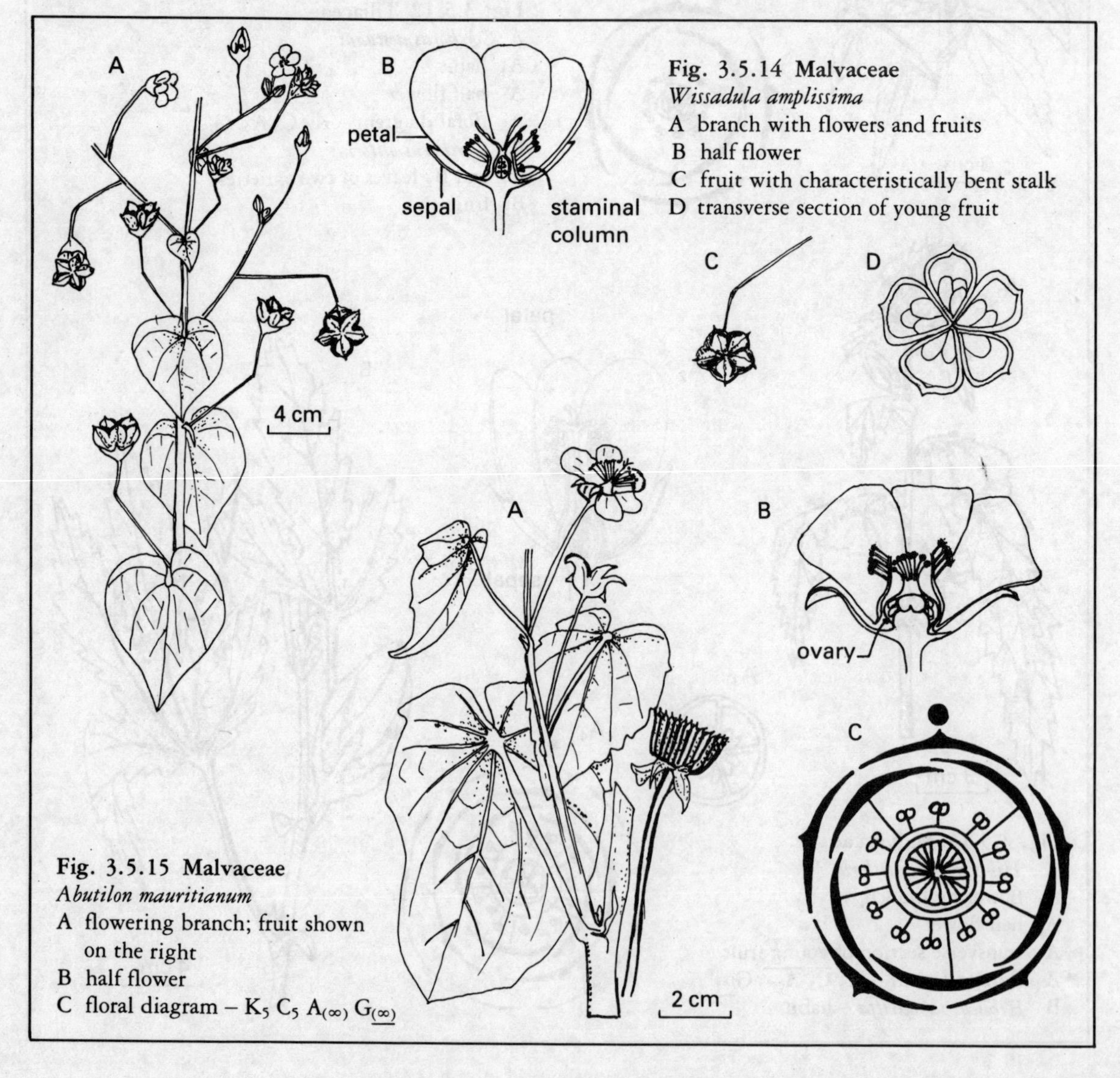

Fig. 3.5.14 Malvaceae
Wissadula amplissima
A branch with flowers and fruits
B half flower
C fruit with characteristically bent stalk
D transverse section of young fruit

Fig. 3.5.15 Malvaceae
Abutilon mauritianum
A flowering branch; fruit shown on the right
B half flower
C floral diagram – $K_5\ C_5\ A_{(\infty)}\ G_{\underline{(\infty)}}$

the staminal column (see Fig. 3.5.13) in such a way that when the corolla falls it does so with the staminal column, leaving the pistil and a persistent calyx behind. The stamens are numerous and monadelphous (forming a staminal column). The pistil is composed of single- or many-ovuled carpels with axile placentation.

The fruit is a capsule or schizocarp, the latter breaking up to form D-shaped, one-seeded, indehiscent mericarps.

The common species are mostly herbs and shrubs of waste places. *Hibiscus congestiflorus* (Figure 3.5.13A) is a conspicuous, although sparsely distributed, species of southern Guinea and derived savanna with a congested terminal inflorescence and a characteristic malvaceous flower. *H. sabdariffa* (Fig. 3.5.13B) grows as a weed or may be cultivated near houses. The yellow flowers are very conspicuous with a dark-purple centre.

Wissadula amplissima (Fig. 3.5.14) is a savanna or forest regrowth species with cordate leaves covered with stellate hairs. The branches are slender and the stalks of the fruits are bent at peculiar angles when mature. The fruits themselves are very characteristic.

Abutilon mauritianum is also a very common regrowth species in disturbed forest areas. The cream–yellow flowers produce characteristic fruits with many hooked mericarps (Fig. 3.5.15).

Species of *Sida* are mostly erect weedy herbs of waste places or prostrate herbs in places such as lawns. They have characteristic fruits which break into D-shaped mericarps. *S. acuta* (Fig. 3.5.16) is a common erect and much-branched woody herbs of waste places. It has pale yellow to cream flowers and prominent stipules. Often growing with *S. acuta* are other common species such as *S. rhombifolia* and *S. corymbosa* which differ from *S. acuta* in their carpel morphology and leaf shapes. *S. veronicifolia*, on the other hand, is a prostrate herb of waste places usually rooting at the nodes. The solitary flowers with long peduncles are cream–yellow (Fig. 3.5.17D).

Hibiscus surattensis (Fig. 3.5.17A) is a common regrowth weed with yellow flowers which have red centres. The epicalyx segments are diagnostic.

A variety of the common okra (*Abelmoschus esculentus*) is illustrated in Fig. 3.5.17C; many varieties of this vegetable are well known to farmers all over Africa.

Urena lobata grows in cleared forest and dervied savanna environments. The purple flowers produce characteristically warty or tuberculate fruits surrounded by a persistent calyx (Fig. 3.5.17B). The leaves of *U. lobata* vary from subentire to deeply digitately lobed.

Cienfuegosia is a savanna genus with angular stems. One species (*C. heteroclada*) sends up branches from woody rootstocks after bush fires and has pinkish or purplish pink flowers. *C. digitata* is a small shrub with digitately-divided leaves and yellow flowers.

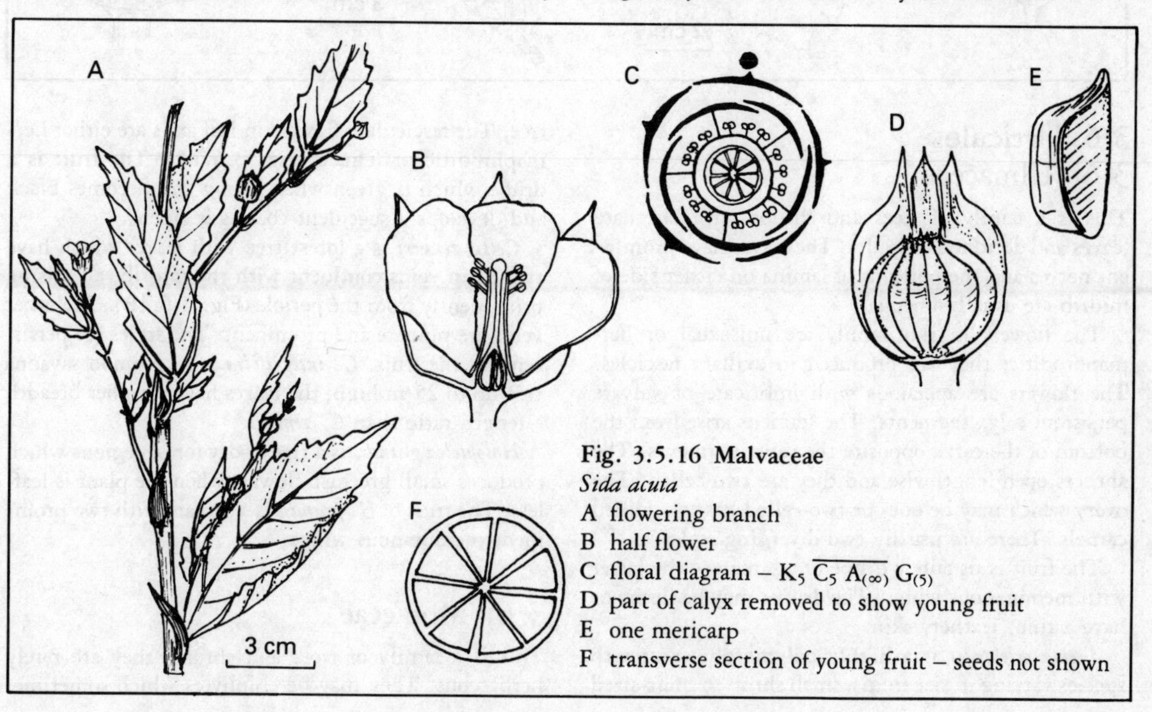

Fig. 3.5.16 Malvaceae
Sida acuta
A flowering branch
B half flower
C floral diagram – $K_5\ \overline{C_5\ A_{(\infty)}}\ G_{(5)}$
D part of calyx removed to show young fruit
E one mericarp
F transverse section of young fruit – seeds not shown

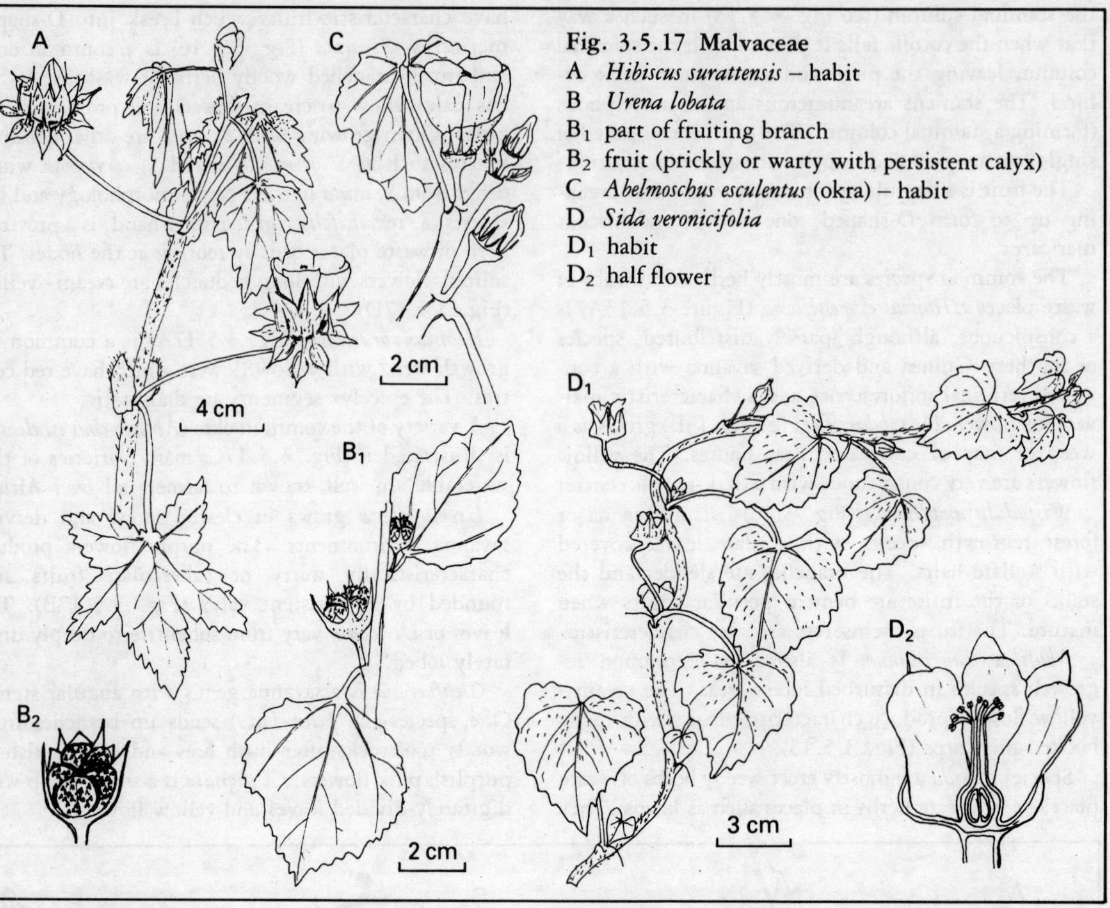

Fig. 3.5.17 Malvaceae
A *Hibiscus surattensis* – habit
B *Urena lobata*
B_1 part of fruiting branch
B_2 fruit (prickly or warty with persistent calyx)
C *Abelmoschus esculentus* (okra) – habit
D *Sida veronicifolia*
D_1 habit
D_2 half flower

3·6 Urticales

3·6i Ulmaceae

This is a family of trees and shrubs with alternate leaves and deciduous stipules. The leaves have prominent nerves and the portions of lamina on either side of midrib are usually unequal.

The flowers in this family are unisexual or hermaphroditic; they are produced in axillary fascicles. The flowers are apetalous with imbricate or valvate persistent calyx segments. The stamens arise from the bottom of the calyx opposite the calyx segments. The anthers open lengthwise and they are two-celled. The ovary which may be one- or two-celled has two joined carpels. There are usually two diverging styles.

The fruit is usually a drupe or a samara – the latter with membranous wings. The drupes may be fleshy or have a thin, leathery skin.

Trema orientalis is a highly polymorphic regrowth species varying in size from a small shrub to a fair-sized tree. The fasciculate flowers in leaf axils are either hermaphroditic or female (Fig. 3.6.1A). The fruit is a drupe which is green when young and becomes black and develops a succulent coat as it ripens.

Celtis zenkeri is a forest tree with leaves which have two main veins confluent with the midrib, all arising prominently from the petiole (Fig. 3.6.1B). The other veins are pinnate and prominent. The styles are persistent on the fruit. *C. integrifolia* is a common savanna tree up to 25 m high; the leaves have a higher breadth –length ratio than *C. zenkeri*.

Holoptelea grandis is a tree of dry forest regions which produces small greenish flowers when the plant is leafless. The fruit of *H. grandis* is a samara with two prominent membranous wings.

3·6j Moraceae

This is a family of trees and shrubs; they are rarely herbaceous. They may be epiphytes which sometimes

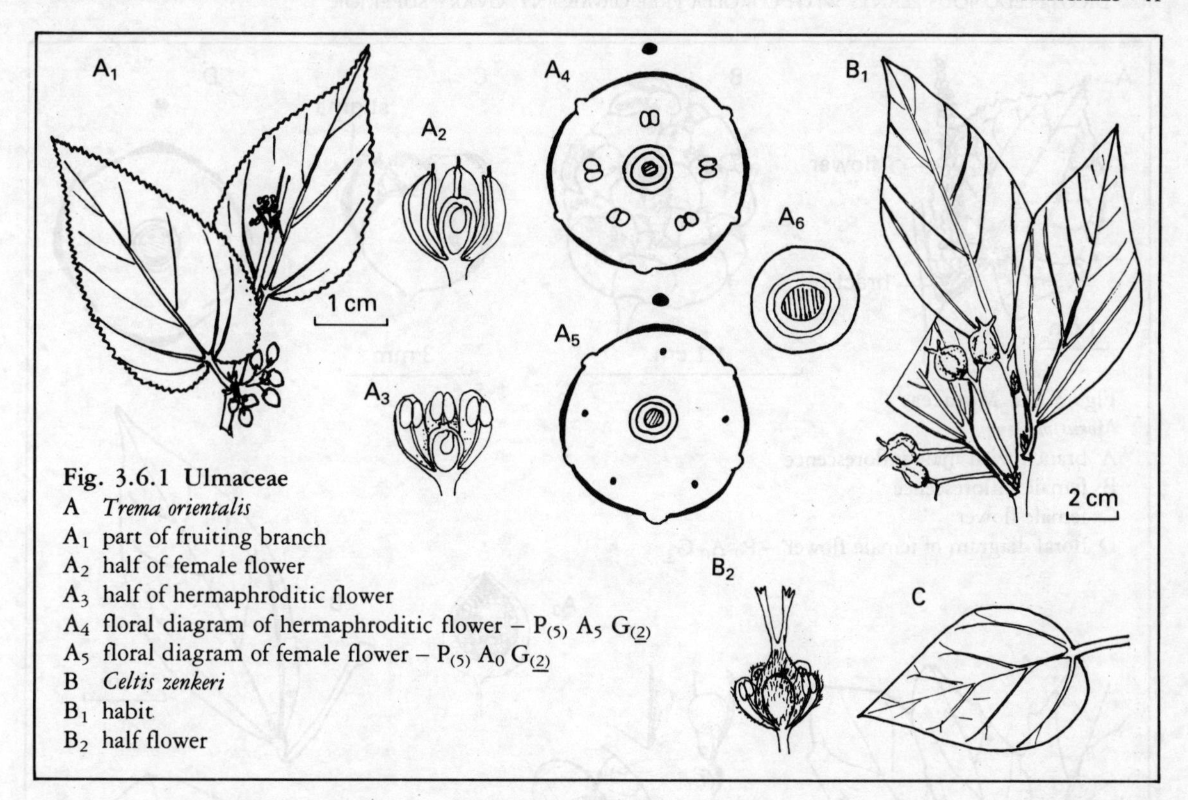

Fig. 3.6.1 Ulmaceae
A *Trema orientalis*
A_1 part of fruiting branch
A_2 half of female flower
A_3 half of hermaphroditic flower
A_4 floral diagram of hermaphroditic flower – $P_{(5)}\ A_5\ G_{\underline{(2)}}$
A_5 floral diagram of female flower – $P_{(5)}\ A_0\ G_{\underline{(2)}}$
B *Celtis zenkeri*
B_1 habit
B_2 half flower

strangle the plants on which they are epiphytic. The leaves are alternate (rarely opposite) and simple (rarely compound). Paired stipules are usually present; the stipules are usually large and brightly-coloured; they enclose the terminal bud. They are deciduous and fall off to leave prominent scars on the branch. The inflorescence, which has a common receptacle, may be in the form of a disc, a head, a hollow bag (with flowers inside) or a fleshy false fruit which is actually the whole inflorescence. Species of this family usually have milky latex.

Economic species in this family include timber species, one ornamental species and a number of species whose fruits are important food items. *Chlorophora excelsa* is an important timber tree, widespread from wet forest to derived savanna environments. *Ficus pumila* is an ornamental plant which attaches itself to walls of houses with its roots. *Ficus carica* is the European fig whose 'fruit' is edible. The genus of breadfruit (*Artocarpus*) is important because of its use as food. *Artocarpus communis* is a monoecious species. In the fertile variety, seeds (the so-called breadnut) are produced, while in the sterile variety seeds are not produced and the 'fruits' (the whole inflorescence) are boiled as a substitute for yams.

Plants may be monoecious (with male and female flowers on separate inflorescences or on the same inflorescence or receptacle) or dioecious. Flowers are apetalous with four or fewer calyx lobes which are often quite reduced and are imbricate or valvate. Stamens are usually one to four in number. When there are the same number of stamens as sepals, the stamens are opposite the sepals. The filaments are erect or inflexed in bud, and anthers are two-celled. The ovary has two carpels or one carpel by abortion. The ovary is superior and has two styles, or one style with two diverging arms and a single pendulous or erect ovule.

Morus mesozygia is a tree attaining various sizes in dry forest environments. The branches are reddish brown and the leaves have three prominent nerves arising from the petiole; the conspicuous reticulation of the other veins and the cordate leaf base are diagnostic. *M. mesozygia* is dioecious, with the male flowers carried on catkin-like inflorescences (Fig. 3.6.2).

Ficus is a large genus with the typical cupular inflorescence called a 'syconium' (Fig. 3.6.3). The male and female flowers are enclosed in the syconium, the latter towards the orifice, which is usually protected by sterile scales. The syconia are produced on leafless stems, usually in large clusters. *F. pumila* is an

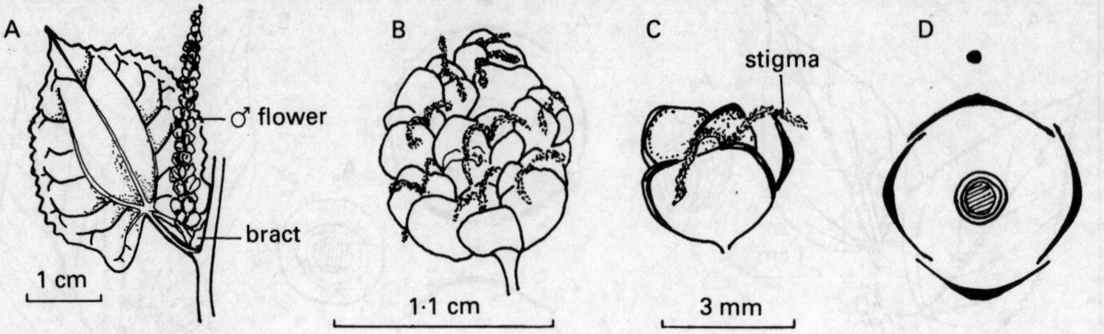

Fig. 3.6.2 Moraceae
Morus mesozygia
A branch with male inflorescence
B female inflorescence
C female flower
D floral diagram of female flower – $P_4 A_0 G_{\underline{1}}$

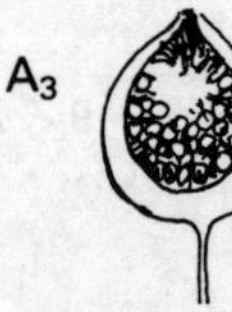

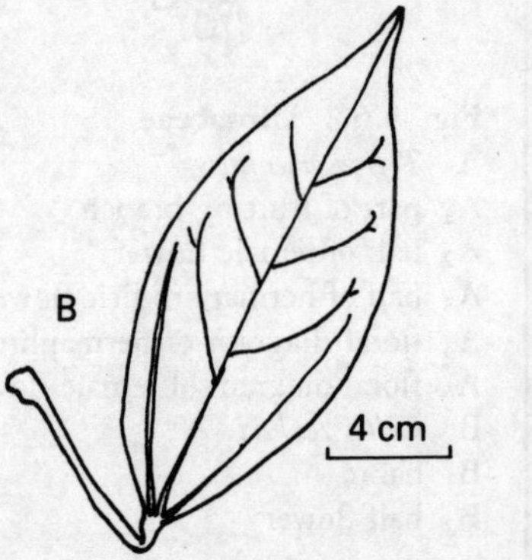

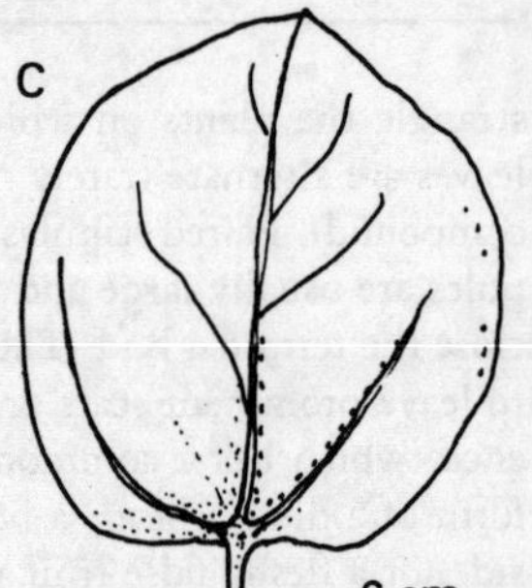

Fig. 3.6.3 Moraceae
A *Ficus capensis*
A_1 part of branch
A_2 stem with false fruits (syconia)
A_3 half of syconium
A_4 female flower
B *Ficus thonningii* – leaf
C *Ficus polita* – leaf

ornamental and *F. carica* is the European fig. *F. thonningii* and *F. polita* are generally planted for the shade they give; the latter has numerous red aerial roots. *F. mucuso* is a conspicuous tree of dry forest zones with a smooth, greenish rust coloured stem. Brownish 'fruits' are produced in large numbers on the bare stem. *F. capensis* is a common small tree of regrowth forest environments. The 'fruits', which are produced in large numbers on the stem, ripen into reddish, edible syconia. *F. exasperata* is the 'sand-paper leaf' tree. The syconium is usually quite small and hard and the rough surface of the leaves is diagnostic.

Bosqueia angolensis is a forest tree, usually with lax branches and smooth greyish bark which produces white latex that turns reddish brown. The inflorescence is similar to that of *Ficus* except that it is subtended by a green bract which protects it when young, and that the receptacle not only carries a single female

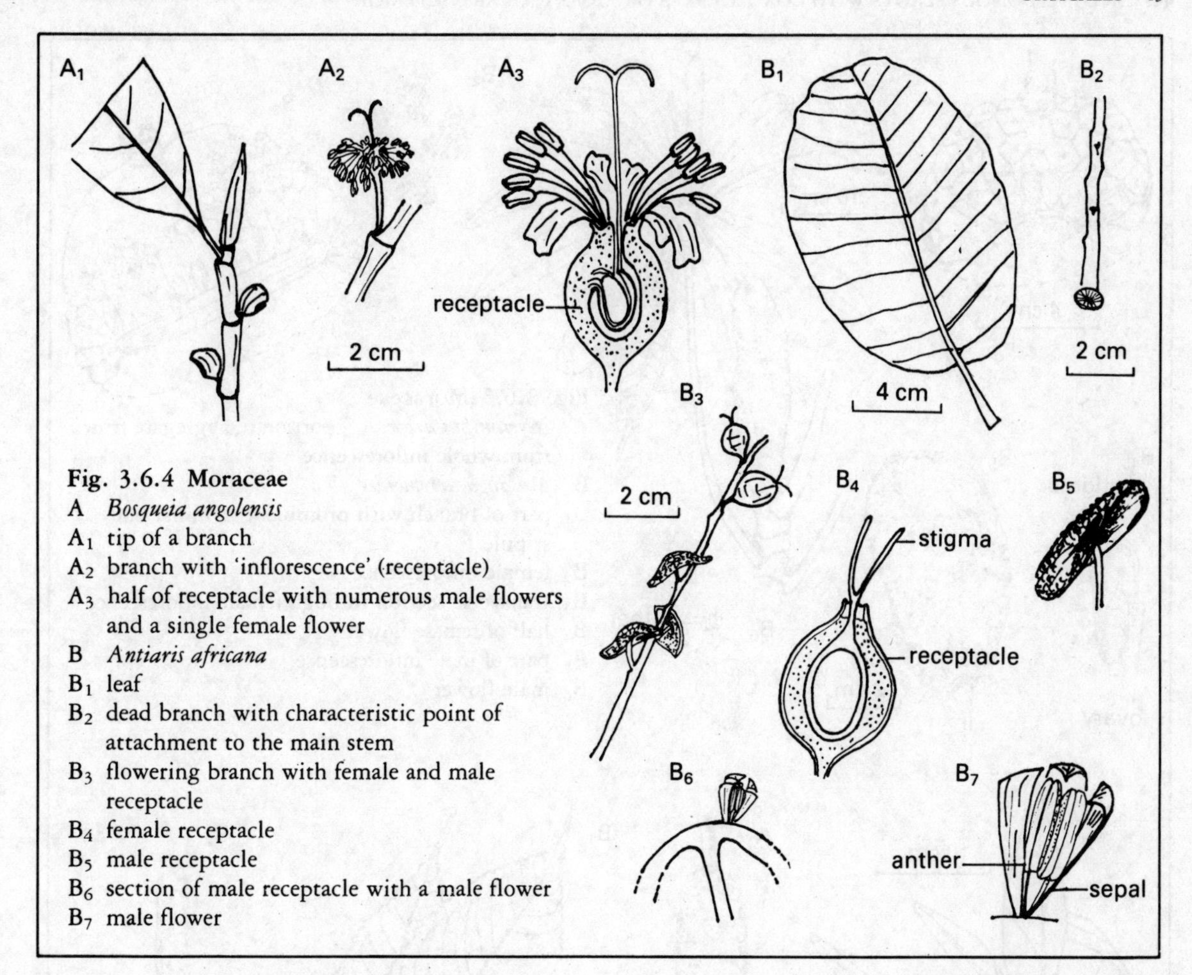

Fig. 3.6.4 Moraceae
A *Bosqueia angolensis*
A_1 tip of a branch
A_2 branch with 'inflorescence' (receptacle)
A_3 half of receptacle with numerous male flowers and a single female flower
B *Antiaris africana*
B_1 leaf
B_2 dead branch with characteristic point of attachment to the main stem
B_3 flowering branch with female and male receptacle
B_4 female receptacle
B_5 male receptacle
B_6 section of male receptacle with a male flower
B_7 male flower

flower, but the stamens are exposed (Fig. 3.6.4A).

Antiaris africana is a tall deciduous tree found in dry forest environments. It has grey bark displaying elliptic transverse patches where branches have fallen off. The bole is usually quite straight because the branches, which are usually at right angles to the main stem, fall off very neatly by some sort of self-pruning mechanism; thus when a 'pruned' branch is observed, the end that was attached to the stem has a neat conical surface (Fig. 3.6.4B). The leaves of *A. africana* are rough and pinnately-nerved. The young branches are often covered with rust-coloured hairs. The male and female receptacles are borne on the same inflorescence which is produced when the tree is leafless.

Digitately-compound or digitately-divided leaves are encountered in *Musanga* and *Myrianthus*. In *Myrianthus arboreus*, the leaves are digitately 5–7 foliolate, although the leaves in the seedlings may be simple. *M. arboreus* is a monoecious forest regrowth tree with a characteristic false fruit which is yellow when ripe (Fig. 3.6.5A). *Musanga cecropioides* is a regrowth tree in wet and dry forests, usually with stilt roots. The leaves are digitately deeply divided and the terminal bud is usually enclosed in large bight red stipules. The leaves and stipules fall off leaving prominent scars. *M. cecropioides* is deciduous – details of inflorescences and flowers are shown in Fig. 3.6.5B.

Artocarpus communis is widely cultivated and it is easily recognised by the pinnately-lobed leaves (Fig. 3.6.6A).

Treculia africana (African breadfruit) is a forest tree whose seeds are eaten.

Chlorophora excelsa is an important timber tree, widespread in environments ranging from wet forest to derived savanna. It is large and usually only shortly buttressed. It has galled leaves which are pinnately nerved. The grey bark is smooth or scaly and produces milky latex. The plant is dioecious with distinct

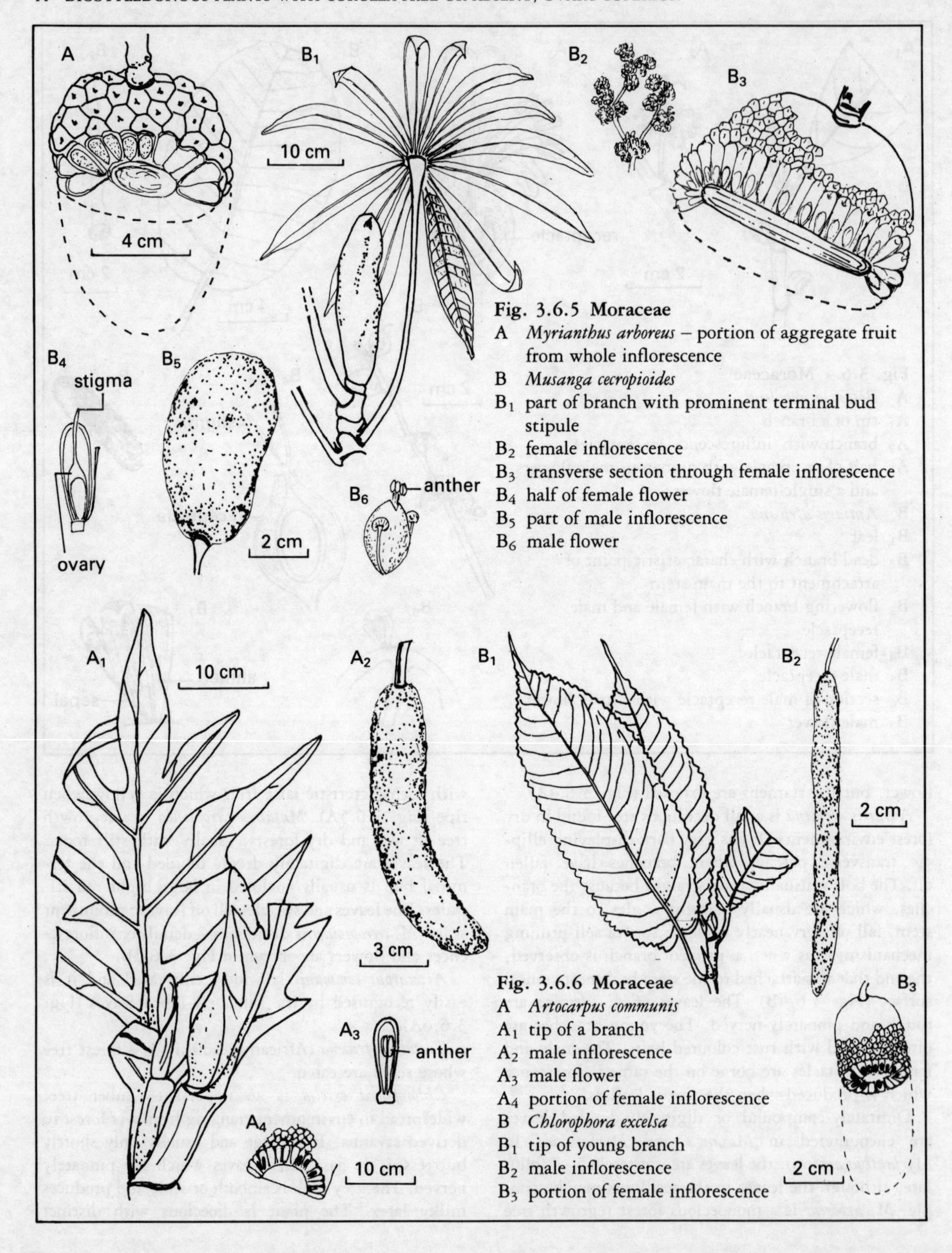

Fig. 3.6.5 Moraceae
A *Myrianthus arboreus* – portion of aggregate fruit from whole inflorescence
B *Musanga cecropioides*
B_1 part of branch with prominent terminal bud stipule
B_2 female inflorescence
B_3 transverse section through female inflorescence
B_4 half of female flower
B_5 part of male inflorescence
B_6 male flower

Fig. 3.6.6 Moraceae
A *Artocarpus communis*
A_1 tip of a branch
A_2 male inflorescence
A_3 male flower
A_4 portion of female inflorescence
B *Chlorophora excelsa*
B_1 tip of young branch
B_2 male inflorescence
B_3 portion of female inflorescence

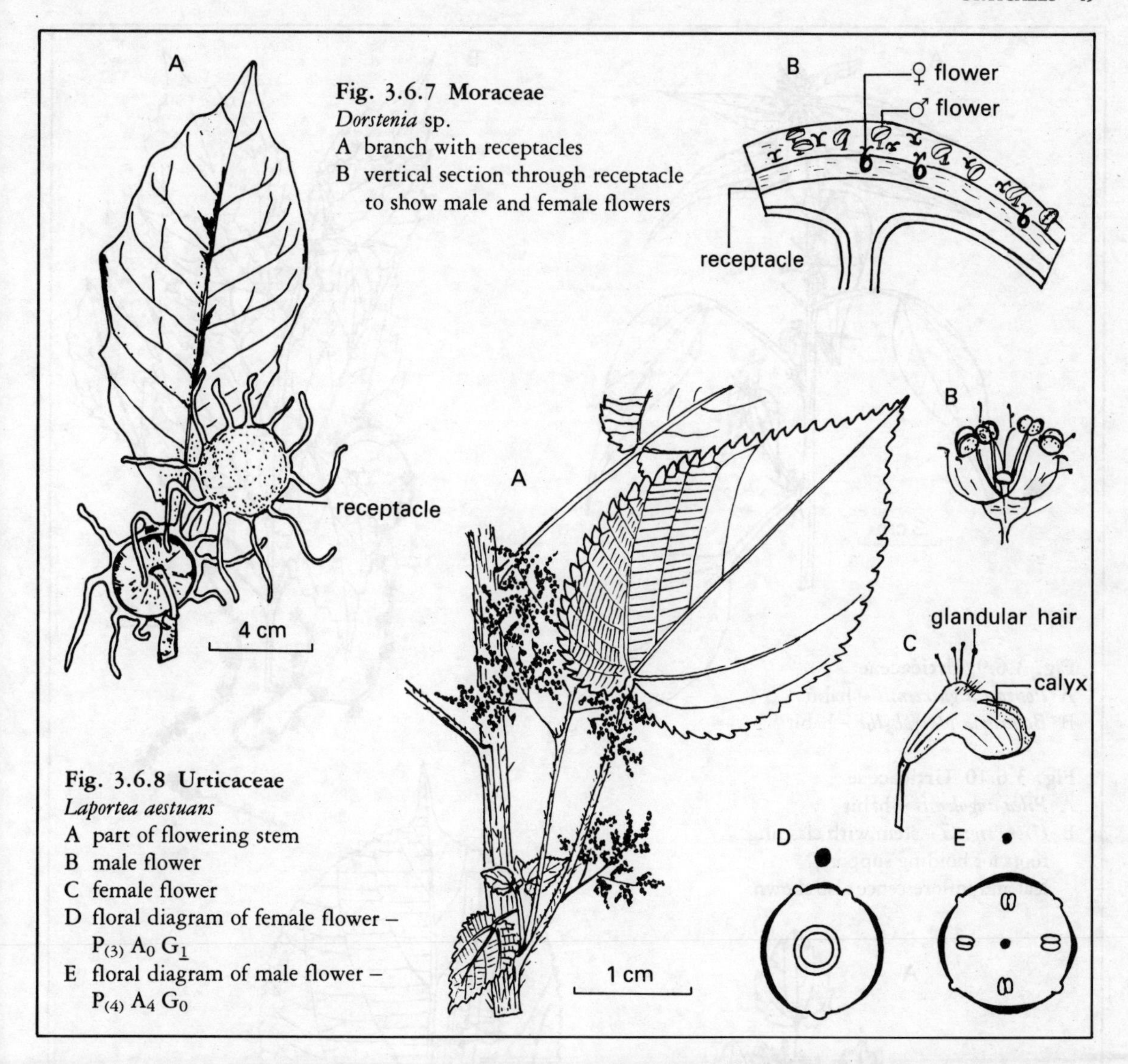

Fig. 3.6.7 Moraceae
Dorstenia sp.
A branch with receptacles
B vertical section through receptacle to show male and female flowers

Fig. 3.6.8 Urticaceae
Laportea aestuans
A part of flowering stem
B male flower
C female flower
D floral diagram of female flower – $P_{(3)}\ A_0\ G_{\underline{1}}$
E floral diagram of male flower – $P_{(4)}\ A_4\ G_0$

inflorescences – the male inflorescence is longer and catkin-like (Fig. 3.6.6B).

Dorstenia is a fairly large genus of herbs and small shrubs mostly in wet forest environments. The receptacles are usually plate-shaped, boat-shaped or linear and with or without elongate lobes at the plate margins; male and female flowers are found on the same receptacle (Fig. 3.6.7).

3·6k Urticaceae

This is a family of undershrubs and (seldom) soft-wooded trees, soft-wooded climbers or terrestial or epiphytic herbs, the latter usually in wet and shaded environments. Species are often armed with stinging hairs; their epidermal cells have cystoliths and the stem may be fibrous. The leaves are opposite or alternate, simple and often with finely-toothed margins. Stipules may be present or absent. The family is cosmopolitan and includes important weed species.

The flowers are usually quite small, unisexual, in cymose inflorescences, crowded together in an enlarged 'receptacle'. The flowers are apetalous. The calyx in male flowers is made of four or five imbricate or valvate segments, while calyx lobes in female flowers are similar in general character to those in male flowers except that some segments become larger when the fruit is mature (Fig. 3.6.8). Stamens are as many

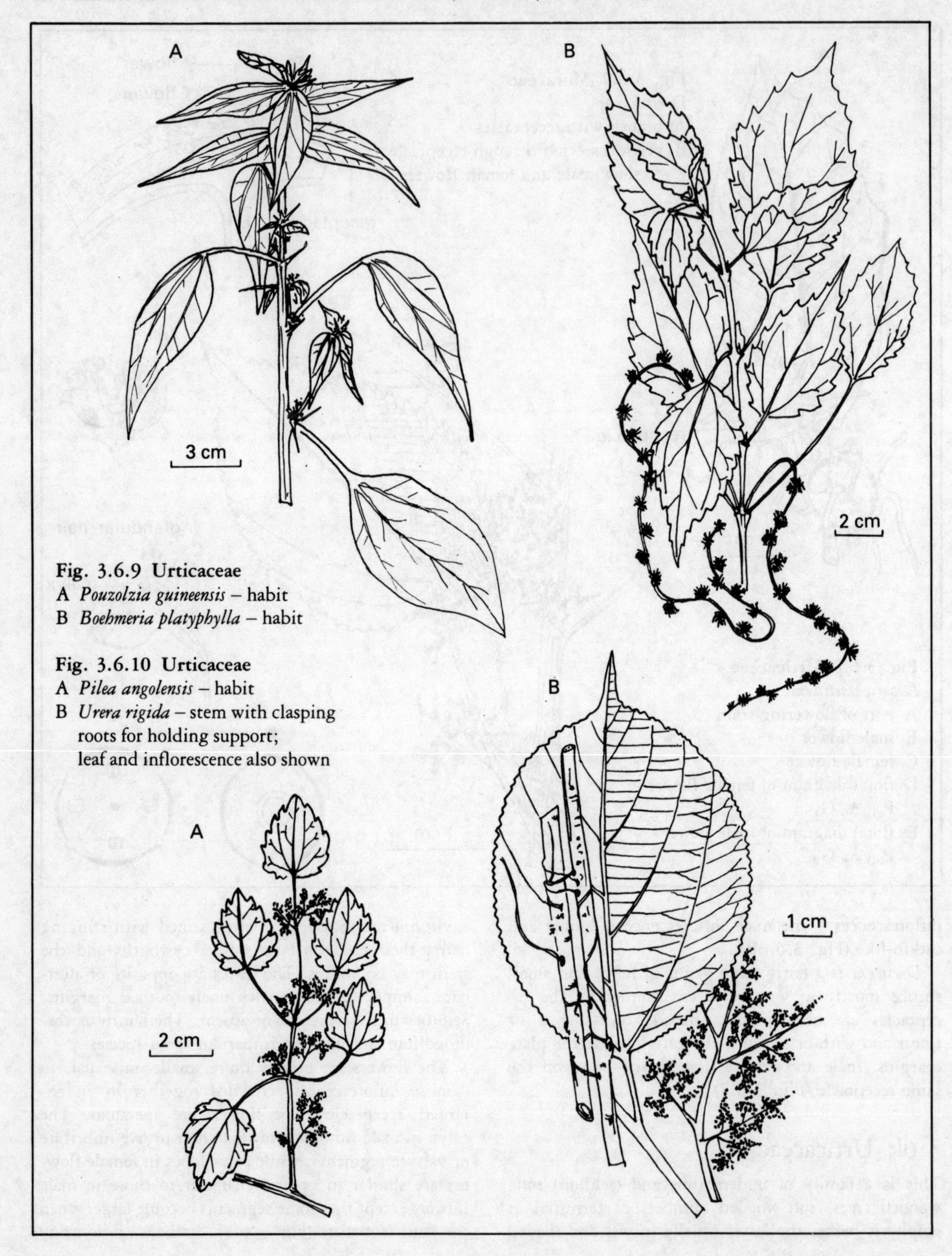

Fig. 3.6.9 Urticaceae
A *Pouzolzia guineensis* – habit
B *Boehmeria platyphylla* – habit

Fig. 3.6.10 Urticaceae
A *Pilea angolensis* – habit
B *Urera rigida* – stem with clasping roots for holding support; leaf and inflorescence also shown

as, and opposite, the calyx segments; staminodes may be present in the female flowers while a rudimentry pistil may be present in the male flowers. The ovary is one-celled; the pistil has a simple style while the ovule is solitary and erect. The fruit is a dry achene or a fleshy drupe.

Laportea aestuans is a common herbaceous weed in forest farms and clearings. It may be up to 1·5 m in height; the stems are weak, more or less flimsy and hairy. The alternate leaves are hairy with finely-toothed margins. The much-branched inflorescences are axillary and they carry unisexual flowers (Fig. 3.6.8).

Pouzolzia guinensis is also a weed of open places and farmland in forest environments. The leaves are alternate with long petioles and three digitate veins. The inflorescences are spike-like and axillary (Fig. 3.6.9A).

Boehmeria platyphylla is mostly a highland species with opposite leaves and typical axillary inflorescences. The inflorescences are long with the flowers clustered at intervals along the inflorescence axes (Fig. 3.6.9B).

Pilea is a genus of herbs found growing on wet rock surfaces and epiphytic on wet tree trunks in wet and dark forest environments. Their stems are weak and transparent. Their leaves are opposite with branched axillary inflorescences carrying minute unisexual flowers. Figure 3.6.10A shows *P. angolensis* – a species which grows on wet rock surfaces in shaded forest environments.

Urera is a fairly large genus of erect, creeping or climbing species, the latter with clasping roots produced all over the stems. *U. rigida* is a common forest climber with a soft, fibrous stem and clasping roots; the stem is also characterised by short prickly protuberances. The much-branched inflorescences (Fig. 3.6.10B) carry flowers which produce red fruits.

3·7 Rosales

3·71 Rosaceae

This is the garden rose family. Although this is a cosmopolitan family of trees, shrubs and herbs, it is poorly represented in the Tropics. The herbaceous genera – *Alchemilla* (Fig. 3.7.1A) and *Neurada* – are plants of highland locations or desert environments, while *Rubus* (Fig. 3.7.1B and C), a genus of scrambling shrubs, is composed almost entirely of species of relatively high elevations. The tree genera of Rosaceae are, however, found in diverse elevations and different vegetational zones. The leaves are simple (in most tree species) or compound and usually with paired stipules which are adnate to the petiole. Of the genera in our area, *Rubus* is the closest to the type genus *Rosa* with respect to both floral and vegetative attributes.

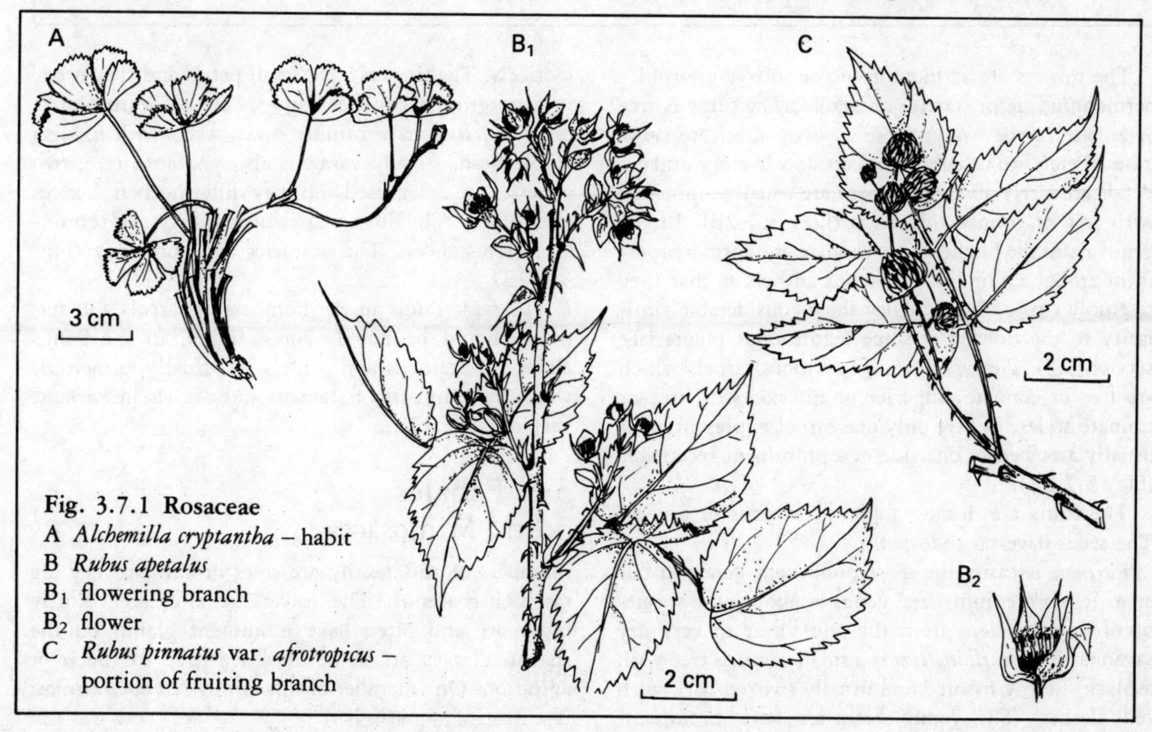

Fig. 3.7.1 Rosaceae
A *Alchemilla cryptantha* – habit
B *Rubus apetalus*
B_1 flowering branch
B_2 flower
C *Rubus pinnatus* var. *afrotropicus* – portion of fruiting branch

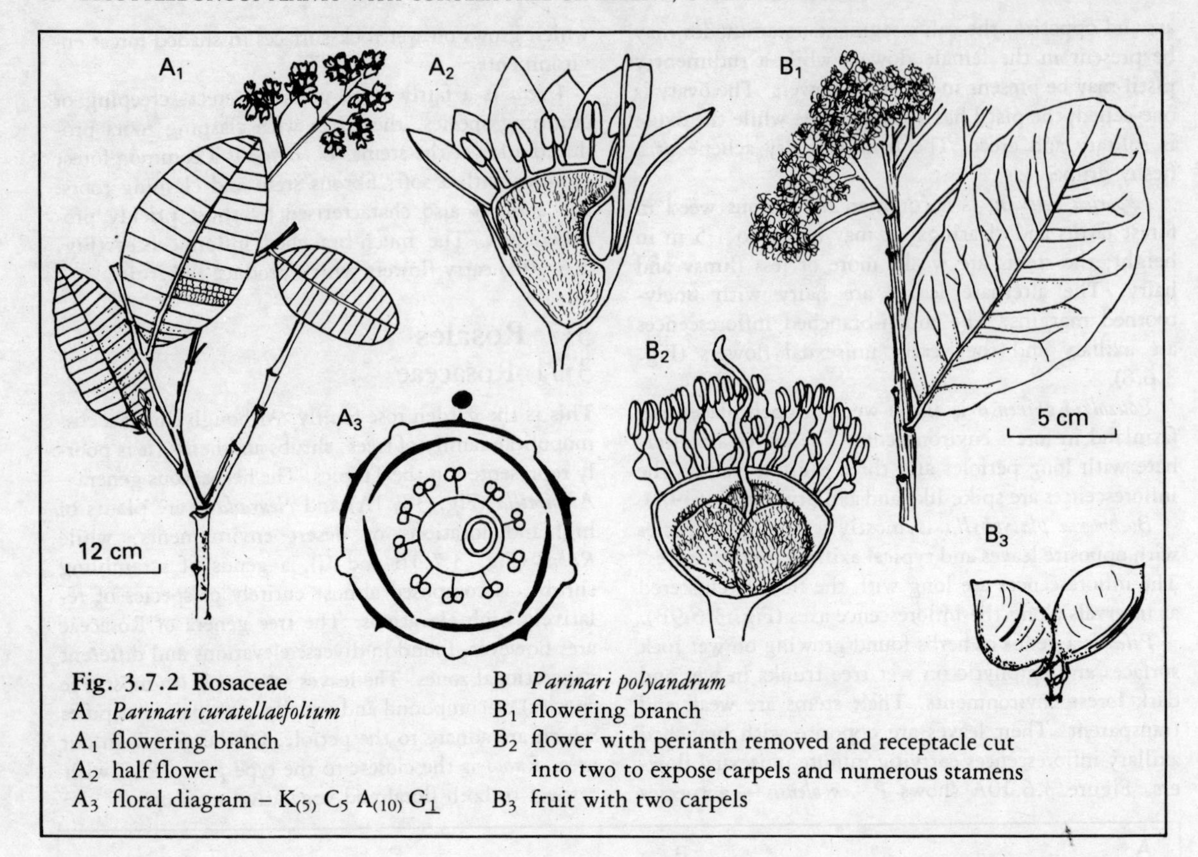

Fig. 3.7.2 Rosaceae
A *Parinari curatellaefolium*
A_1 flowering branch
A_2 half flower
A_3 floral diagram – $K_{(5)}\ C_5\ A_{(10)}\ G_{\underline{1}}$
B *Parinari polyandrum*
B_1 flowering branch
B_2 flower with perianth removed and receptacle cut into two to expose carpels and numerous stamens
B_3 fruit with two carpels

The flowers are actinomorphic or subzygomorphic, hermaphroditic or (rarely) unisexual. The calyx is free or adnate to the ovary; there are five imbricate calyx lobes. Petals are usually imbricate and five in number. Petals are rarely absent. Stamens are usually numerous with free or connate filaments (Fig. 3.7.2B). In the genus *Acioa* the filaments are connate and form a prominent trough ending in numerous anthers so that they resemble a paint brush. They show considerable similarity to the flowers of some leguminous plants (see section 3.8). Flowers have one or more carpels which are free or connate, superior or inferior with free or connate styles. Where only one carpel is present, it is usually attached to one side of a prominent receptacle (Fig. 3.7.2A).

The fruits are drupes, follicles, achenes or pomes. The seeds have no endosperm.

Parinari is easily the most prominent genus in our area. It is a medium-sized genus of about nine species occupying locations from the rain forest to very dry savanna. *P. curatellaefolium* is a small savanna tree with a black, deeply fissured and usually twisted bark with red slashes. The young branches have prominent lenticels. The flowers have small petals and they occur in large terminal panicles (Fig. 3.7.2A). The fruits of *P. curatellaefolium* have minute corky warts with reddish edible flesh. *P. polyandrum* is also a savanna tree; two varieties are recognised and they differ in their degrees of hairiness. The flowers are white in flattened terminal inflorescences. The stamens are numerous (Fig. 3.7.2B).

The wide range in the numbers of carpels and stamens should be noted. For example, in the genus *Rubus* the stamens and carpels are usually numerous, while the numbers of stamens and carpels in *Parinari* vary considerably.

3·8 Fabales

3·8m Mimosaceae

Members of this family are trees or shrubs; they are rarely herbaceous. The leaves are alternate, mostly bipinnate and often have prominent glands on the rhachis; leaves are sometimes sensitive to touch or agitation. One member of this family *Neptunia oleracea* is a floating aquatic herb.

Species of economic importance include those cultivated for ornamental purposes, those that yield timber, those that are used for medicinal and culinary purposes and one species, *Acacia senegal*, which yields gum – the so-called gum arabic. Species commonly planted as ornamentals include *Mimosa pudica, Acacia farnesiana, Calliandra haematocephala, Samanea saman* and *Leucaena leucocephala*. The timber species of commercial consequence are *Piptadeniastrum africanum, Cyclodiscus gabunensis, Albizia adianthifolia* and *Aubrevillea platycarpa*. *Tetrapleura tetraptera* yields a characteristic fruit illustrated in Fig. 3.8.4 which is used as a condiment and for various medicinal purposes. *Parkia clappertoniana* is a fair-sized savanna tree that yields the locust bean commonly used as a condiment.

The flowers in this family are small, actinomorphic and hermaphroditic. They occur in spicate, racemose or capitate (headlike, globose) inflorescences. The flowers are usually white, yellow or pink and pentamerous. The calyx is five-lobed or toothed, tubular, valvate or (very rarely) imbricate. The petals are valvate and free or joined into a tube. The stamens are as many or twice as many as the petals or they are numerous; they are free or monadelphous (joined into a staminal tube around the pistil). The anthers are small, often with a deciduous apical gland; they open lengthwise. The ovary is of one carpel and is superior.

The fruit is usually a pod (dehiscent). Apart from indehiscent fruit forms encountered in this family, some genera also produce lomenta: fruits that break into one-seeded segments.

Acacia is a large genus of trees and shrubs of the drier parts of tropical and subtropical regions. The thorny stipules are characteristic. The stamens are numerous and they are usually joined into a tube. *A. farnesiana* is a spiny shrub with scented yellow flowers in globular heads (Fig. 3.8.1A). The plant is commonly cultivated as an ornamental. The numerous anthers are free and the fruits are characteristically produced in fasciculate bunches. *A. senegal* is a shrub or small tree with spines in groups of three, below the nodes. The flowers are in spikes about 12 cm long and the flowers have monadelphous stamens (Fig. 3.8.2B and C); the fruits are flat and they are about 10 cm long. Many other species of *Acacia* are widespread, often forming pure thickets, in the dry savanna.

Mimosa is a genus of plants with sensitive leaves. *M. pudica* is a straggling perennial with one or two pairs of pinnae and globular axillary inflorescences of pink flowers (Fig. 3.8.1B). The fruits are jointed (lomenta). *M. pigra* is a prickly shrub of river banks with three or more pairs of pinnae and larger fruits than in *M. pudica*. In both species of *Mimosa* the fruits are prickly.

Schrankia leptocarpa is a straggling or climbing perennial with a prickly and ridged stem and two or three pairs of pinnae. The leaves are sensitive to touch as in *Mimosa*, but the fruits are terete and not jointed.

Dichrostachys cinerea is a small tree or shub in the savanna. There are about ten pairs of pinnae and the leaves have axillary spines. The inflorescence is a

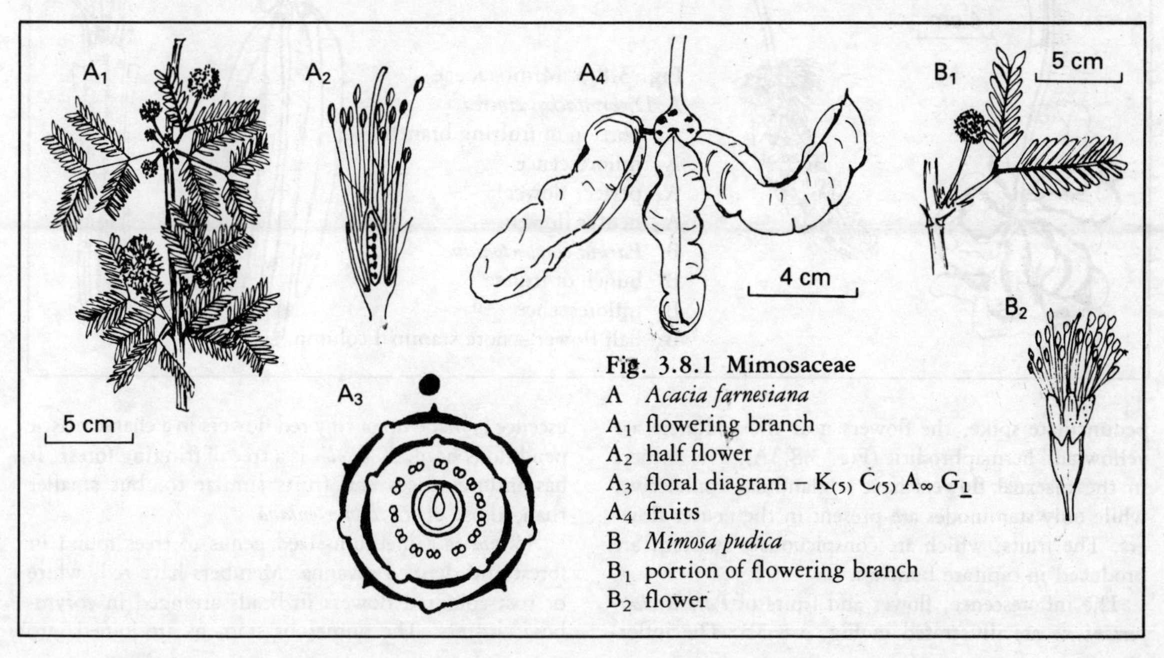

Fig. 3.8.1 Mimosaceae

A *Acacia farnesiana*

A_1 flowering branch

A_2 half flower

A_3 floral diagram – $K_{(5)}\ \dot{C}_{(5)}\ A_{\infty}\ G_{\underline{1}}$

A_4 fruits

B *Mimosa pudica*

B_1 portion of flowering branch

B_2 flower

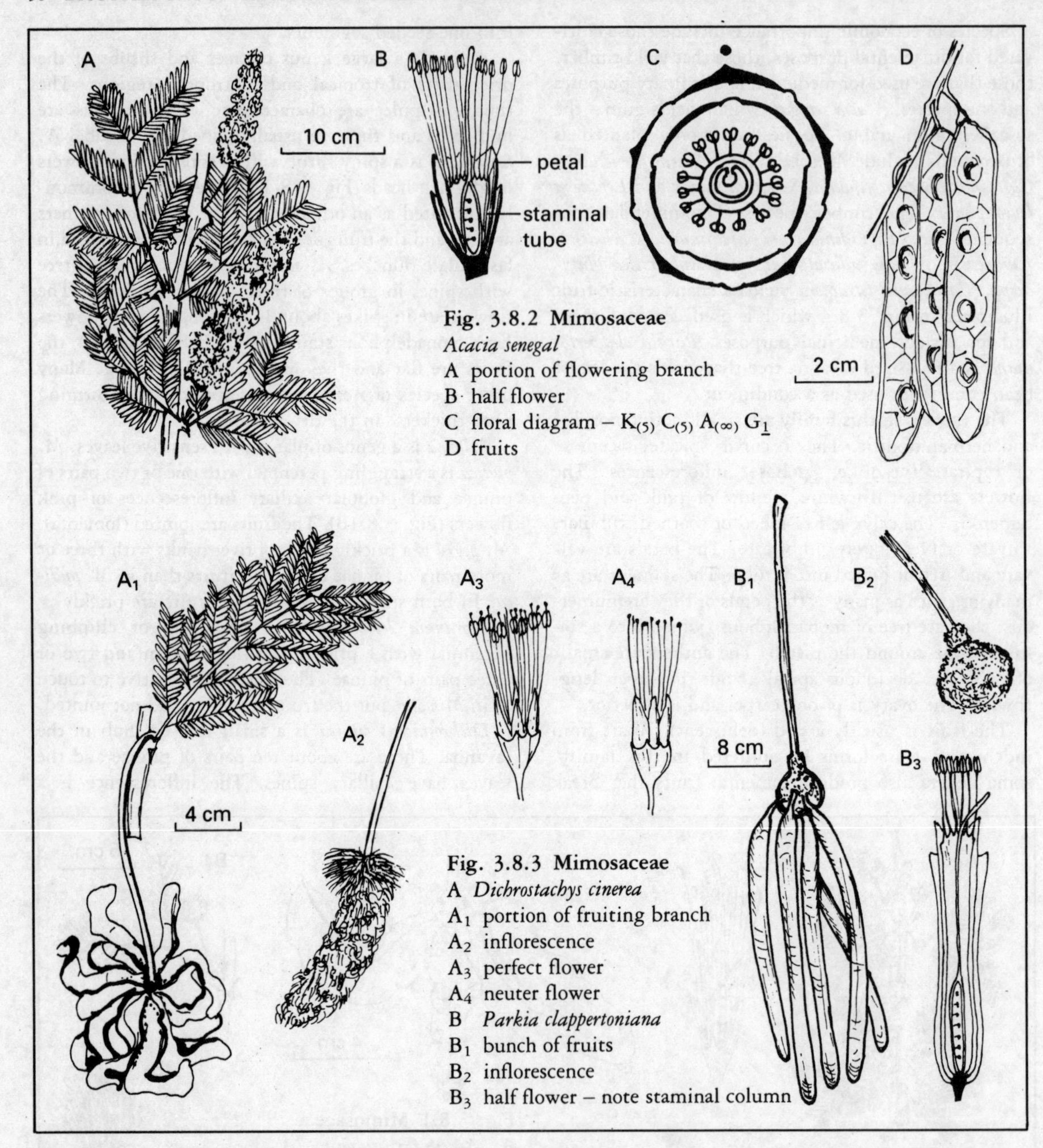

Fig. 3.8.2 Mimosaceae
Acacia senegal
A portion of flowering branch
B half flower
C floral diagram – $K_{(5)}\ C_{(5)}\ A_{(\infty)}\ G_{\underline{1}}$
D fruits

Fig. 3.8.3 Mimosaceae
A *Dichrostachys cinerea*
A_1 portion of fruiting branch
A_2 inflorescence
A_3 perfect flower
A_4 neuter flower
B *Parkia clappertoniana*
B_1 bunch of fruits
B_2 inflorescence
B_3 half flower – note staminal column

pedunculate spike; the flowers near the peduncle are yellow and hermaphroditic (Fig. 3.8.3A). The anthers in the bisexual flowers have a glandular connective, while only staminodes are present in the neuter flowers. The fruits, which are conspicuously twisted, are produced in capitate bunches.

The inflorescence, flower and fruits of *Parkia clappertoniana* are illustrated in Fig. 3.8.3B. The inflorescence is made up of tiny red flowers in a characteristic pendulous head. *P. bicolor* is a tree of fringing forest. It has inflorescences and fruits similar to, but smaller than, those of *P. clappertoniana*.

Albizia is a medium-sized genus of trees found in forest and derived savanna. Members have red, white or rust-coloured flowers in heads arranged in corymbose clusters. The numerous stamens are joined into

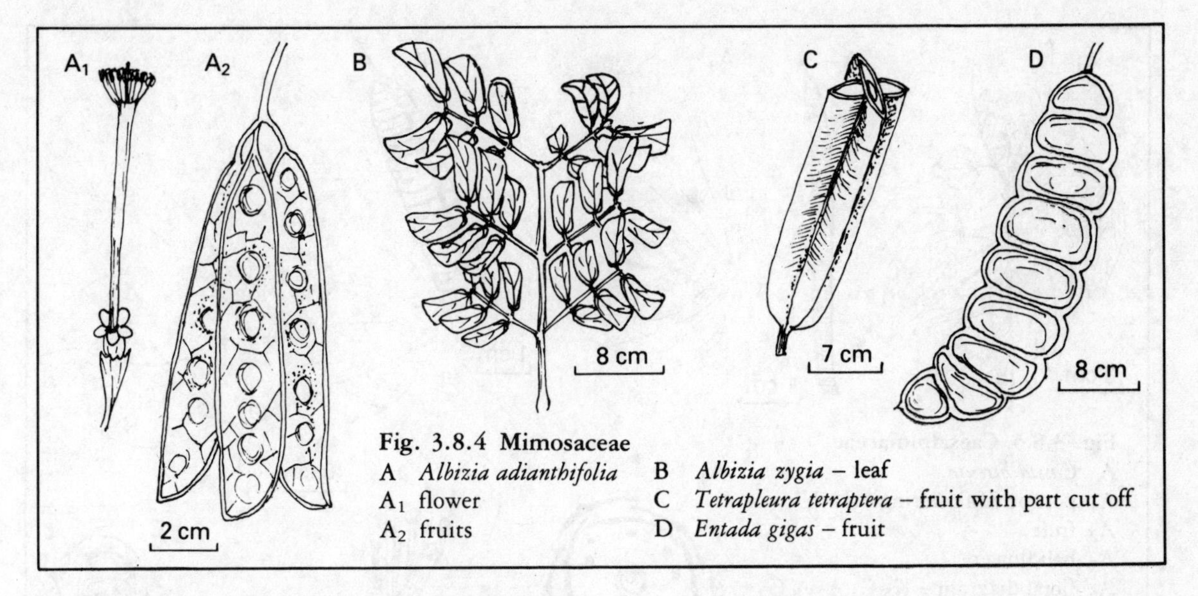

Fig. 3.8.4 Mimosaceae
A *Albizia adianthifolia*
A_1 flower
A_2 fruits
B *Albizia zygia* – leaf
C *Tetrapleura tetraptera* – fruit with part cut off
D *Entada gigas* – fruit

a prominent staminal tube and their fruits are thin and flat (Fig. 3.8.4A). The leaflets are conspicuously asymmetrical. *A. adianthifolia* is a large forest tree with a spreading rust-coloured crown during the flowering season; a flower is illustrated in Fig. 3.8.4A. *A. zygia* is a common regrowth tree species in dry forest and derived savanna. The spreading crown produces pinkish red flowers. The leaf (Fig. 3.8.4B) is very diagnostic, the pinnae and pinnules increasing progressively in size from the base upwards.

Aubrevillea platycarpa is a large, buttressed forest tree. The spikes of creamy flowers produce bunches of thin and papery fruits which are rough white.

Prosopis africana is an acacia-like savanna tree often planted in towns. The ten stamens are free with apical glands on the anthers. The flowers are yellow; the inflorescence is a spike. The fruit is indehiscent, cylindrical and edible.

Tetrapleura tetraptera is a forest tree with fern-like foliage and creamy flowers in axillary spikes. The fruit is about 30 cm long with four characteristic wings (Fig. 3.8.4C).

The genus *Entada* is a genus of scandent shrubs and small trees with flowers in axillary spikes. *E. pursaetha* is a woody climber with a thick stem and large woody, segmented fruits up to almost 1 m in length (Fig. 3.8.4D). It is found in regrowth thickets in hilly locations. *E. gigas* is also a vigorous climber with smaller subwoody fruits. *E. africana* is a small savanna tree with thin and glossy 'entada-type' fruits that appear in the dry season. The fruits are green at first and then mature into reddish brown; the flowers are creamy.

3·8n Caesalpiniaceae

This is the *Cassia* family – a family of trees, shrubs and, rarely, stragglers or herbs with very showy flowers in showy and large inflorescences. As in Mimosaceae and the next family we will consider (Papilionaceae), the pod is the usual form of fruit. Indeed, Mimosaceae, Caesalpiniaceae and Papilionaceae have been variously treated as an order or family (Leguminales or Leguminosae) – the legumes in popular conception. Leaves are pinnate or bipinnate, stipulate and (rarely) one-foliolate or simple.

Members of this family are important as ornamental plants while some yield wood for poles and others are important timber species. Important ornamental species include *Caesalpinia pulcherrima* ('Pride of Barbados'), various species of *Cassia* such as *C. alata* and *C. sieberiana*, *Bauhinia* spp. and *Delonix regia* ('Flamboyant' or 'Flame of the Forest'). *Cassia siamea* is a large tree, commonly planted for shade and often planted in drier areas in afforestation programmes: the straight boles are used as poles for various purposes. The important timber species in this family are *Distemonanthus benthamianus* (commercial name: Ayan), *Daniellia ogea*, *Gosswielerodendron balsamiferum* (commerical name: Agba), *Brachystegia eurycoma* (commercial name: Okwen) and some species of *Afzelia*. Apart from all these species of economic importance *Dialium guineense* produces black velvety fruits the inner pulp of which is eaten; this fruit is sold in urban centres.

The plants are monoecious, dioecious or have hermaphroditic flowers. The flowers are very showy and

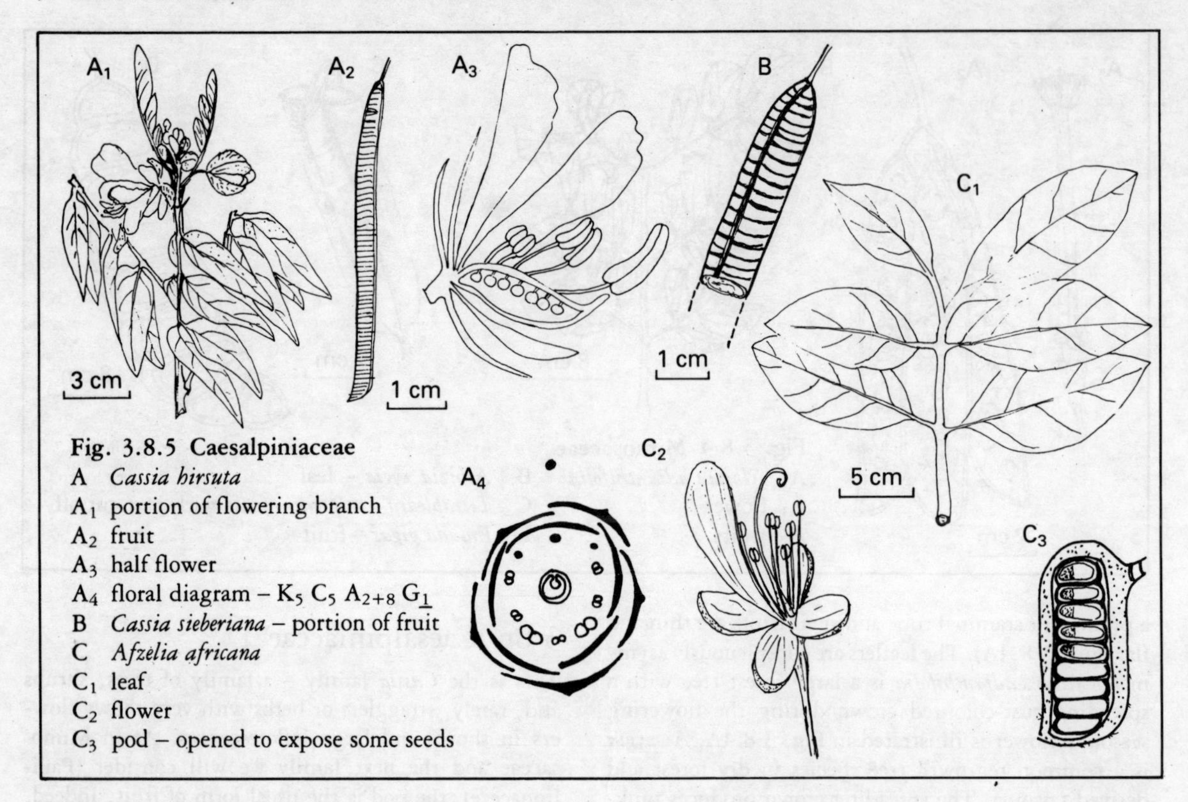

Fig. 3.8.5 Caesalpiniaceae
A *Cassia hirsuta*
A_1 portion of flowering branch
A_2 fruit
A_3 half flower
A_4 floral diagram – $K_5\ C_5\ A_{2+8}\ G_{\underline{1}}$
B *Cassia sieberiana* – portion of fruit
C *Afzelia africana*
C_1 leaf
C_2 flower
C_3 pod – opened to expose some seeds

zygomorphic (rarely subactinomorphic). They are arranged in large, racemose, spicate or (rarely) cymose inflorescences. Bracteoles are usually present and large and they may enclose the flowers in bud. There are five sepals and they are mostly free (the upper two may be connate), imbricate or valvate; they are sometimes reduced but replaced by conspicuous bracteoles that may be mistaken for sepals. Petals are usually five or fewer in number; they may be completely absent; the adaxial petal is inside the others in bud, while the others are variously imbricate (Fig. 3.8.5A). Stamens are ten in number or (rarely) numerous; they are free or variously joined. The anthers are of various sizes, sometimes opening by apical pores. The ovary is of one superior carpel which is also one-celled.

The fruit is usually a woody pod; samaras, drupes, 'nuts' and inflated and prickly pods are also encountered. Seeds are usually many in the pod although some species have one-seeded fruits.

Cassia is a large genus of trees, shrubs, undershrubs and woody herbs with paripinnate leaves and yellow flowers. The flowers have deciduous bracteoles and they are arranged in large conspicuous racemes. *C. rotundifolia* is a subwoody, locally abundant prostrate weed with stems arising from woody rootstocks. Each leaf has only one pair of leaflets, which have distinctly unequal bases. *C. hirsuta* is a hairy undershrub about 2 m tall often found around habitations. The flowers are yellow (Fig. 3.8.5A): this plant may be confused with *C. tora* and *C. occidentalis*. *C. mimosoides* is a woody herb or small shrub of disturbed locations and poor soil. The flowers are yellow and the leaflets are very small (about 1 cm long). Members of the *Cassia* genus have a vast array of pod types – some are flat, some are cylindrical (Fig. 3.8.5B), while some are variously winged as in *C. alata*.

Afzelia africana (Fig. 3.8.5C) is a tree about 30 m tall, found in savanna and dry forest. The sepals are green, while the single white petal has red markings. The fruit matures into a black woody pod carried conspicuously when the plant is leafless. The seeds are black and shining with a red aril at one end.

Piliostigma and *Bauhinia* have characteristic bilobed leaves (Fig. 3.8.6A). *Piliostigma thonningii* is a large savanna shrub with white unisexual flowers produced in large panicles during the dry season. The fruits, which are rust-coloured, are about 4 cm broad and 20 cm long. Various species of *Bauhinia* are grown as ornamental plants. These include *B. monandra* with only one stamen (Fig. 3.8.6A and C) and *B. tomentosa*

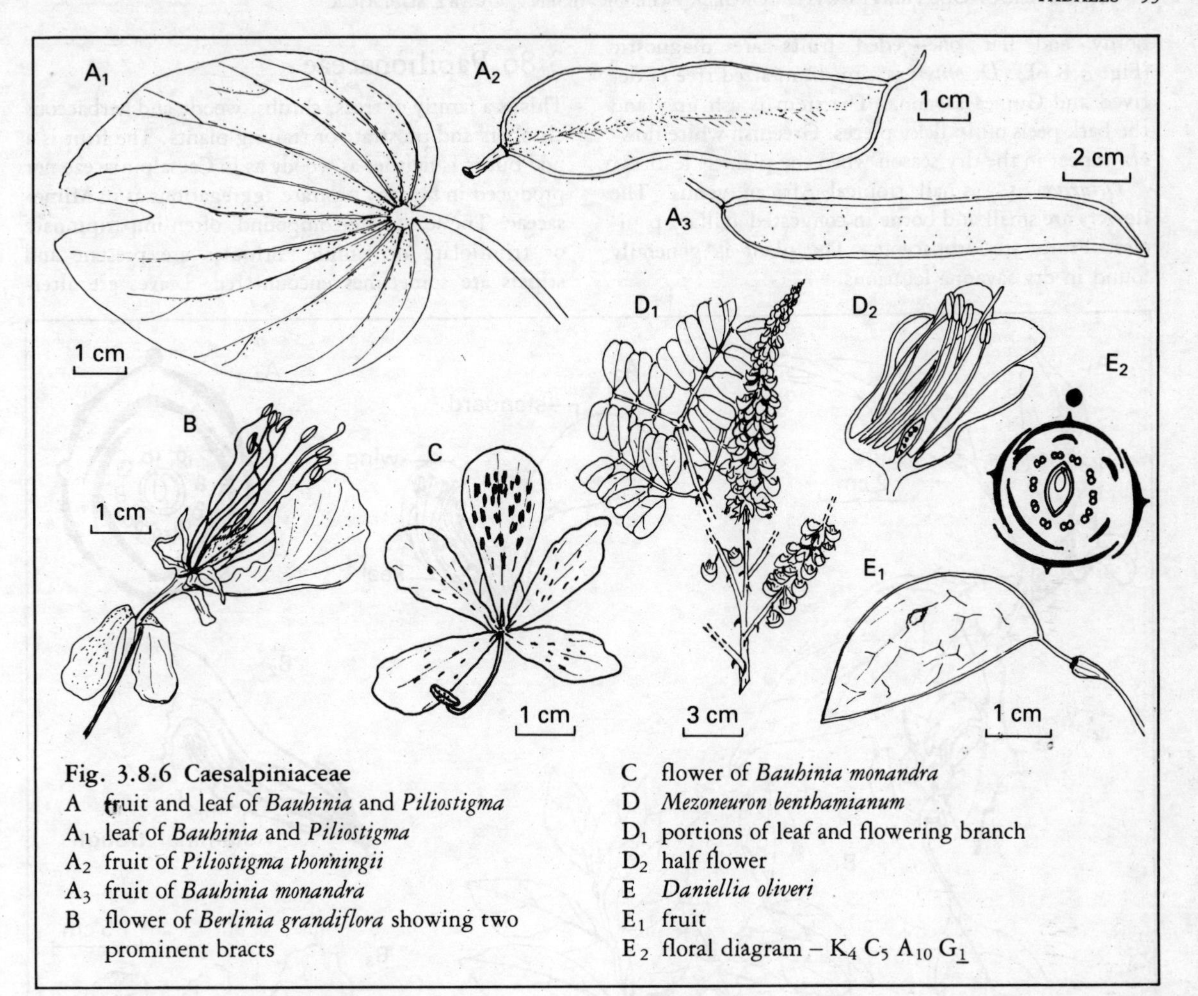

Fig. 3.8.6 Caesalpiniaceae

A fruit and leaf of *Bauhinia* and *Piliostigma*
A_1 leaf of *Bauhinia* and *Piliostigma*
A_2 fruit of *Piliostigma thonningii*
A_3 fruit of *Bauhinia monandra*
B flower of *Berlinia grandiflora* showing two prominent bracts
C flower of *Bauhinia monandra*
D *Mezoneuron benthamianum*
D_1 portions of leaf and flowering branch
D_2 half flower
E *Daniellia oliveri*
E_1 fruit
E_2 floral diagram – $K_4\ C_5\ A_{10}\ G_{\underline{1}}$

with small leaves and subactinomorphic, light yellow flowers.

Berlinia is a genus of medium to large trees found in habitats ranging from coastal swamps to derived savanna. There are up to five pairs of leaflets and the flowers have valvate bracteoles, which are velutinous on the inner side. The sepals are much smaller, subtending a large adaxial petal. Nine of the stamens in this genus are joined, while one is free. The fruits are velutinous and brown and up to 40 cm long. *B. grandiflora* is usually a decumbent tree in forest regions or in river valleys in the savanna. The white flowers are usually quite conspicuous (Fig. 3.8.6B).

Isoberlinia is a small genus closely related to *Berlinia* and distinguished from it by the smaller flowers; species of *Isoberlinia* are also mostly savanna plants. *I. doka* is a medium-sized tree in the southerly regions of the savanna; it has glabrous leaves and fruits. *I. dalzielii* (= *I. tomentosa*) is a more northerly species found on poor stony soils and with ochraceous, tomentose leaves and branches.

Mezoneuron benthamianum and *Caesalpinia bonduc* are prickly forest climbers with bipinnate leaves; the former has recurved prickles on the leaves and stems, and the flowers are on branched racemes (Fig. 3.8.6D). The fruits are prickly, brown and one- or two-seeded in *C. bonduc*, while the fruits of *M. benthamianum* are flat, reddish, glabrous and indehiscent and are produced in conspicuous bunches in the dry season. *C. bonduc* is found near houses; the seeds are used for the 'ayo' game.

Daniellia has representatives in forest and savanna environments. The paripinnate leaves have four to eleven pairs of leaflets and their flowers are white or purple, in large spreading panicles. The flowers have imbricate, sepaloid bracteoles and a regular imbricate calyx. Some of the petals are quite reduced in size. The horny and flat one-seeded fruits are diagnostic (Fig. 3.8.6E). *D. oliveri* is a medium-sized tree of derived and Guinea savanna. The stem is ash grey and the bark peels off in flaky pieces. Greenish white flowers appear in the dry season when the plant is leafless.

Detarium is a small tropical African genus. The flowers are small and borne in congested axillary panicles. Fruits are drupaceous. The plant is generally found in dry savanna locations.

3·80 Papilionaceae

This is a family of trees, shrubs, woody and herbaceous climbers and prostrate or trailing plants. The fruit is a pod but it is neither as woody as in Caesalpiniaceae nor produced in bunchy capitate aggregations as in Mimosaceae. The leaves are compound, often imparipinnate or trifoliolate or simple. Stipules are present and stipels are sometimes encountered. Leaves are alter-

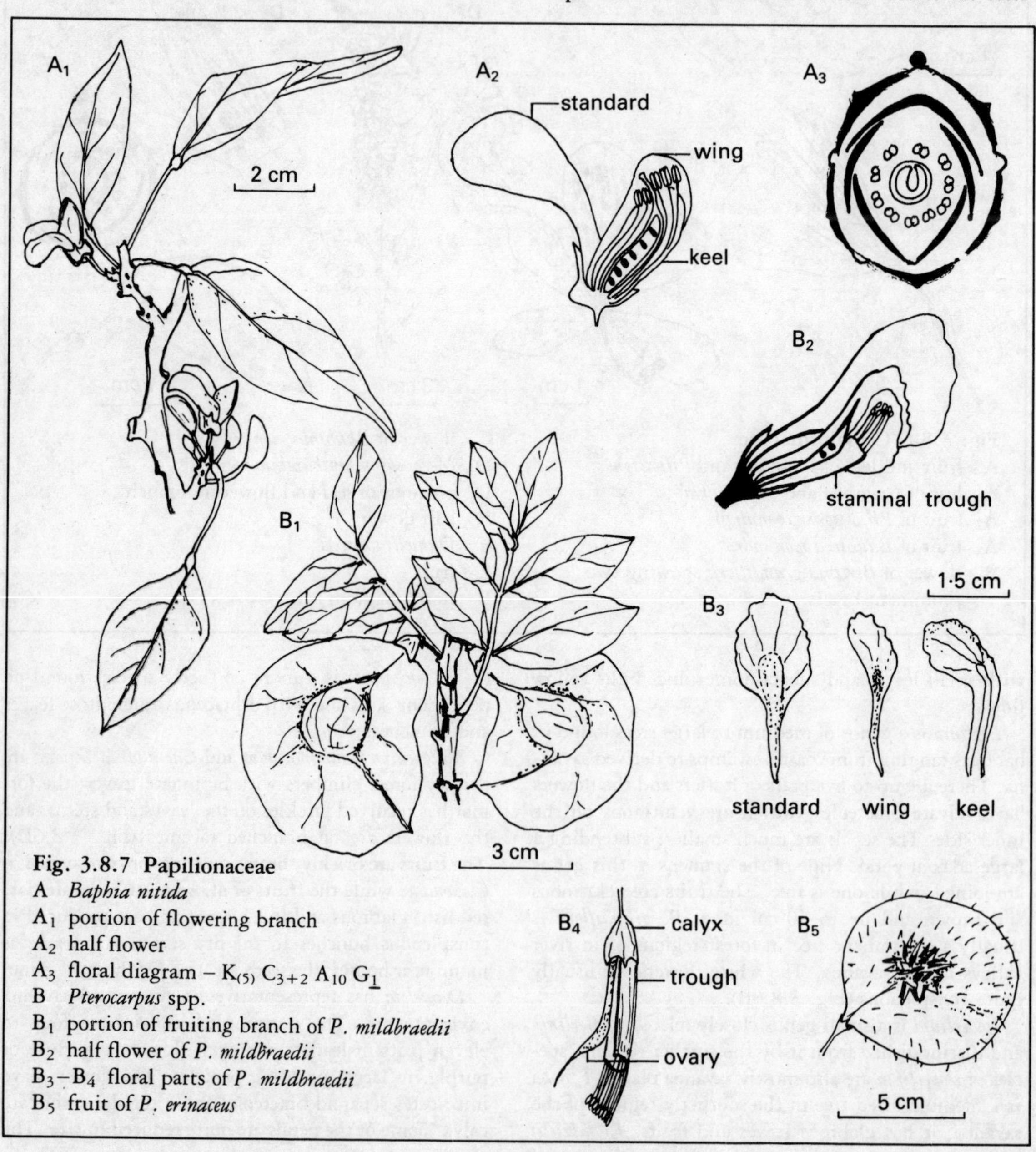

Fig. 3.8.7 Papilionaceae
A *Baphia nitida*
A_1 portion of flowering branch
A_2 half flower
A_3 floral diagram – $K_{(5)}\ C_{3+2}\ A_{10}\ G_{\underline{1}}$
B *Pterocarpus* spp.
B_1 portion of fruiting branch of *P. mildbraedii*
B_2 half flower of *P. mildbraedii*
B_3–B_4 floral parts of *P. mildbraedii*
B_5 fruit of *P. erinaceus*

nate. The butterfly-like flowers are diagnostic.

This family comprises some of the most important crop plants (the beans), cover crops and forage crops. The West African indigo used for dyeing clothes is also a member of this family. Many members are also important ornamental plants. The important food crops include the regular kinds of beans such as *Vigna unguiculata* (common cowpea or black-eye pea), *Phaseolus lunatus* (Lima bean), *Cajanus cajan* (pigeon pea) and *Canavalia ensiformis* (sword bean). Another group of food crop species in this family are those that bury their fertilized ovaries in the soil; these include *Arachis hypogaea* (groundnut or peanut), *Kerstingiella geocarpa* and *Voandzeia geocarpa* (Bambara groundnut). The important cover crops are *Mucuna* spp. and *Calopogonium mucunoides*. *Centrosema pubescens, Vigna unguiculata, Stylosanthes* spp., *Pueraria phaseoloides* and *Arachis hypogaea* are some of the fodder species commonly encountered. *Lonchocarpus cyanescens* is the shrub whose young leaves are used to make preparations for dyeing clothes in southwestern Nigeria. Some of the few decorative species are *Erythrina senegalensis, Clitoria ternatea* and *Gliricidia sepium*.

The flowers are zygomorphic, butterfly-shaped and mostly hermaphroditic. There are usually five sepals which are more or less joined. There are usually five imbricate petals with the adaxial one (the standard) quite large and outside all the others; the two lateral petals (the wings) are parallel and outside the two lower ones, which are connate to form the keel which encloses the stamens. The petals are free or partially joined at the base (Fig. 3.8.7A). There are usually ten stamens which are monadelphous, diadelphous or (rarely) free and with anthers that open by longitudinal slits. Floral characteristics (such as joining of stamens) and the structure of fruits are important in the tribal delimitation of this family, but such delimitation is not essential to this work.

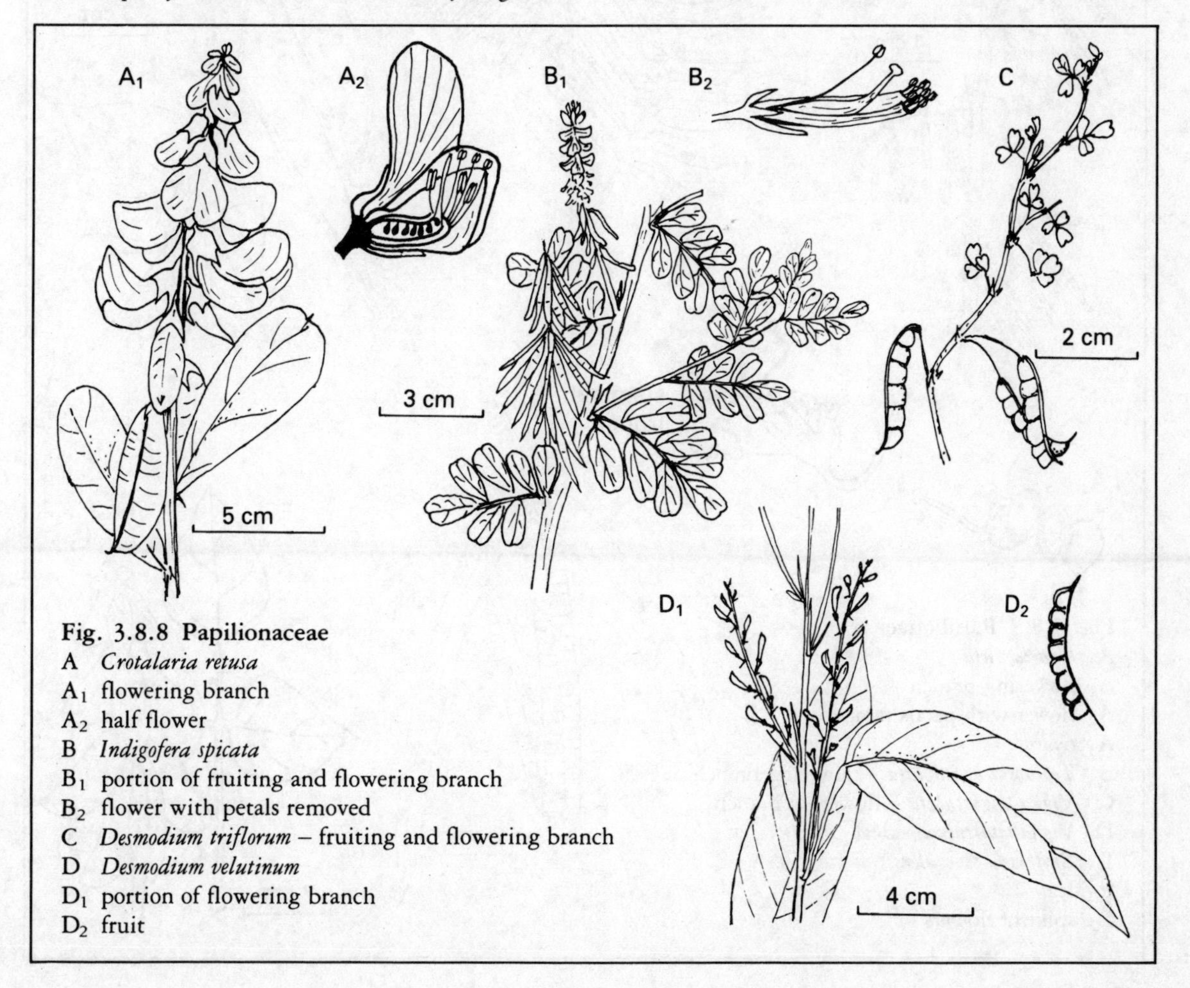

Fig. 3.8.8 Papilionaceae
A *Crotalaria retusa*
A_1 flowering branch
A_2 half flower
B *Indigofera spicata*
B_1 portion of fruiting and flowering branch
B_2 flower with petals removed
C *Desmodium triflorum* – fruiting and flowering branch
D *Desmodium velutinum*
D_1 portion of flowering branch
D_2 fruit

Baphia nitida is a small tree or shrub with glabrous grey branchlets and is found in relatively open secondary forest. The leaves are unifoliolate. The flowers are white with a yellow centre; they may be fasciculate in axillary clusters. As in other members of the tribe Sophoreae, the stamens are free in *B. nitida* (Fig. 3.8.7A).

Pterocarpus is a small genus of trees with winged fruits. *P. mildbraedii* is a tree about 15 m tall with a straight, grey bole and is found in forest environments. The flowers are yellow and the young fruits are green with foliaceous wings. The fruits are smooth and brown when mature (Fig. 3.8.7B). *P. erinaceus* is a small savanna tree with green prickly fruits produced when plants are leafless, while *P. osun* is a dry-forest species with prickly fruits about twice the size of those of *P. erinaceus*. *P. santalinoides* is a shrub or small tree of river banks with shining foliage and rugose fruits; the wing of the fruit is short and the fruits float easily on water. The samara found in *Pterocarpus*, and indehiscent

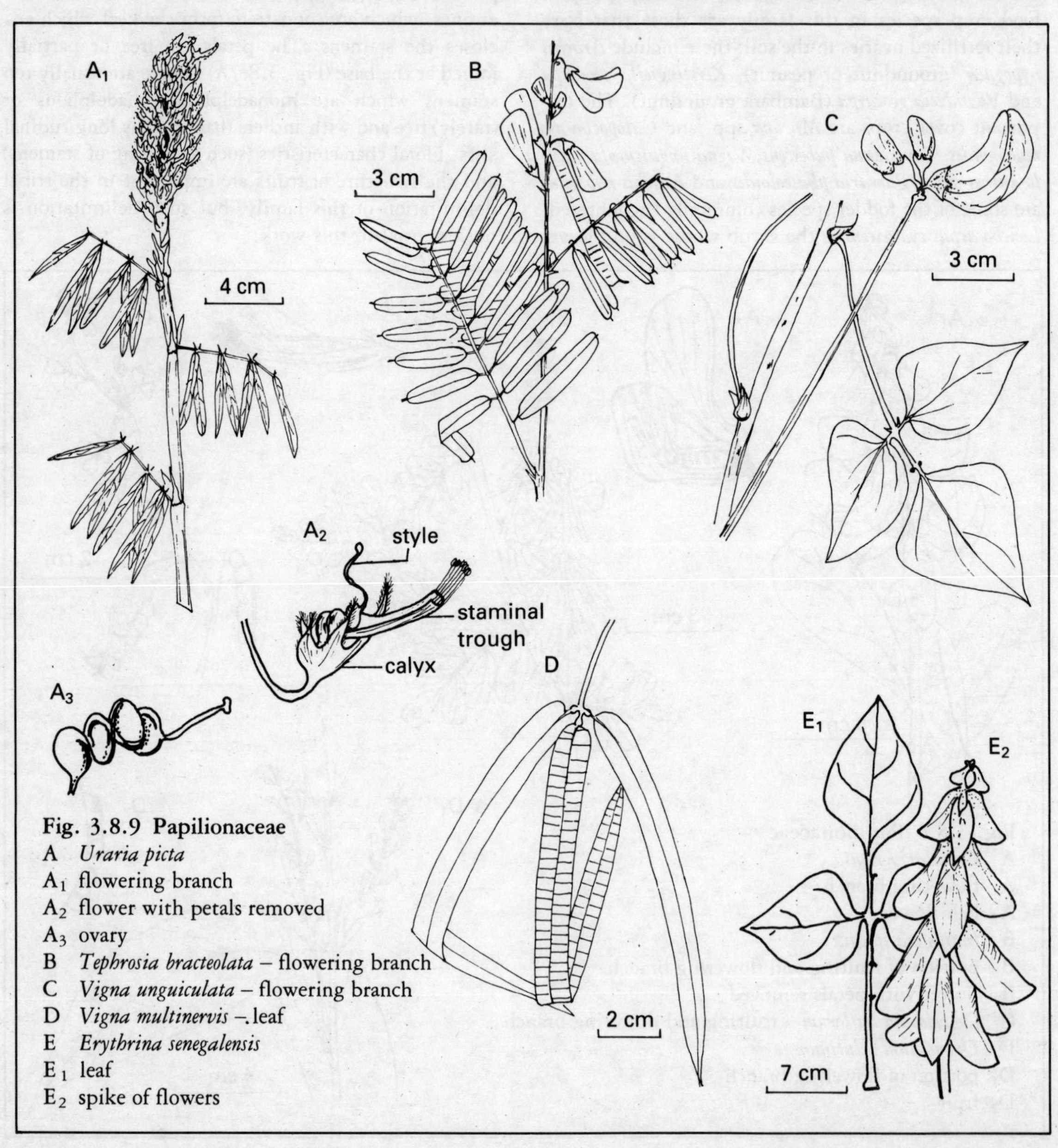

Fig. 3.8.9 Papilionaceae
A *Uraria picta*
A_1 flowering branch
A_2 flower with petals removed
A_3 ovary
B *Tephrosia bracteolata* – flowering branch
C *Vigna unguiculata* – flowering branch
D *Vigna multinervis* – leaf
E *Erythrina senegalensis*
E_1 leaf
E_2 spike of flowers

fruits in general, are characteristic of the tribe Dalbergieae.

Foliage of the genus *Millettia* may be easily confused with that of *Pterocarpus*, but members of the former have purple or blue flowers and their pods are generally many-seeded.

Crotalaria is a large genus of erect, rarely prostrate, plants. Inflated fruits which rattle are produced from yellow flowers which are borne in axillary or terminal spikes. The plants have simple, trifoliolate or pentafoliolate leaves. Five of the anthers are round and carried on long filaments alternating with, and joined to, five other filaments carrying fairly long anthers; this dimorphism of anthers in *Crotalaria* is characteristic of the tribe Genisteae. *C. retusa* is illustrated in Fig. 3.8.8A.

Indigofera is a very large genus of erect or prostrate woody herbs. *I. spicata* is a highly variable woody herb of disturbed or open locations. The axillary spikes of pink flowers produce a bunch of many-seeded fruits (Fig. 3.8.8.B).

Desmodium is a fairly large genus of mostly woody herbs and undershrubs. Their flowers are pink or purple and their leaves are trifoliolate or simple. The jointed fruits, which break into lomenta that are easily dispersed by animals, are characteristic of the tribe Hedysareae. *D. triflorum* is a tiny herb on lawns often forming a thick mat. The lomenta stick very easily on clothes (Fig. 3.8.8C). *D. velutinum* is a hairy and woody undershrub with simple leaves and large branched racemes of purple flowers (Fig. 3.8.8D).

Uraria picta is a distinctive undershrub (often cultivated for medicinal purposes) found in the savanna. The pink or purplish flowers are in dense terminal spikes which are very hairy and which produce twisted pearl grey fruits (Fig. 3.8.9A).

Tephrosia is a fairly large genus of woody undershrubs with procumbent or erect stems. The leaves are mostly pinnately five- or more-foliolate. The flowers are solitary or racemose, pink or purple producing more or less flat fruits. *T. bracteolata* is illustrated in Fig. 3.8.9B. Various species of *Tephrosia* are widespread in the savanna where they may form pure and extensive stands and are used as fodder.

Vigna is also a large genus of herbaceous trailing or climbing species with blue, pink, purple, cream, reddish or white flowers widespread in various vegetational belts. *V. unguiculata* is cultivated; it also has many wild varieties in open forest and derived savanna locations (Fig. 3.8.9C). *V. multinervis* is a distinctive climber in grassland; the stem is hairy with trifoliolate leaves, the secondary veins of which are perpendicular to the main vein (Fig. 3.8.9D).

Erythrina senegalensis is a common savanna tree or shrub. The branchlets and leaf petioles are armed with recurved spines; the flowers are bright red and conspicuous when the plant is leafless. The bark is usually grey and older ones are deeply fissured (Fig. 3.8.9E).

Abrus precatorius is a climbing herb or woody twining shrub with stout racemes of pinkish yellow flowers and pods that split to expose red seeds with black ends.

3·9 Euphorbiales

3·9p Euphorbiaceae

This is a large family of trees, shrubs and herbs which is mostly tropical. They are rarely climbing (or twining) herbs with stinging hairs as in *Tragia*, or they may be of peculiar habits as in *Euphorbia kamerunica* which is often wrongly called cactus and is used for hedges. The leaves are alternate, simple, undivided or digitately lobed or (rarely) digitately compound as in *Ricinodendron heudelotii*. They are quite often stipulate. The leaves are rarely opposite as in *Mallotus oppositifolius*. Milky juice is encountered in some genera such as *Euphorbia* and *Manihot* (the cassava genus).

The family includes many introduced ornamental species such as the so-called 'ice plant' or 'snow bush' (*Breynia nivosal*), *Acalypha, Jatropha* spp., the plant often wrongly called 'croton' (*Codiaeum variegatum*), Poinsettia (*Euphorbia pulcherrima*) and *Hura crepitans*, the fruits of which children use to make toy tyres. Important economic species include the rubber tree (*Hevea brasiliensis*), cassava (*Manihot esculentus*), castor oil (*Ricinus communis*) and awusa, asala (*Tetracarpidium conophorum*).

The flowers in this family are unisexual and mostly monoecious; dioecism is found, for example in *Mallotus*. The ovary is superior and the flowers are radially symmetrical and usually apetalous. Perianth segments are usually pentamerous and valvate or imbricate. In the specialised inflorescence of *Euphorbia* spp. (called a 'cyathium'), the sepals are very much reduced or absent. Stamens range from one to numerous and they are free with two- to four-celled anthers which generally open lengthwise. A rudimentary pistil may be present in the centre of the male flower, while staminodes may be present in the female flower. A disc is usually present in the flower. The ovary is usually three-celled with solitary or paired pendulous ovules attached to the inner side of the ovary cells.

The fruit is a capsule, a drupe or (rarely) a berry. The seeds may have a conspicuous caruncle and they usually have copious endosperm.

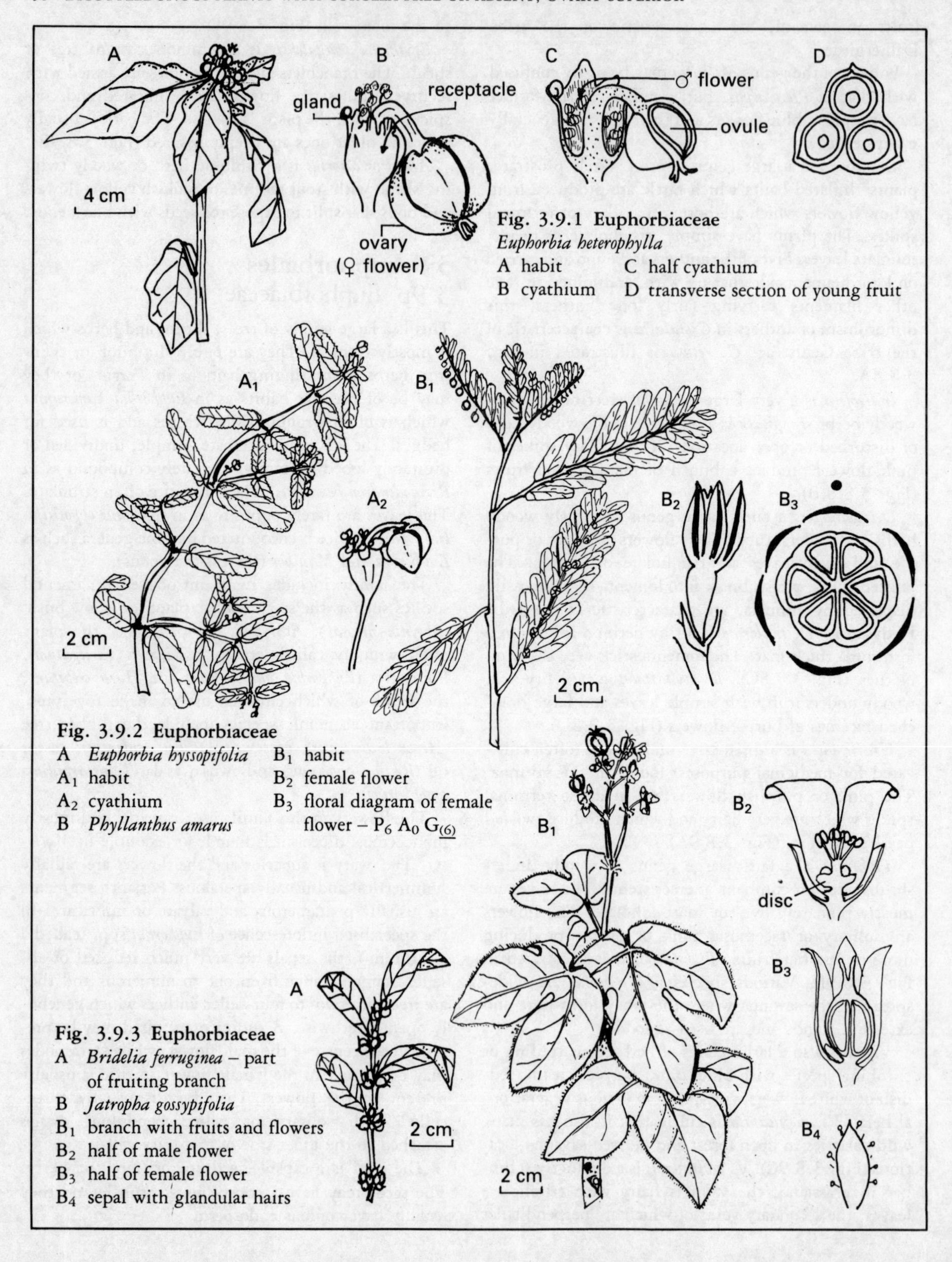

Fig. 3.9.1 Euphorbiaceae
Euphorbia heterophylla
A habit
B cyathium
C half cyathium
D transverse section of young fruit

Fig. 3.9.2 Euphorbiaceae
A *Euphorbia hyssopifolia*
A_1 habit
A_2 cyathium
B *Phyllanthus amarus*
B_1 habit
B_2 female flower
B_3 floral diagram of female flower – $P_6\ A_0\ \underline{G_{(6)}}$

Fig. 3.9.3 Euphorbiaceae
A *Bridelia ferruginea* – part of fruiting branch
B *Jatropha gossypifolia*
B_1 branch with fruits and flowers
B_2 half of male flower
B_3 half of female flower
B_4 sepal with glandular hairs

Figures 3.9.1 and 3.9.2A show two representative species of *Euphorbia*: the characteristic inflorescence of the genus (cyathium) is illustrated. *E. heterophylla* is a common weed of abandoned farmland in wet and dry forest regions. The leaves around the flowers turn bright red towards the dry season. *E. hyssopifolia* is a common glabrous, erect or decumbent herb of lawns and waste places in forest and derived savanna regions. It is found growing with other related species of *Euphorbia* such as *E. hirta*, which is also prostrate, but more hairy and with rather stout stems.

Phyllanthus is a genus of herbs, shrubs and small trees with simple alternate leaves, the flowers appearing as if they are borne on the leaves (hence the name of the genus). Species of the genus are monoecious with inconspicuous flowers. *P. amarus* (Figure 3.9.2B) is a common weed of waste places in forest and derived savanna environments. The flowers are arranged linearly on the underside of branches which appear as compound leaves. The fruit is a berry with six cells. *P. muellerianus* is a straggling regrowth species of secondary forest with fasciculate flowering branchlets. The plants are leafless when in flower and the recurved stipular spines are characteristic. *P. discoideus* is a deciduous shrub or tree, a regrowth species flowering when leafless. The female plants produce green, three-lobed fruits.

Bridelia is a genus of trees and shrubs usually with marginal nerves in the leaves. They are widespread in forest and savanna environments. The plants, which are monoecious, have pinnately-nerved alternate leaves. The flowers and fruits occur in characteristic axillary clusters (Fig. 3.9.3A). The flowers have five sepals and five small petals. The two-celled ovary usually ripens into a one-or two-stoned drupe. The species are often armed with stout or fairly long spines on their stems.

Jatropha is represented by many cultivated species. They are monoecious with prominent glandular discs present in their flowers. The fruits are three-celled with one seed per cell. *J. gossypifolia* is the common red species (Fig. 3.9.3B) which is planted around houses.

Alchornea is a genus of regrowth species of secondary forests. They are monoecious. The genus is represented

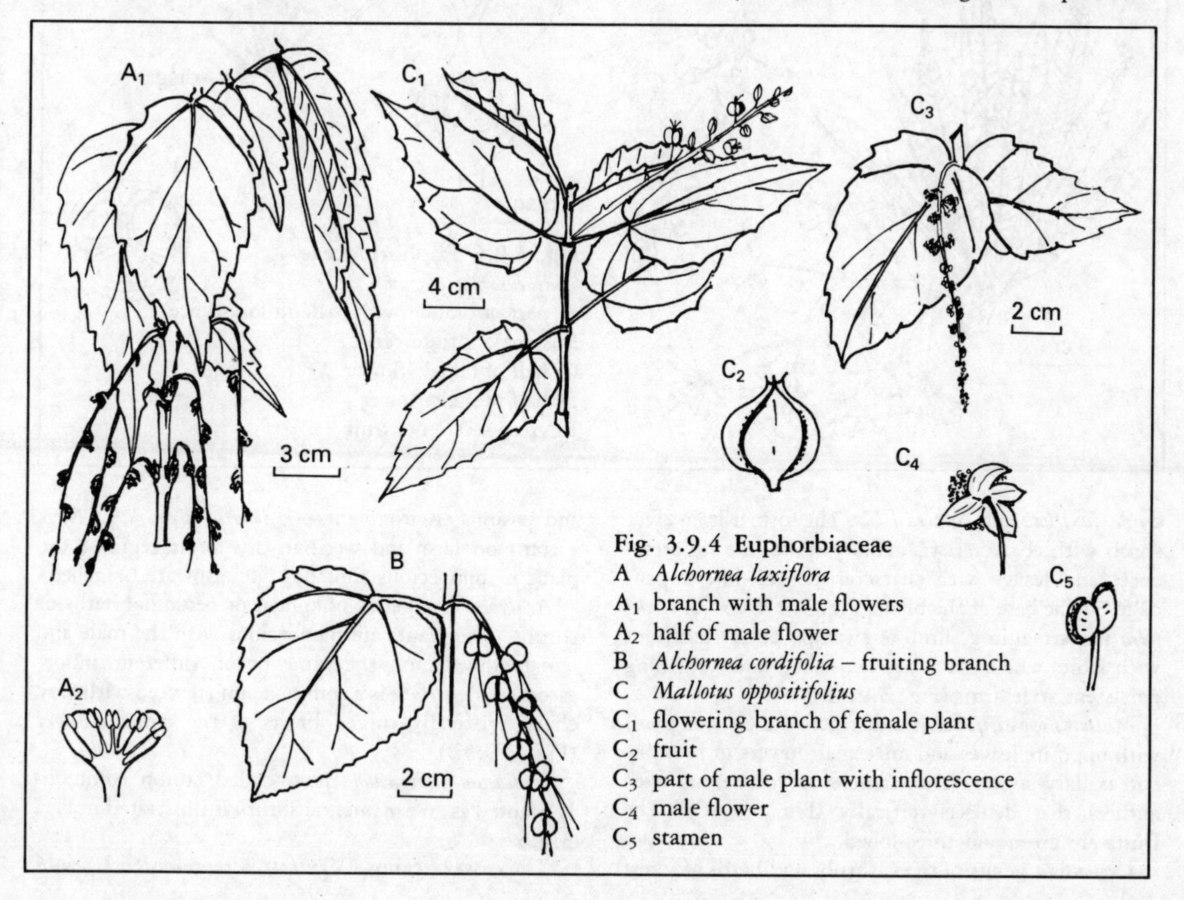

Fig. 3.9.4 Euphorbiaceae
A *Alchornea laxiflora*
A_1 branch with male flowers
A_2 half of male flower
B *Alchornea cordifolia* – fruiting branch
C *Mallotus oppositifolius*
C_1 flowering branch of female plant
C_2 fruit
C_3 part of male plant with inflorescence
C_4 male flower
C_5 stamen

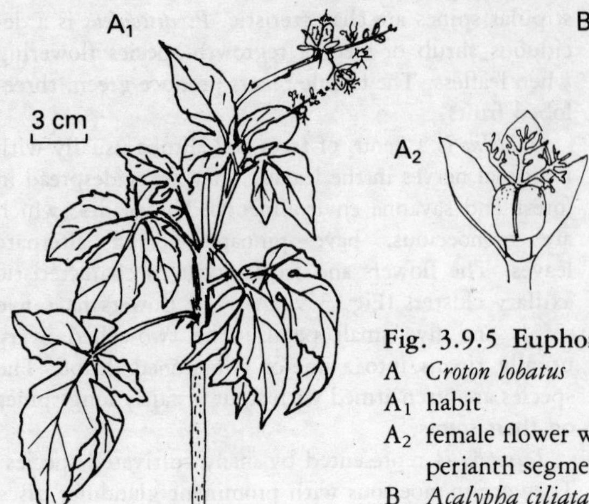

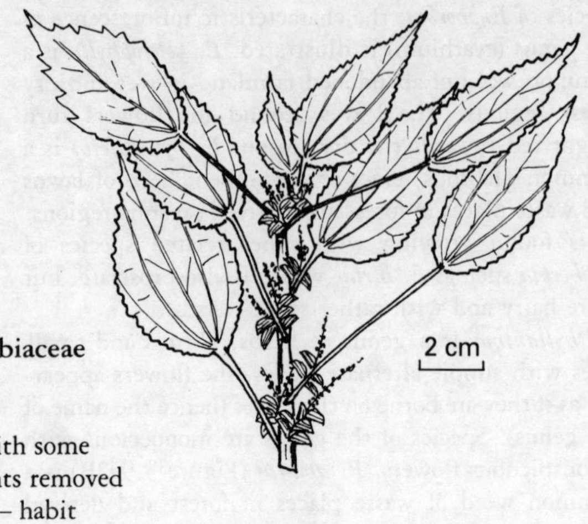

Fig. 3.9.5 Euphorbiaceae
A *Croton lobatus*
A_1 habit
A_2 female flower with some perianth segments removed
B *Acalypha ciliata* – habit

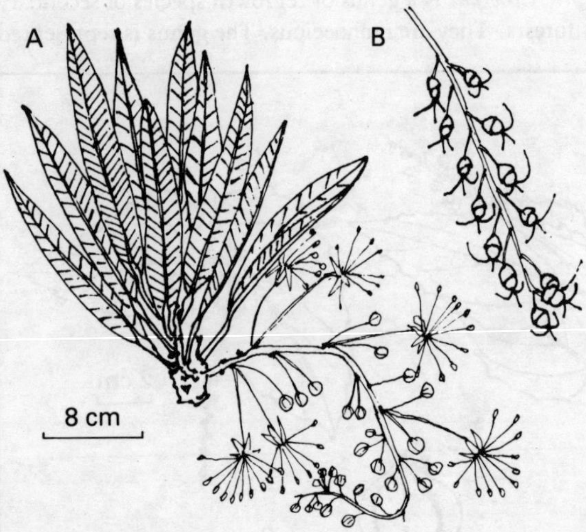

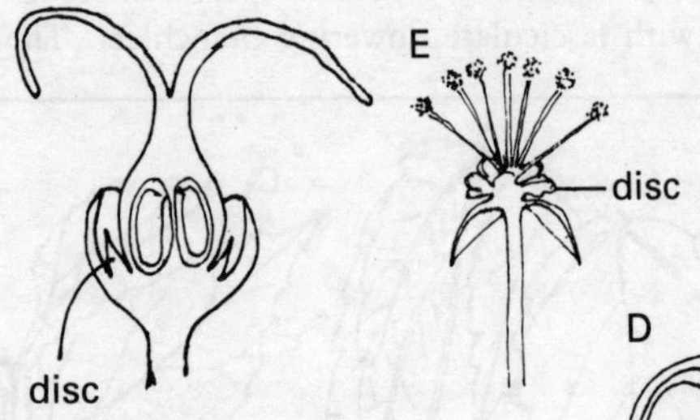

Fig. 3.9.6 Euphorbiaceae
Codiaeum variegatum
A part of branch with male inflorescence
B female inflorescence
C half of female flower
D half of male flower
E cross-section of fruit

by *A. laxiflora* and *A. cordifolia*. The former is an erect shrub with characteristic axillary spike-like inflorescences and leaves with characteristic tail-like appendages at the base of the blade (Fig. 3.9.4A.). *A. cordifolia* is a straggling shrub of swampy or dry locations with fruits which are often two-lobed and with long persistent style arms (Fig. 3.9.4B).

Mallotus oppositifolius is a distinctive dioecious plant with opposite leaves and unisexual flowers in prominent axillary spikes. The stamens are numerous with anthers that dehisce vertically (Fig. 3.9.4C). The fruits are green and three-lobed.

Croton is a genus of trees, shrubs and herbs of forest and savanna environments. *C. lobatus* (Fig. 3.9.5A) is a common farmland weed in dry forest regions; the plant is monoecious and the style arms are branched.

Acalypha is a genus of annual or perennial herbs or shrubs. The plants are monoecious with the male and female flowers on the same or on different inflorescences. *A. ciliata* is a common annual weed with very characteristically ridged bracts of the female flower (Fig. 3.9.5B).

Codiaeum variegatum (the so-called 'croton' commonly planted as an ornamental shrub) is illustrated in Fig. 3.9.6.

The cassava genus (*Manihot*) is represented by wild

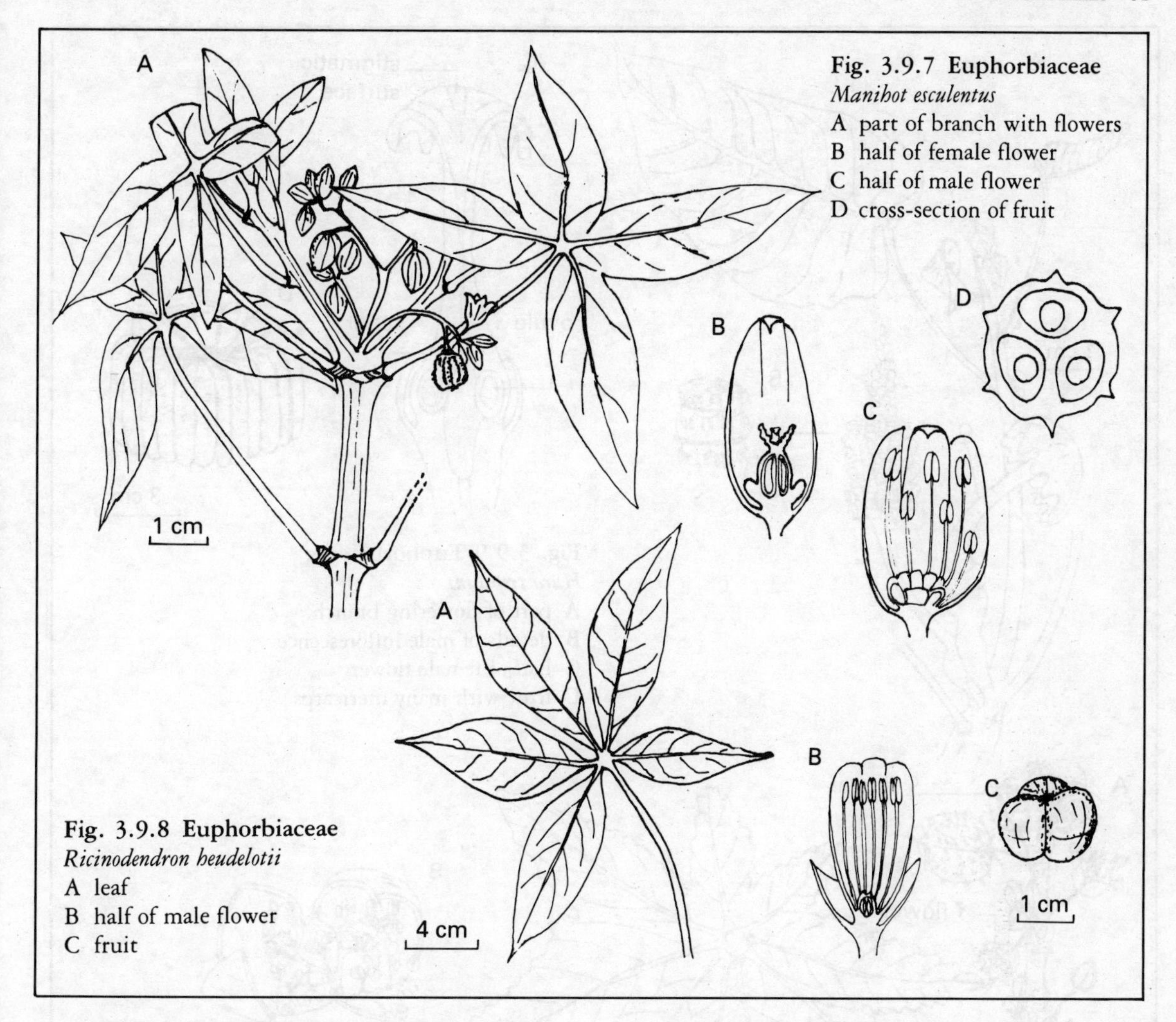

Fig. 3.9.7 Euphorbiaceae
Manihot esculentus
A part of branch with flowers
B half of female flower
C half of male flower
D cross-section of fruit

Fig. 3.9.8 Euphorbiaceae
Ricinodendron heudelotii
A leaf
B half of male flower
C fruit

(*M. glaziovii*) and cultivated (*M. esculentus*) species. They are monoecious and their flowers have prominent glandular discs (Fig. 3.9.7). *M. glaziovii* grows abundantly, almost like a weed, in forest openings and forest road-sides.

Ricinodendron heudelotii (Fig. 3.9.8) is a common forest tree with digitate compound leaves subtended by prominent foliaceous stipules. The flowers are produced in large panicles. The fruits are two- or three-seeded with a glabrous coat. The fruits, which are produced in large numbers, germinate very readily.

Hura crepitans is illustrated in Figure 3.9.9. The male flowers are in a characteristic inflorescence, while the female flowers are solitary on the same plant. The fruits are characteristic.

Ricinus communis is monoecious with male and female flowers on the same inflorescence. The male flowers have peculiar stamens which are branched (Fig. 3.9.10).

There are many other species of ecological significance in the family. *Hymenocardia acida* is a common savanna shrub or small tree with a stem that becomes rust-coloured when the bark peels off. The pale green leaves are simple and the fruit is a samara. *Securinega virosa* is a common regrowth woody shrub with glabrous stems and elliptic or obovate leaves. The tiny greenish white flowers are usually clustered in the leaf axils. The plants are dioecious. The fruit is a three-celled berry with two seeds per cell.

Uapaca is a genus of trees often with stilt roots, usually found in swampy locations from the coast to the savanna. They are monoecious or dioecious with characteristic male inflorescences, which appear as if they are a single flower with numerous stamens.

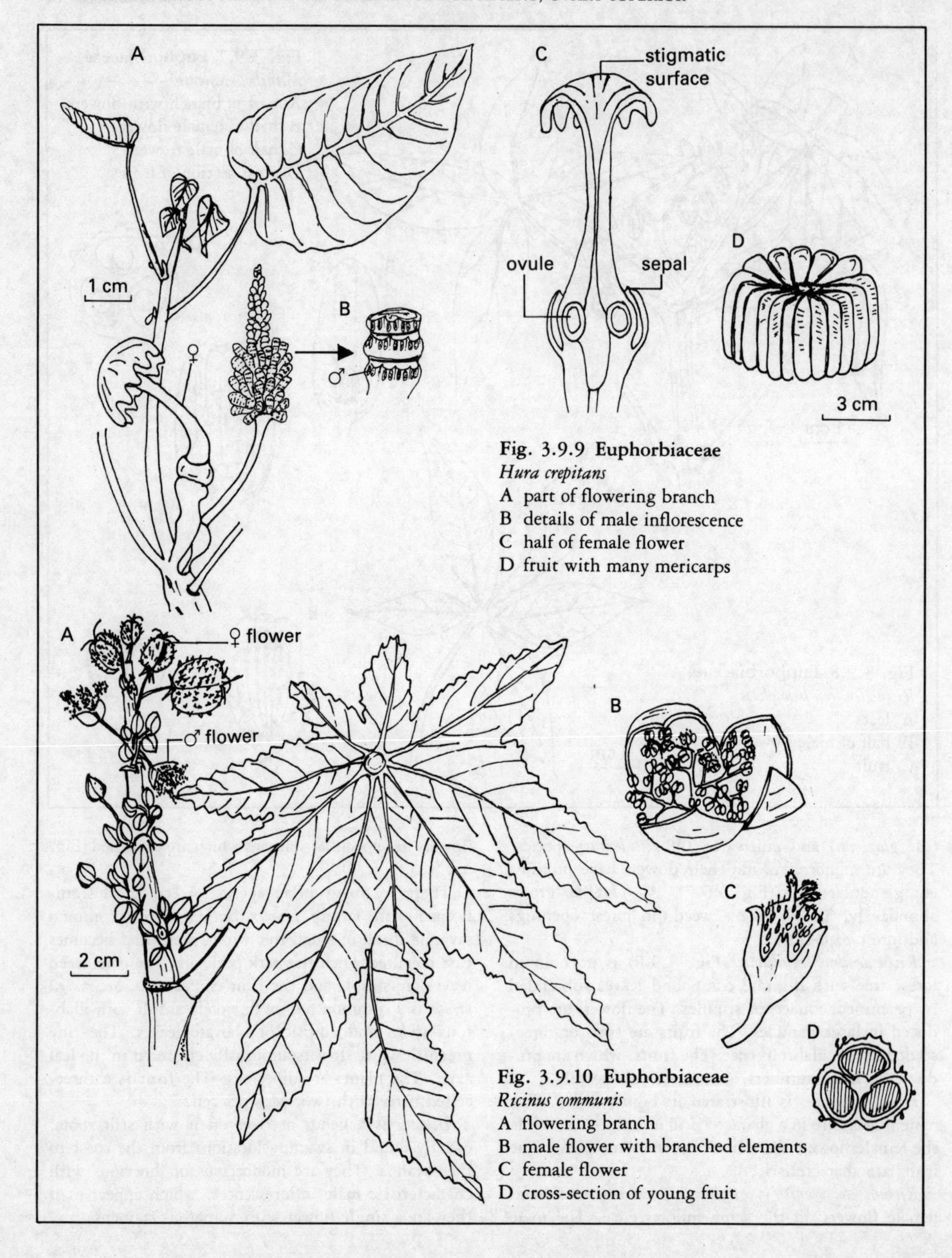

Fig. 3.9.9 Euphorbiaceae
Hura crepitans
A part of flowering branch
B details of male inflorescence
C half of female flower
D fruit with many mericarps

Fig. 3.9.10 Euphorbiaceae
Ricinus communis
A flowering branch
B male flower with branched elements
C female flower
D cross-section of young fruit

3·10 Sapindales

3·10q Meliaceae

This family is composed of trees and shrubs with alternate, pinnate and exstipulate, eglandular leaves. The wood usually emits a strong odour and the small (usually whitish) flowers are borne in loose axillary panicles.

Azadirachta indica is an introduced shade tree often planted, especially in the drier parts of West Africa. Some members of this family are important timber species, e.g. *Khaya grandifoliola, K. ivorensis, Lovoa trichilioides* and *Entandrophragma* spp. Species of *Khaya* are generally called 'African mahogany' while *Entandrophragma* spp. are variously referred to as 'Utile', 'Omu' or 'Sapele'.

The flowers of Meliaceae are actinomorphic, apparently hermahphroditic, but usually unisexual. They are tetramerous or pentamerous and they have a superior ovary. The calyx is short and mostly imbricate. The petals are free, rarely united; they are either imbricate or contorted – they are rarely valvate. There are usually twice as many stamens as there are petals and they are joined basally in varying degrees to form a petaloid staminal tube. A disc is present in the flower. There are two to five carpels with between one to twelve pendulous ovules per cell.

The fruit is usually woody and dehiscent (often splitting explosively), usually with many winged seeds attached to a woody central axis. The valves of the woody fruits may split from the top down, from the base up or both ways. When the valves are leathery, the seeds are few and arillate as in *Trichilia heudelotii* (Fig. 3.10.1). Fruits are rarely baccate or drupaceous.

Trichilia heudelotii is illustrated in Fig. 3.10.1. It is a small tree which is very common in secondary forest. The yellowish white flowers are borne on a loose axillary inflorescence and develop to produce many three-valved fruits which are purplish when immature and

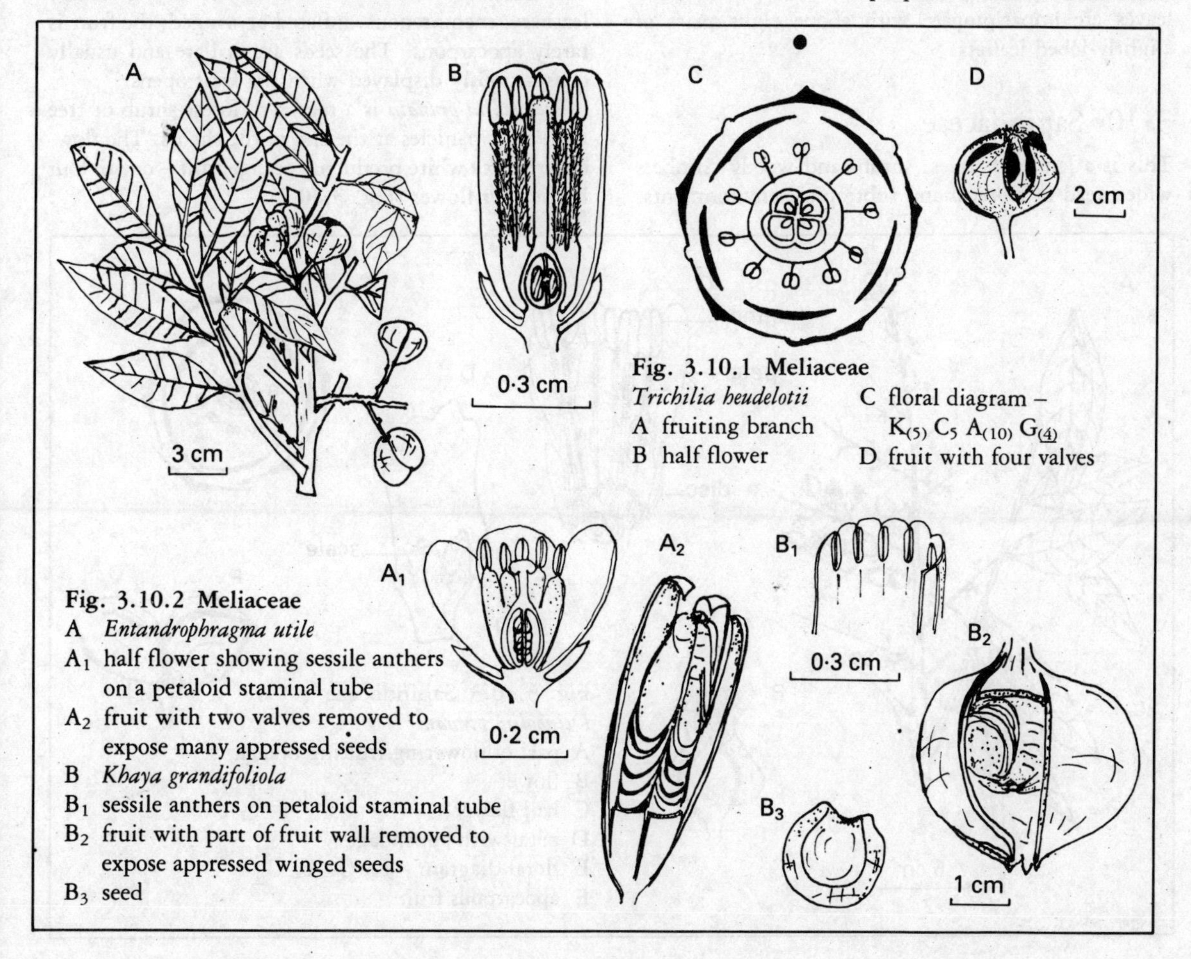

Fig. 3.10.1 Meliaceae
Trichilia heudelotii
A fruiting branch
B half flower
C floral diagram – $K_{(5)}\ C_5\ A_{(10)}\ G_{\underline{(4)}}$
D fruit with four valves

Fig. 3.10.2 Meliaceae
A *Entandrophragma utile*
A_1 half flower showing sessile anthers on a petaloid staminal tube
A_2 fruit with two valves removed to expose many appressed seeds
B *Khaya grandifoliola*
B_1 sessile anthers on petaloid staminal tube
B_2 fruit with part of fruit wall removed to expose appressed winged seeds
B_3 seed

velvety brown when ripe. The seeds are arillate and have a characteristic red and black coat.

Khaya grandifoliola is a buttressed tree of dry forests. The imparipinnate leaves have about four pairs of leaflets which are bright red when they are young. *Entandrophragma utile* is also a buttressed forest species with paripinnate leaves crowded at the end of branches. Certain fruit and floral characteristics of *Entandrophragma utile* and *Khaya grandifoliola* are illustrated in Fig. 3.10.2. Note that the attachment of the anthers to the staminal tube can be a useful taxonomic criterion.

Azadirachta indica had imparipinnate leaves, with leaflets characterised by serrated margins. The flowers, which have conspicuous staminal tubes, are greenish white and borne in lax axillary panicles. The wood and fruits have a strong and characteristic odour.

Savanna species in this family are represented in genera such as *Trichilia, Khaya* and *Pseudocedrela*. In *P. kotschyi,* which is a small savanna tree with grey and fissured bark producing white flowers in panicles, the leaves are imparipinnate with about eight pairs of slightly-lobed leaflets.

3·10r Sapindaceae

This is a family of trees, shrubs and woody climbers widespread in tropical and subtropical environments. The leaves are alternate or (rarely) opposite and exstipulate. The trees and shrubs have paripinnate leaves which lack tendrils. The climbers have characteristic tendrils subtending the racemose inflorescences of white flowers. *Blighia sapida* is the edible akee apple: this is the only species of economic importance; it grows wild most of the time. The flowers are actinomorphic or zygomorphic, unisexual or hermaphroditic and tetramerous or pentamerous. The sepals are free and valvate or imbricate. The petals, when present, are free and imbricate; one of the petals may have a scale or some other appendage associated with it. A disc is usually present. The stamens are usually free and there are twice as many stamens as petals or they are numerous; they may be all drawn towards one side of the flower. The ovary is superior with one to eight carpels (usually three). It is syncarpous or rarely apocarpous. The ovules are usually in axile placentation with one or two of them per ovary cell.

The fruit is usually a capsule which may be fleshy, leathery, membranous, inflated or winged; the fruit is rarely apocarpous. The seeds are arillate and usually conspicuously displayed when the fruit opens.

Deinbollia pinnata is a regrowth forest shrub or tree with erect panicles at the end of the shoots. The flowers are dirty white producing orange fruit – one to four berries per flower (Fig. 3.10.3).

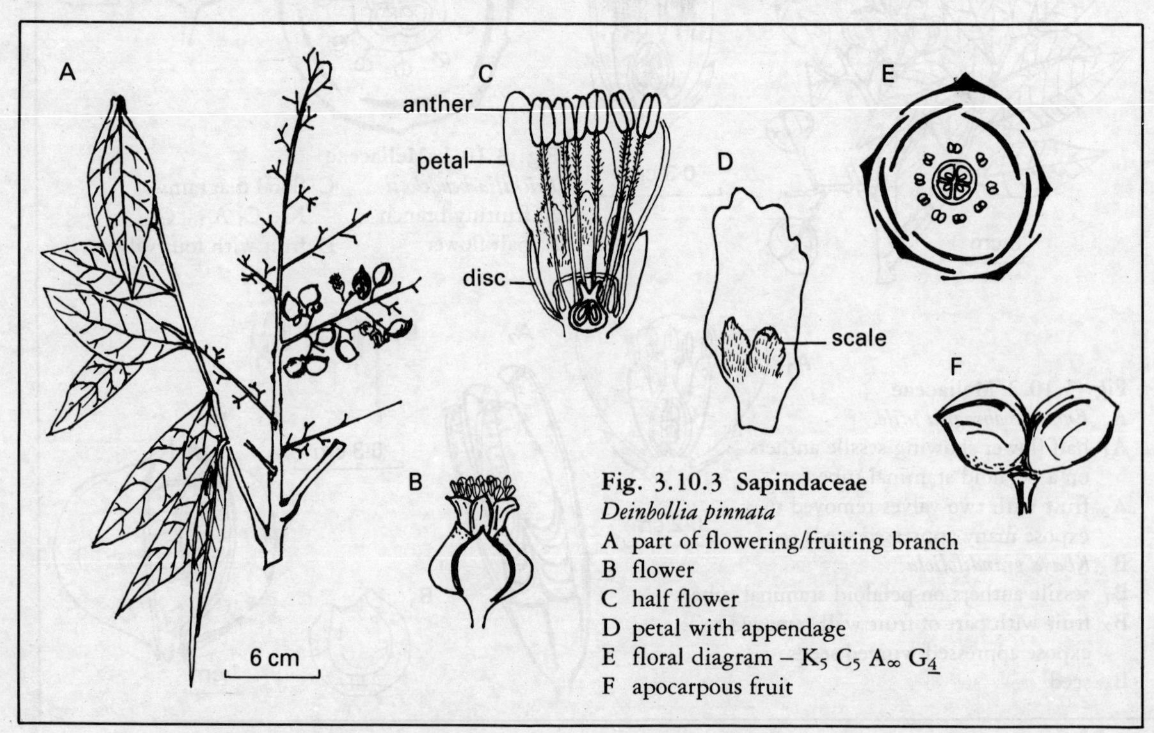

Fig. 3.10.3 Sapindaceae
Deinbollia pinnata
A part of flowering/fruiting branch
B flower
C half flower
D petal with appendage
E floral diagram – $K_5\ C_5\ A_\infty\ G_{\underline{4}}$
F apocarpous fruit

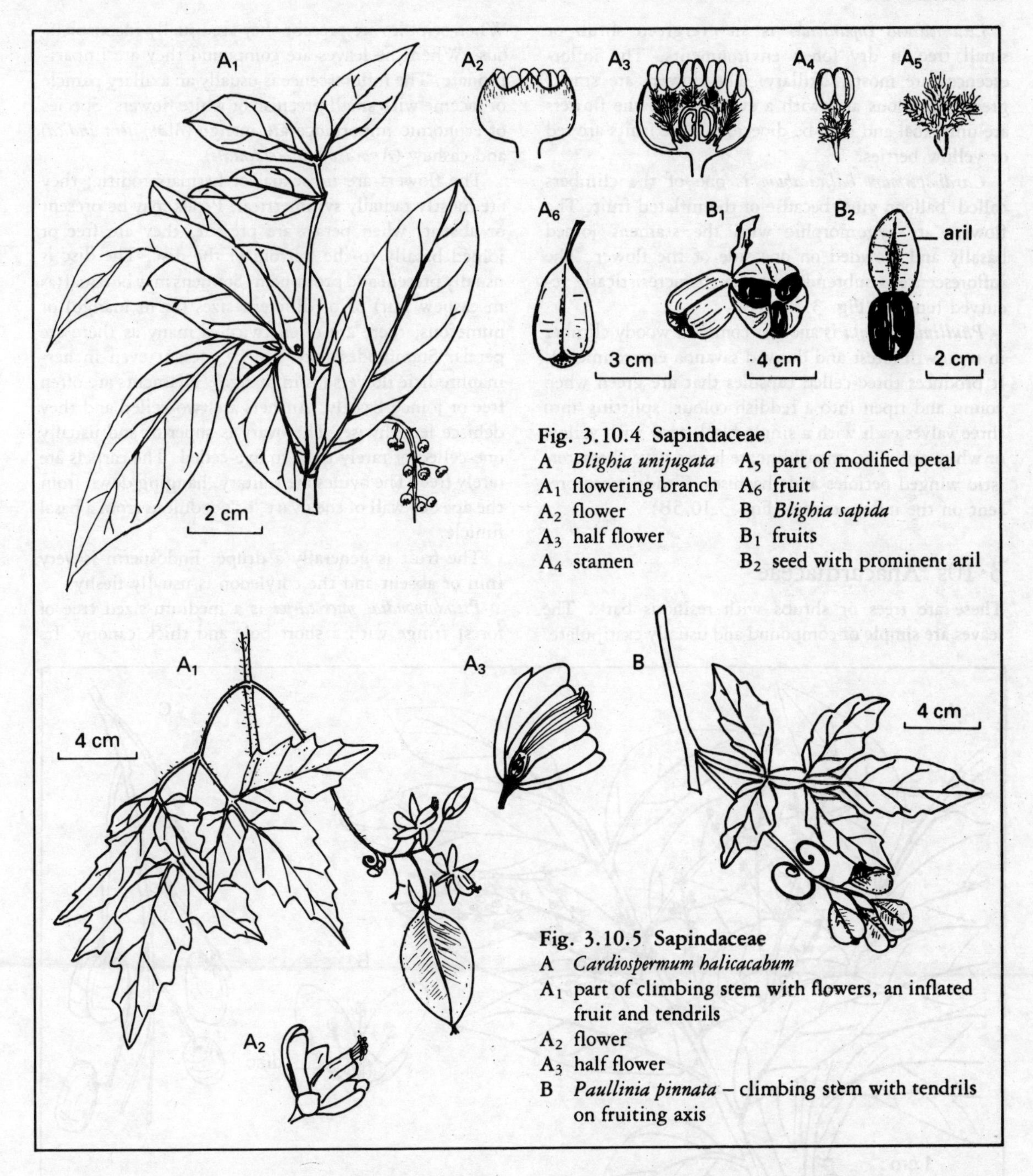

Fig. 3.10.4 Sapindaceae

A *Blighia unijugata*
A_1 flowering branch
A_2 flower
A_3 half flower
A_4 stamen
A_5 part of modified petal
A_6 fruit
B *Blighia sapida*
B_1 fruits
B_2 seed with prominent aril

Fig. 3.10.5 Sapindaceae

A *Cardiospermum halicacabum*
A_1 part of climbing stem with flowers, an inflated fruit and tendrils
A_2 flower
A_3 half flower
B *Paullinia pinnata* – climbing stem with tendrils on fruiting axis

Blighia unijugata is a dioecious forest tree, usually with two pairs of leaflets. The white fragrant flowers produce red capsules which split to expose shining black seeds with prominent yellow arils (Fig. 3.10.4). *B. sapida*, like *B. unijugata*, is a medium-sized tree, but is found also in the savanna. It is similar to the latter in many respects except for the larger fruits and more leaflets per leaf (Fig. 3.10.4B).

Sapindus abyssinicus is a forest tree with woody inflorescences producing red one-seeded berries, one other carpel having aborted but still remaining at the base of the fruit.

Lecaniodiscus cupanioides is an evergreen shrub or small tree in dry forest environments. The inflorescences are mostly axillary. The flowers are small, green, apetalous and with a sweet smell; the flowers are unisexual and may be dioecious. The fruits are red or yellow berries.

Cardiospermum halicacabum is one of the climbers called 'balloon vine' because of the inflated fruit. The flowers are zygomorphic with the stamens joined basally and crowded on one side of the flower. The inflorescence is subtended by the characteristically recurved tendrils (Fig. 3.10.5A).

Paullinia pinnata is another common woody climber in regrowth forest and derived savanna environments. It produces three-celled capsules that are green when young and ripen into a reddish colour, splitting into three valves each with a single black seed with yellow or white aril. The imparipinnate leaves have characteristic winged petioles and rhachises. Tendrils are present on the inflorescences (Fig. 3.10.5B).

3·10s Anacardiaceae

These are trees or shrubs with resinous bark. The leaves are simple or compound and usually exstipulate. When stipules are present they are usually inconspicuous. When the leaves are compound they are imparipinnate. The inflorescence is usually an axillary panicle or raceme with small greenish or white flowers. Species of economic importance are mango (*Mangifera indica*) and cashew (*Anacardium occidentale*).

The flowers are unisexual or hermaphroditic; they are mostly radially symmetrical. Petals may be present or absent; when petals are present, they are free or joined basally to the bottom of the disc. The disc is usually present and prominent. Stamens may be equal (as in cashew nut) or of different sizes (as in mango) or numerous; there are often twice as many as there are petals. Staminodes may also be present even in hermaphroditic flowers (as in mango). Filaments are often free or joined basally. Anthers are two-celled and they dehisce lengthwise. The ovary is superior and usually one-celled or rarely two- to five-celled. The carpels are rarely free. The ovules are solitary, hanging down from the apex or wall of the ovary, or pendulous from a basal funicle.

The fruit is generally a drupe. Endosperm is very thin or absent and the cotyledon is usually fleshy.

Pseudospondias microcarpa is a medium-sized tree of forest fringe with a short bole and thick canopy. Its

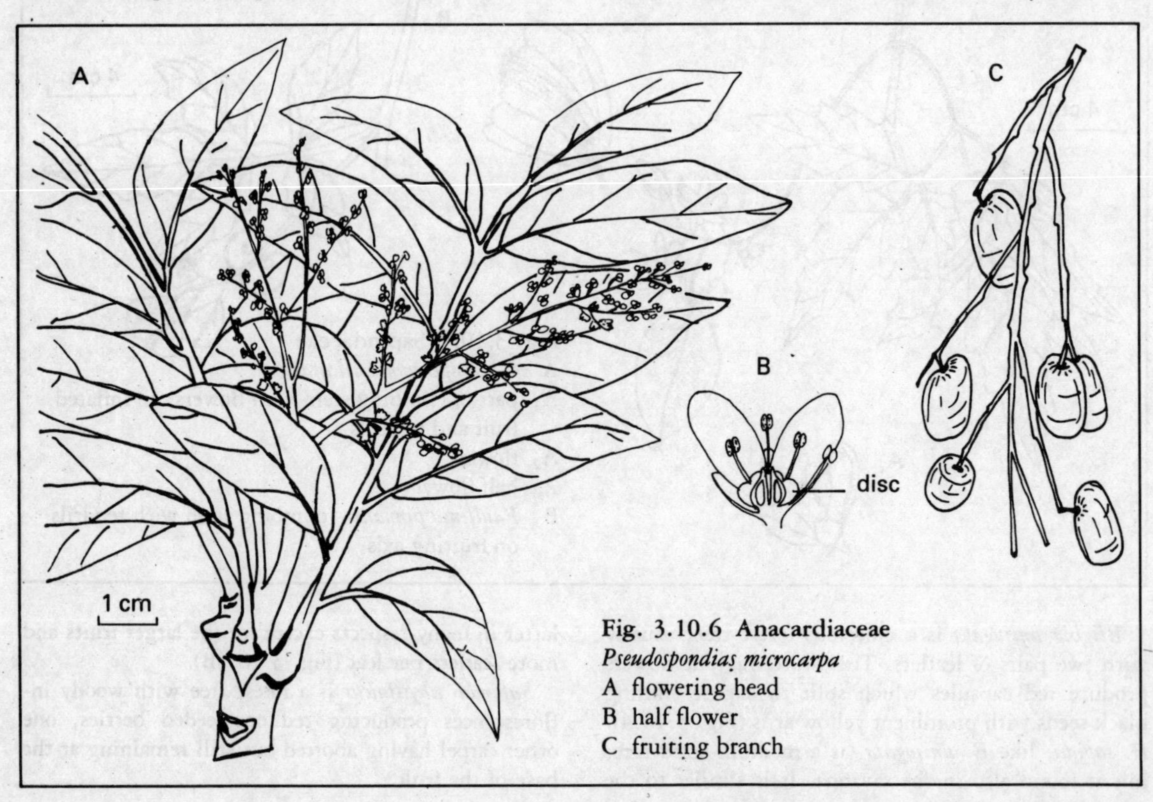

Fig. 3.10.6 Anacardiaceae
Pseudospondias microcarpa
A flowering head
B half flower
C fruiting branch

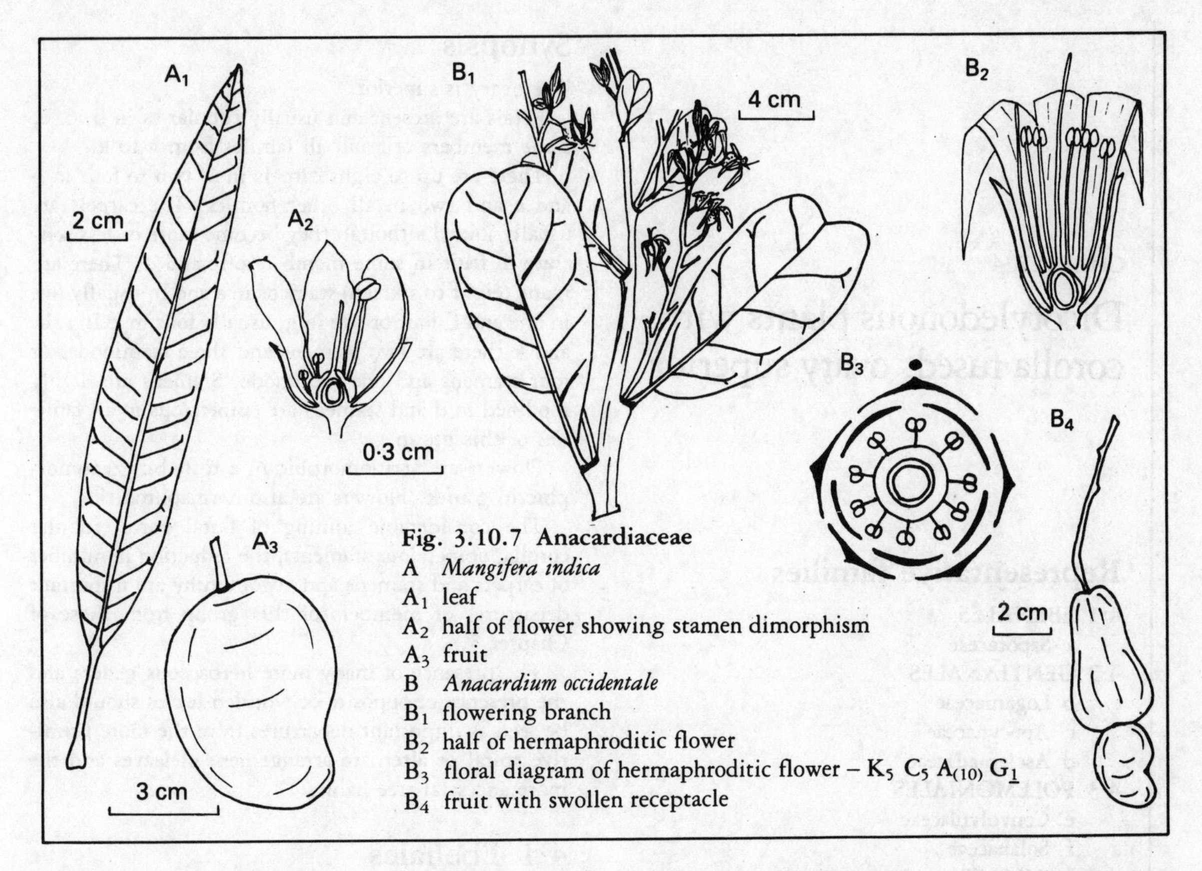

Fig. 3.10.7 Anacardiaceae
A *Mangifera indica*
A_1 leaf
A_2 half of flower showing stamen dimorphism
A_3 fruit
B *Anacardium occidentale*
B_1 flowering branch
B_2 half of hermaphroditic flower
B_3 floral diagram of hermaphroditic flower – $K_5\ C_5\ A_{(10)}\ G_{\underline{1}}$
B_4 fruit with swollen receptacle

pale green flowers are borne on large axillary panicles. The leaves are imparipinnate, with up to seventeen alternate leaflets having unequal sides of the lamina (Fig. 3.10.6). The flowers are male, female or hermaphroditic on the same inflorescence. The drupaceous fruits are produced in lax bunches.

Spondias mombin grows into a large tree in the forest zone. Generally known as the hog plum because of the yellow edible drupaceous fruits, the tree is usually planted near habitations; it grows very readily from cuttings.

Lannea and *Sclerocarya* are other prominent genera with compound leaves; the latter is represented by a single savanna species – *S. birrea* – while the former has forest and savanna representatives. The stamens are numerous in *Sclerocarya* (12-16), but fewer (about eight) in *Lannea*. *Lannea* is a genus of deciduous trees and shrubs with characteristic pinkish-reddish woolly hairs on the young shoots. The panicles or racemes of yellow flowers are produced before the emergence of new leaves in the dry season. *L. schimperi* is a savanna tree with pinkish or reddish indumentum on the young parts. The leaves have two to four pairs of leaflets. Other prominent savanna species are *L. fruticosa*, *L. humilis*, *L. kerstingii* and *L. microcarpa*; the last two have spirally-twisted bark.

Both *Mangifera indica* and *Anacardium occidentale* are introduced species. Their leaves are simple. In both of them, male, female and hermaphrodite flowers are produced on the same inflorescence. In mango, one or two of the stamens are large and fertile and the others are very small: staminodes are present (Fig. 3.10.7A). The fruit in *A. occidentale* is characteristic with a nut subtended by the swollen receptacle (Fig. 3.10.7B).

CHAPTER 4

Dicotyledonous plants with corolla fused, ovary superior

Representative families

Synopsis

The ovary is superior.

Petals are present and usually tubular as in b, c, e, some members of f and all familes from g to k.

There are up to eight carpels in a, two to four in b and e and two in all other familes. The carpels are usually joined although they become more or less separate in fruit in some members of c and d. There are many (eight to sixteen) stamens in a and b, usually five in c, e and f, four or five in g; usually four in j. In i, h, and k there are two stamens and three staminodes or four stamens and one staminode. Stamens are highly modified in d and stamens are epipetalous in all families of this group.

Flowers are actinomorphic in a to f, but zygomorphic in g to k. Flowers are also hermaphroditic.

The considerable joining of floral parts (tubular corolla, epipetalous stamens), the reduction in number of carpels and stamens and zygomorphy are important departures of members of this group from those of Chapter 3.

The presence of many more herbaceous genera and the presence of opposite or whorled leaves should also be seen as important departures from the more primitive spiral or alternate arrangement of leaves and the more ancestral tree habit.

4·1 Ebenales

4·1a Sapotaceae

This family is composed of trees, shrubs, and (rarely) climbers. They are mostly rainforest trees. Members of the family often have milky latex. Their leaves are alternate, entire, leathery and usually exstipulate. The widespread savanna species (*Butyrospermum paradoxum*), from which shea butter is made, belongs in this family. *Chrysophyllum albidum* also belongs in this family; it produces edible fruits.

The flowers are actinomorphic and usually hermaphroditic. The calyx is usually four- to eight-lobed and the corolla is also four- to eight-lobed, the lobes being in one or two whorls. Corolla lobes are imbricate and sometimes with petaloid external appendages. The stamens are epipetalous. The fertile stamens are as many as, and opposite, the petals; or they may be more numerous and in two or more whorls; staminodes are usually present. The anthers open by longituidinal slits. The flowers usually occur in leaf-axil clusters or on older and leafless parts of branches or (rarely) on the main trunk (Fig. 4.1.1A). The ovary is superior and many-celled; the style is simple with a small stigma;

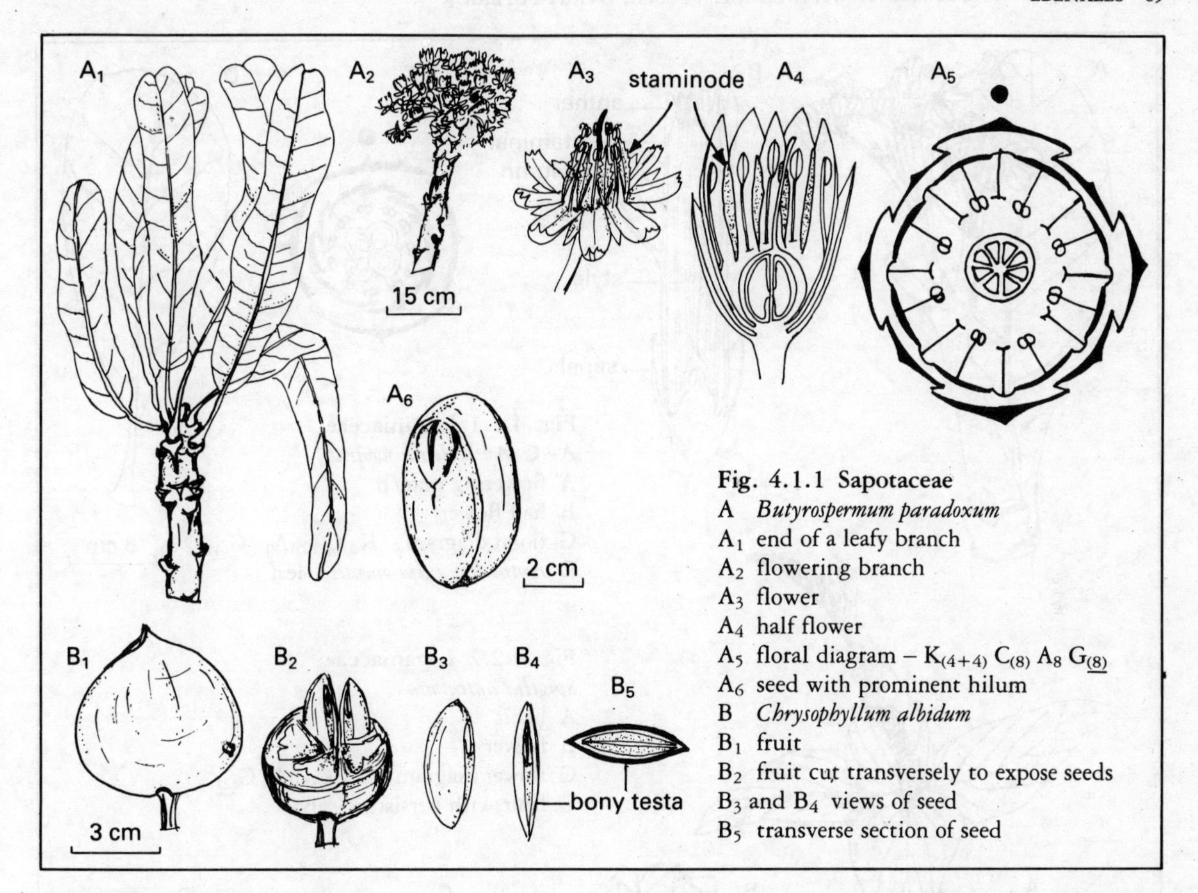

Fig. 4.1.1 Sapotaceae
A *Butyrospermum paradoxum*
A_1 end of a leafy branch
A_2 flowering branch
A_3 flower
A_4 half flower
A_5 floral diagram – $K_{(4+4)}\ C_{(8)}\ A_8\ G_{\underline{(8)}}$
A_6 seed with prominent hilum
B *Chrysophyllum albidum*
B_1 fruit
B_2 fruit cut transversely to expose seeds
B_3 and B_4 views of seed
B_5 transverse section of seed

the ovules are solitary in each cell.

The fruit is usually a fleshy berry with one to many cells. The seeds have hard bony shells; they have rather broad hilums and sparse endosperm but a large embryo and foliaceous cotyledons.

As we noted above, *Butyrospermum paradoxum* (the shea butter tree) is a prominent and abundant savanna species occurring within a wide latitudinal range (Fig. 4.1.1A). The shea butter tree has prominently fissured bark and the stem has a flesh-coloured slash which produces copious, gummy latex. The leaves, which are usually crowded at the end of the branch, are broadly obovate with prominent pinnate nerves. The fleshy skin of the fruit is edible when ripe and soft and the seeds are processed for shea butter. The fruit is ovoid and is about 5 cm long.

Manilkara obovata is a tree of varied habitat. It is widespread from edges of lagoons, through rainforest locations to outlying areas of forest and riverside woodlands in the savanna. The leaves have pinnate nerves and flowers are usually carried in axillary clusters. There are six hairy sepals and the corolla has undivided appendages. The fruits are globose, single-seeded, pale yellow and about 2 cm in diameter.

Mimusops is a small genus of trees with corolla appendages, deeply three-lobed – this character distinguishes members of the genus from related genera in the family. They are plants of rainforest and fringe-forest locations.

Omphalocarpum is a distinctive genus characterised by having more than one stamen opposite each petal. The species has distinctive button-shaped fruits which are about 10 cm in diameter and 9 cm thick in *O. elatum*, and about 30 cm in diameter and 12 cm thick in *O. procerum*. In both species the flowers occur in clusters on the trunk.

Chrysophyllum albidum grows wild or it is cultivated. The edible fruits are olive-coloured and ripen into bright orange. The flowers occur in axillary clusters (Fig. 4.1.1B). The leaves have prominent pinnate veins; they are olive green on the upper surface and whitish green on the lower surface. The flattened seeds with brown bony testa (Fig. 4.1.1B), are characteristic of the species.

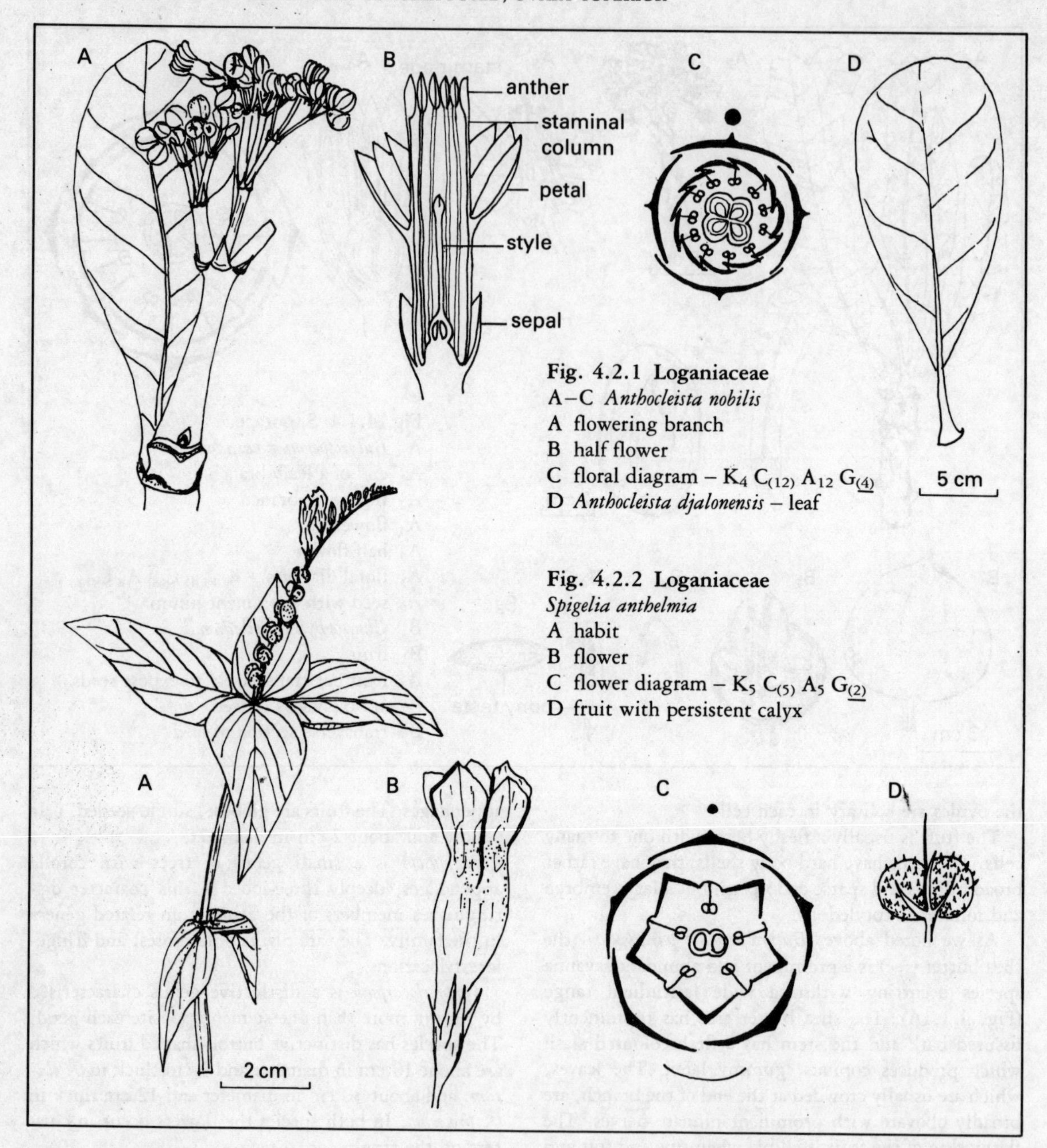

Fig. 4.2.1 Loganiaceae
A–C *Anthocleista nobilis*
A flowering branch
B half flower
C floral diagram – $K_4\ C_{(12)}\ A_{12}\ G_{\underline{(4)}}$
D *Anthocleista djalonensis* – leaf

Fig. 4.2.2 Loganiaceae
Spigelia anthelmia
A habit
B flower
C flower diagram – $K_5\ C_{(5)}\ A_5\ G_{\underline{(2)}}$
D fruit with persistent calyx

4·2 Gentianales

4·2b Loganiaceae

This is a family of trees, shrubs and herbs. The leaves are simple and opposite. Representative species are found in forest and savanna environments.

The flowers are hermaphroditic and actinomorphic; the inflorescence may be paniculate, corymbose, cymose or in globose heads. Calyx segments are imbricate or valvate, while the petals are valvate, imbricate or contorted and are joined into a tubular corolla, the latter with four to sixteen segments. There are as many stamens as corolla segments and the stamens are epipetalous. The ovary has two or four cells; the style is single. Ovules are numerous (rarely solitary), usually in axile placentation.

The fruit is usually a capsule, a berry or a drupe and the seeds may be winged.

Anthocleista is a fairly large genus of trees with opposite, usually large leaves; the stems usually carry spines and the flowers are diagnostic. *A. nobilis* is a large tree usually with a relatively thin, greyish bole branched conspicuously at the top. The flowers are white and the fruits yellowish. *A. nobilis* is common in secondary forest and riverine environments in derived savanna (Fig. 4.2.1).

Strychnos is a large genus of small trees or erect or climbing shrubs with three to seven digitate nerves arising from the base of the leaves. A small and common savanna tree, *S. spinosa*, has opposite leaves with axillary spines; it produces greenish white flowers in short compound cymes; the flowers produce orange-like berries (with hard coats) which are green when immature and ripen into yellow fruits, the inner pulp of which is edible.

Spigelia anthelmia is a common herbaceous weed of waste places in forest and derived savanna locations usually growing in very large populations; the habit (Fig. 4.2.2), is diagnostic.

4·2c Apocynaceae

This is a family of trees, shrubs and woody climbers or (rarely) perennial herbs. Milky latex is diagnostic. The leaves are opposite or whorled, simple, entire and mostly exstipulate; leaves are rarely alternate. The family is cosmopolitan.

This family has no member of major economic consequence. The soft wood of *Alstonia boonei* is used for carving and may yield timber too. The latex of *Funtumia elastica* yields some kind of 'rubber'. Some members of this family may also contain important medicinal compounds and poisonous substances. Preparations of *Rauvolfia vomitoria* are used locally as a sedative, while strophanthin from *Strophantus* spp. is also an important drug. Many introduced species in this family are ornamental plants. They include *Allamanda* spp., milk bush (*Thevetia neriifolia*), oleander (*Nerium oleander*), rose periwinkle (*Vinca rosea*) and frangipani (*Plumeria* spp.).

The flowers are hermaphroditic and radially symmetrical. There are five (rarely four) calyx segments which are imbricate and the calyx is often glandular

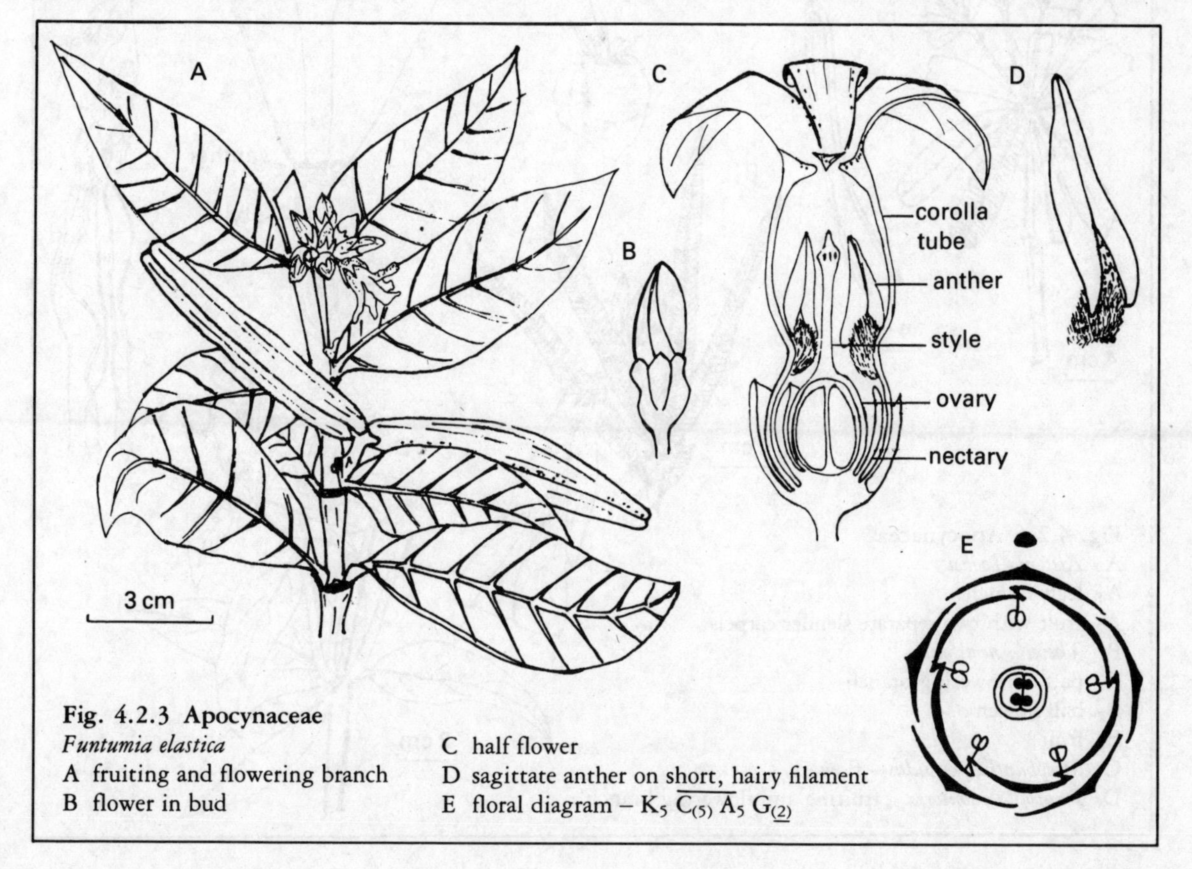

Fig. 4.2.3 Apocynaceae
Funtumia elastica
A fruiting and flowering branch
B flower in bud
C half flower
D sagittate anther on short, hairy filament
E floral diagram – $K_5\ \overline{C_{(5)}\ A_5}\ G_{\underline{(2)}}$

inside. The corolla is tubular with contorted or imbricate segments; it is rarely valvate. There are four or five epipetalous stamens inserted in the tube. Filaments are mostly free. The anthers are sagittate and more or less clasp the stigma and the connective is often produced at the apex (Fig. 4.2.3). The ovary is superior and may be one-celled with two parietal placentae, or it may be two-celled with placentae on the septum, or the two carpels may be free, or only basally joined, with ventral placentae in each carpel. There is one style which is entire or split at the base; the style is usually thick at the apex. There are two or more ovules in each carpel. A disc is present; it may be cupular or divided into segments.

The fruits are entire and indehiscent or of two separate carpels, baccate, drupaceous or follicular, the latter splitting on one side. The seeds often have wings or an appendage carrying long silky hairs.

Funtumia elastica is a large forest tree with opposite leaves and axillary clusters of white flowers with contorted petals. The fruits with two woody follicles are diagnostic (Fig. 4.2.3). *Holarrhena floribunda* is another tree of forest and derived savanna environments whose vegetative features are similar to those of *F. elastica*; the follicles are, however, slender and flexible (about 12 cm long).

Alstonia boonei has a characteristic whorled branching and leaf arrangement (Fig. 4.2.4A). It is a large tree of forest and derived savanna locations. The fruits (Fig. 4.2.4B) which are produced from a congested inflorescence of white flowers are like the fruits of *H. floribunda* described above.

The branch, fruit and details of a flower of *Thevetia neriifolia* are illustrated in Fig. 4.2.4B.

Rauvolfia vomitoria is a shrub or a small tree in forest and savanna locations with leaves in whorls of four and terminal cymes of small white flowers. The fruit is a pair of drupes or a single two-stoned drupe which is green when immature and ripens into red (Fig. 4.2.4D).

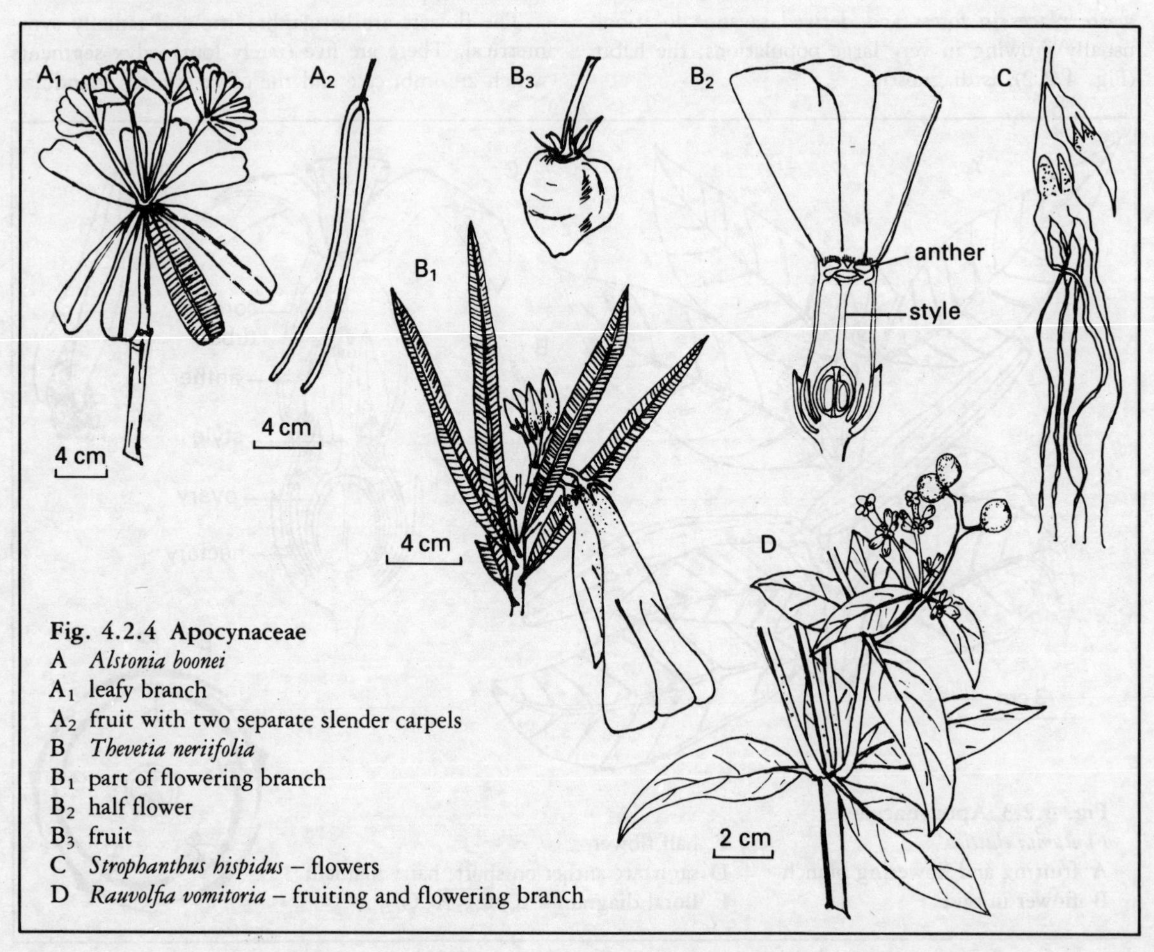

Fig. 4.2.4 Apocynaceae
A *Alstonia boonei*
A_1 leafy branch
A_2 fruit with two separate slender carpels
B *Thevetia neriifolia*
B_1 part of flowering branch
B_2 half flower
B_3 fruit
C *Strophanthus hispidus* – flowers
D *Rauvolfia vomitoria* – fruiting and flowering branch

Voacanga africana is a common regrowth shrub or small tree of forest environments with grey bark and opposite leaves 6–30 cm long. The fruit is a pair of berries with white and green patches; each of the berries is about 5 cm in diameter.

Tabernaemontana pachysiphon is a glabrous shrub or small tree in regrowth forest. The large and glabrous leaves are oppostie. The fleshy white flowers have contorted corolla lobes and they are fragrant. The fruit is a pair of berries joined basally, and each of the berries is about 12 cm in diameter.

Another distinct habit group in this family is that of woody climbers represented prominently by species of *Alafia* and *Strophanthus*. In both genera, the leaves are opposite. *Alafia barteri* is a high-climbing shrub with a glabrous stem and white or pinkish flowers in lax corymbose inflorescences. Its fruit is a pair of slender follicles each of them about 30 cm long. *Strophanthus* is a genus of woody climbers characterised by the twisted petal tails of their conspicuous flowers. It is found in wet and dry forest environments. The fruits are also diagnostic with a pair of follicles each having warty lenticels on the skin. The flowers of *S. hispidus* are shown in Fig. 4.2.4C.

4·2d Asclepiadaceae

This is a large, mostly tropical, genus of mesic and arid or sub-arid environments. Members are twining or erect shrubs or perennial herbs with entire, opposite, linear or orbicular leaves with cordate bases. The plants have milky latex. The inflorescence is cymose, umbelliform or more or less racemose.

Asclepias curassavica is often planted as an ornamental while *Calotropis procera* is planted around villages for medicinal purposes. *C. procera* is also an important additive in the preparation of Nigerian (Fulani) cheese.

The flowers are actinomorphic, hermaphroditic, pentamerous, petalous and have a short calyx. The five stamens are epipetalous on the five joined petals. A petaloid appendage grows from the back of each stamen and five such appendages constitutes the corona which is characteristic of the specialised flowers of Asclepiadaceae (Fig. 4.2.5). The ovary is made up of two superior carpels joined by the top of the styles only, which gives a solid, more or less peltate, structure. The anthers are joined together around the stigma in such a way that an anther cell from one anther is joined to another anther cell from the neighbouring

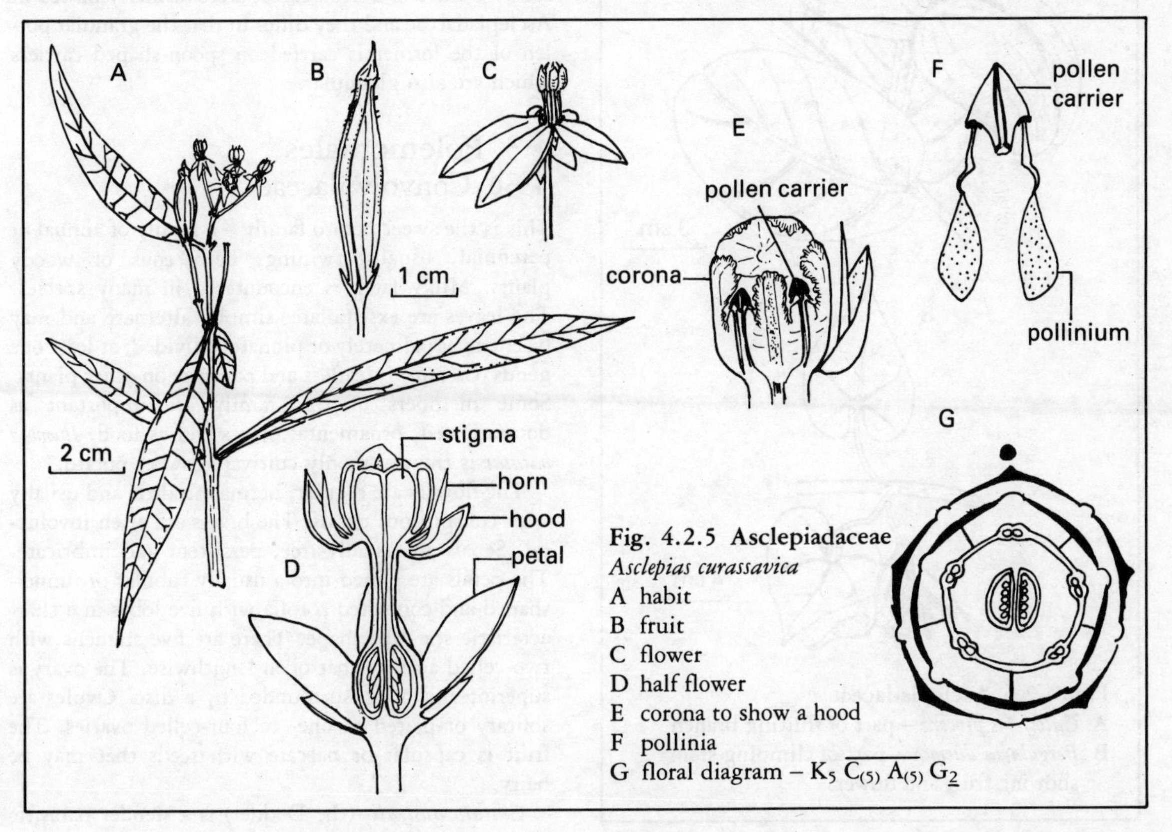

Fig. 4.2.5 Asclepiadaceae
Asclepias curassavica
A habit
B fruit
C flower
D half flower
E corona to show a hood
F pollinia
G floral diagram – $K_5\ \overline{C_{(5)}\ A_{(5)}}\ G_{\underline{2}}$

anther by means of a pollen carrier (Fig. 4.2.5E, F and G) known as a 'translator'. In most members of this family the pollen exists in a mass called a 'pollinium'. Pollen vectors transport pollen from one plant to another quite easily because the translator is sticky and it therefore readily adheres to the legs of insects, which thus carry pollen (pollinia) from two neighbouring anthers as a unit. Figure 4.2.5E and F show that the petaloid appendages of the stamens are developed into specialised structures (hood and horn) which are joined to the petals.

The fruit is usually a pair of follicles. The seeds are numerous and usually have a crown of long silky hairs which aid dispersal.

Asclepias curassavica is a perennial herb or shrub often grown as an ornamental. The flowers are red and yellow and produce many-seeded capsules (Fig. 4.2.5).

Calotropis procera is a soft-wooded shrub of savanna and semi-arid regions. It has opposite leaves which have a greyish powdery bloom. The greyish flowers with purplish centres produce fruits which are a pair of inflated follicles and which dry to produce many flat seeds with long, silky hairs (Fig. 4.2.6A).

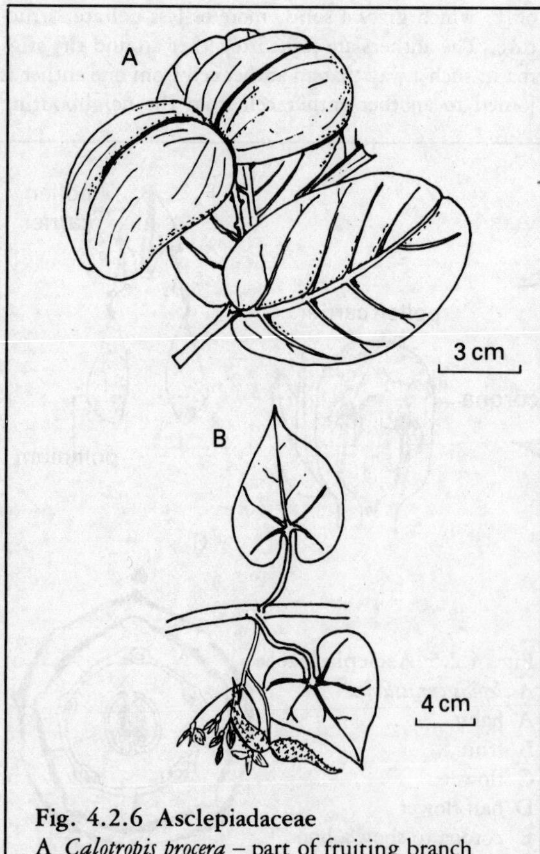

Fig. 4.2.6 Asclepiadaceae
A *Calotropis procera* – part of fruiting branch
B *Pergularia daemia* – part of climbing stem showing fruit and flowers

Pachycarpus is a genus of perennial species found in savanna environments. They have erect stems. The plant is hispid and produces inflated or fusiform follicles. The leaves are oblong and rounded at the base. The flowers are greenish with purple streaks and they are carried in axillary clusters.

Pergularia daemia is a hairy climber found in forest regrowth environments. The green–white flowers are carried on stalked, lax inflorescences. The fruit is a pair of follicles each of them with soft prickles (Fig. 4.2.6B).

Sarcostemma viminale is a straggling or climbing plant with leafless, round branches which are hairy at the nodes. It is mostly a species of savanna or semi-arid locations with greenish white flowers which produce pairs of follicles up to about 10 cm long.

Members of this family are similar to members of Periplocaceae in vegetative morphology, in fruit form, in gross floral morphology and in the structure of the seeds. Genera of Periplocaceae were formerly placed in Asclepiadaceae and they differ in that the granular pollen of the former is carried on spoon-shaped carriers which are also glandular.

4·3 Polemoniales

4·3e Convolvulaceae

This is the sweet potato family – a family of annual or perennial, usually twining, herbaceous or woody plants. Milky latex is encountered in many species. The leaves are exstipulate, simple, alternate and may be entire or palmately or pinnately divided; at least one genus (*Cuscuta*) is leafless and parasitic on other plants. Some members of this family are important as domesticated, ornamental plants and as food. *Ipomoea batatas* is the commonly cultivated sweet potato.

The flowers are regular, hermaphroditic and usually with conspicuous petals. The bracts are often involucral. Sepals are usually free, persistent and imbricate. The petals are joined into a usually tubular or funnel-shaped and contorted corolla with five lobes in a characteristic star-like shape. There are five stamens with two-celled anthers that open lengthwise. The ovary is superior and often surrounded by a disc. Ovules are solitary or paired in one- to four-celled ovaries. The fruit is capsular or baccate with seeds that may be hairy.

Cuscuta australis (the Dodder) is a slender parasitic

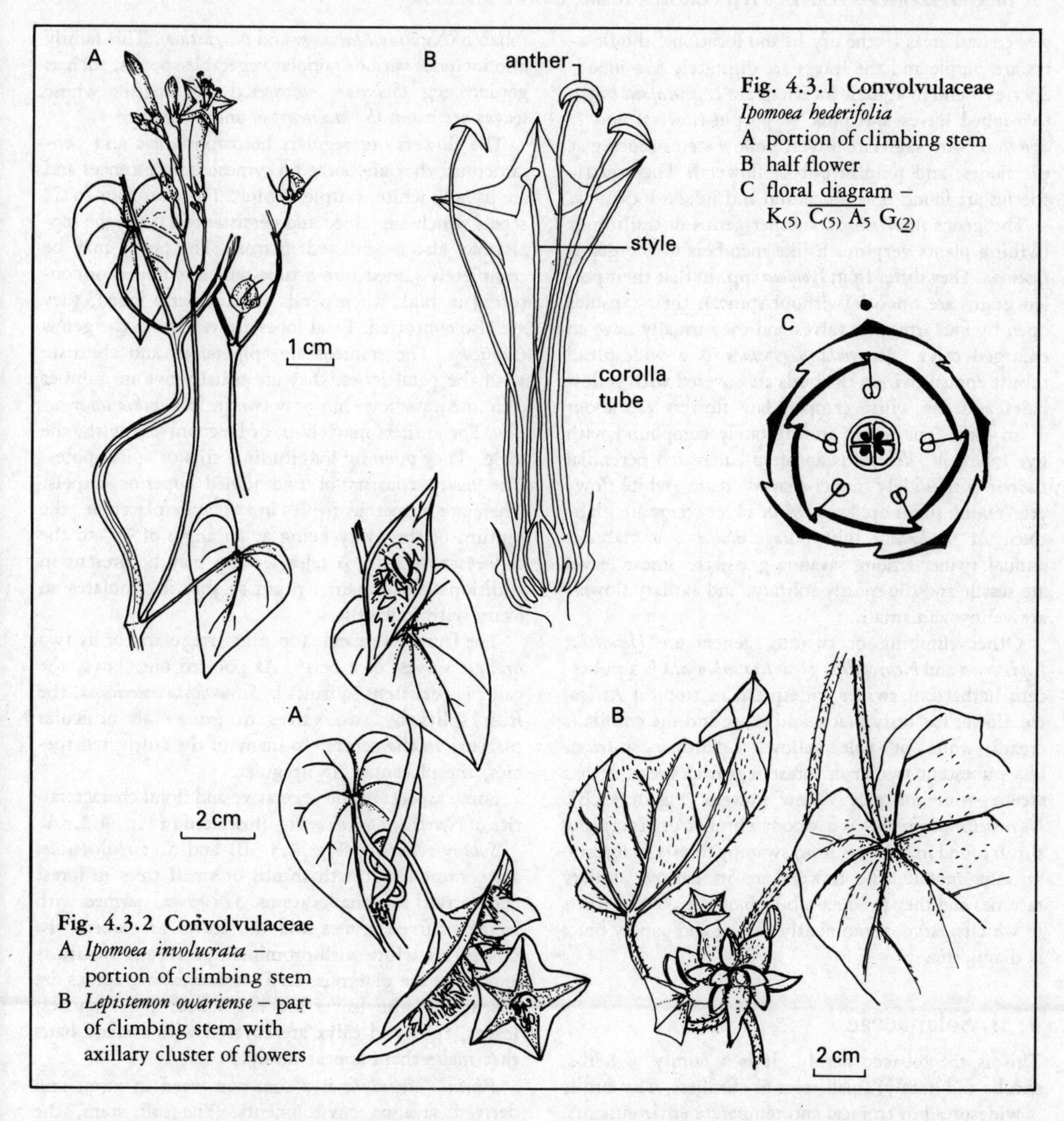

Fig. 4.3.1 Convolvulaceae
Ipomoea hederifolia
A portion of climbing stem
B half flower
C floral diagram – $K_{(5)}\ C_{(5)}\ A_5\ G_{(2)}$

Fig. 4.3.2 Convolvulaceae
A *Ipomoea involucrata* – portion of climbing stem
B *Lepistemon owariense* – part of climbing stem with axillary cluster of flowers

climber twining around plants and producing 'haustoria' (which connect its vascular system with the phloem of the host plant) at intervals along the pale yellow stem. The small white flowers are produced in clusters along the stem. A similar but greener and much commoner parasite – *Cassytha filiformis* – belongs to a different family (Lauraceae).

Ipomoea is a large genus of herbaceous or woody twiners or creepers. They are found in varying habitats from very dry to aquatic. The flowers are funnel-shaped and may be blue, pink, yellow, scarlet or purple with a dark centre. The pollen grains have spines and the capsules open regularly by four valves. *I. hederifolia* (Fig. 4.3.1) is a common regrowth climber in secondary forest with scarlet salverform flowers. *I. quamoclit* (Johnny Walker) is a common weed often cultivated; the featherlike leaves and scarlet flowers are characteristic. *I. involucrata* (Fig. 4.3.2A) is a common creeping weed rooting at the nodes; the boat-shaped involucre is diagnostic. *I. mauritiana* is found from the wet coastal areas to the dry inland locations; the flowers are purple and the leaves are digitately five-lobed. Species found in aquatic locations are *I. pes-caprae* (with two-lobed leaves and pink or purple flowers) and *I. aquatica* (with sagittate leaves, hollow stems rooting at the nodes, and reddish purple flowers). The aquatic species are found both in coastal and inland locations.

The genus *Merremia* is another genus of trailing or twining plants very much like members of the genus *Ipomoea*. They differ from *Ipomoea* spp. in that their pollen grains are smooth (without spines), their capsules open by four irregular valves and they usually have an enlarged calyx. *Merremia aegyptiaca* is a widespread robust annual twiner; the buds are covered with yellow hairs and the white campanulate flowers are about 3 cm long. The leaves are digitately compound with five lobes. *M. dissecta* is an often-cultivated perennial twiner with widely funnel-shaped cream–white flowers. Young stems are hairy while older stems are glabrous. *M. tridentata* subsp. *angustifolia* is a glabrous annual twiner among savanna grass; the linear leaves are sessile and the mostly solitary, and axillary flowers are yellow and small.

Other climbing or twining genera are *Hewittia*, *Lepistemon* and *Neuropeltis*. *Hewittia sublobata* is a pubescent herbaceous twiner widespread in tropical Africa; the flower has leafy bracts and calyx and the corolla is cream–white or pale yellow. *Lepistemon owariense* is a pubescent twiner of forest regrowth areas. It has cream–white or pale yellow flowers (Fig. 4.3.2B). *Neuropeltis acuminata* is a woody climber or straggling shrub found in habitats from swampy forests to forest–savanna mosaic; the flowers are in narrow axillary racemes and they produce a bunch of samara-like fruits in which a large, prominently-veined and papery bract is diagnostic.

4·3f Solanaceae

This is the tobacco family. It is a family of herbs, shrubs and (rarely) small trees or climbers. The family is widespread in tropical and temperate environments. Members of this family have simple, alternate and exstipulate leaves; the leaves are sometimes lobed. Milky latex is lacking.

This family is of considerable economic importance, some of the members being ornamentals and food crops. Among the important ornamentals are *Datura candida* and *Solanum wrightii*, which are shrubs or small trees. The cash crop species include the 'Irish' potato (*Solanum tuberosum*), the tomato (*Lycopersicon lycopersicum*), various species of pepper (*Capsicum* spp.) and tobacco (*Nicotiana tabacum* and *N. rustica*). This family also includes various popular vegetable species, such as garden egg (*Solanum melongena*) and species whose leaves are eaten (*S. macrocarpon* and *S. nigrum*).

The flowers are regular, hermaphroditic and pentamerous; they are borne on cymose inflorescences and are usually white, purple or blue. There are four to six sepals which are joined and persistent in fruit; the sepals may also be inflated in fruit. The petals may be completely joined into a tube which is folded or contorted in bud; when petals are distinctly lobed, they are also contorted. Petal lobes are valvate in the genus *Capsicum*. The stamens are epipetalous and alternate with the petal lobes; they are usually five in number (ten in *Lycopersicon*, but only two in *Schwenckia americana*). The anthers may clasp, or be connivent with, the style. They open by longitudinal slits or apical pores. The ovary consists of two joined superior carpels. There are numerous ovules in two axile placentae, the septum of the ovary being at an angle of 45° to the inflorescence axis. A false septum may be present in addition to the main septum so that it simulates an ovary with four cells.

The fruit is a capsule (opening irregularly or by two or four valves) or a berry. As pointed out above, the calyx is persistent in fruit. In *Schwenckia americana*, the fruit splits by two valves to expose an orbicular placenta in the centre. In many of the cultivated species, the placentae are irregular.

Some aspects of the vegetative and floral charactersitics of *Nicotiana tabacum* are illustrated in Fig. 4.3.3A.

Solanum torvum (Fig. 4.3.3B) and *S. verbascifolium* are common regrowth shrubs or small trees in forest and derived savanna locations. *S. torvum* is armed with stout recurved spines and the leaves are lobed; the flowers are white with prominent petals and the fruits and calyx are glabrous. In *S. verbascifolium*, spines are absent and the leaves are not lobed; the branches, leaves, fruit and calyx are covered with stellate hairs that make them appear mealy.

Physalis angulata is a common weed in forest and derived savanna environments. The soft stem, the flowers and the inflated calyx of *P. angulata* are diagnostic for the genus (Fig. 4.3.4D).

Mention has been made of the fruit characteristics of *Schwenckia americana* above; the plant itself is an erect herb about 0·5 m tall. The stem and flowers have a purplish tinge and the corolla is tubular. The androecium is of only two stamens.

Withania somnifera is a species of drier areas of tropical Africa. It is a branched undershrub (about 2 m tall) with small greenish flowers in axillary clusters.

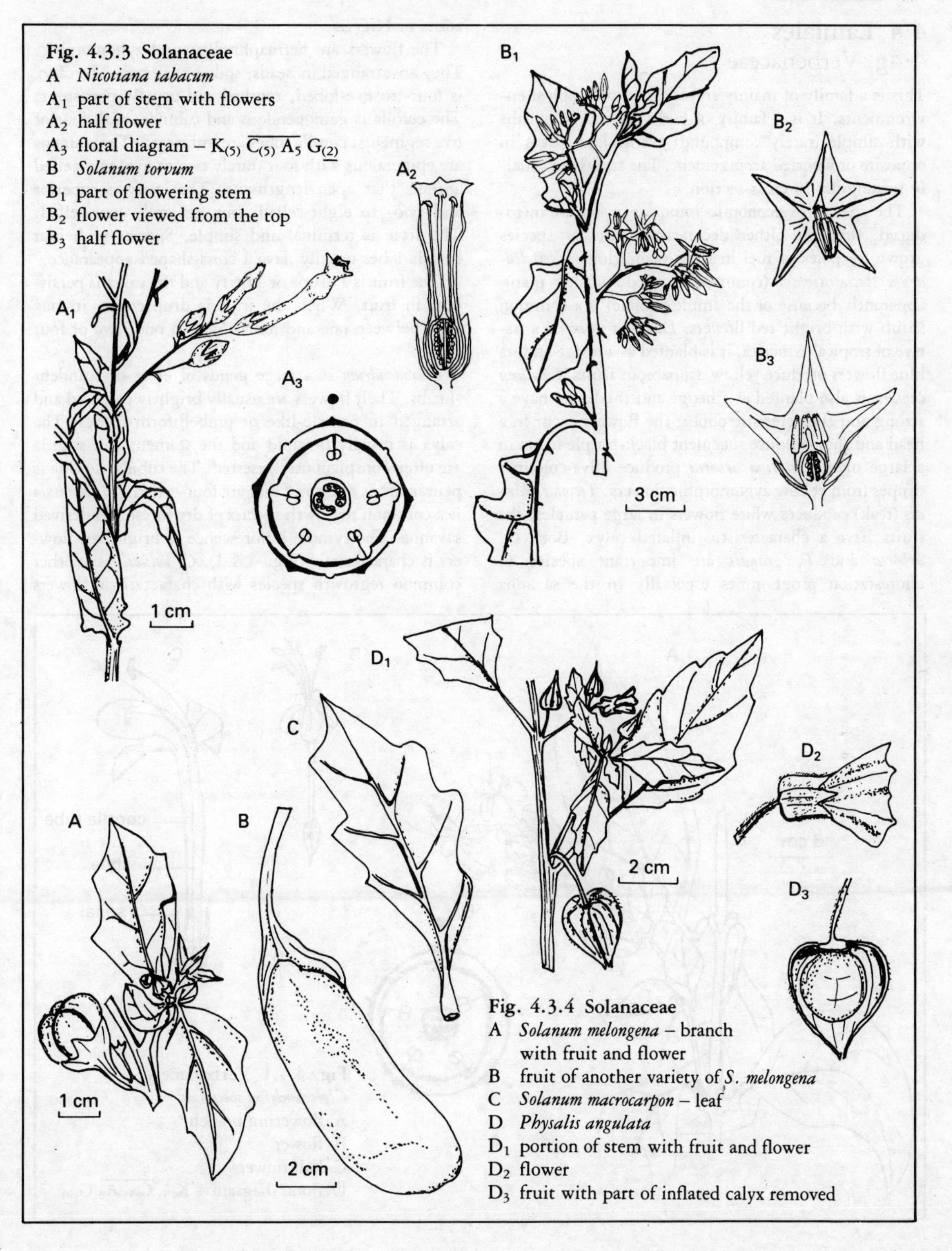

Fig. 4.3.3 Solanaceae
A *Nicotiana tabacum*
A_1 part of stem with flowers
A_2 half flower
A_3 floral diagram – $K_{(5)}\ \overline{C_{(5)}\ A_5}\ \underline{G}_{(2)}$
B *Solanum torvum*
B_1 part of flowering stem
B_2 flower viewed from the top
B_3 half flower

Fig. 4.3.4 Solanaceae
A *Solanum melongena* – branch with fruit and flower
B fruit of another variety of *S. melongena*
C *Solanum macrocarpon* – leaf
D *Physalis angulata*
D_1 portion of stem with fruit and flower
D_2 flower
D_3 fruit with part of inflated calyx removed

4·4 Lamiales
4·4g Verbenaceae

This is a family of mainly tropical and subtropical environments. It is a family of herbs, trees and shrubs with simple (rarely compound) exstipulate leaves in opposite or whorled arrangement. The stems are usually rectangular in cross-section.

The species of economic importance are all introduced: they are either decorative plants or species grown for poles or fuel in large plantations. *Clerodendrum speciosissimum* (commonly called 'pagoda plant' apparently because of the ample panicle) is a common shrub with bright red flowers. *Duranta repens* is a native of tropical America; it is planted as a hedge and its blue flowers produce yellow drupaceous fruits. *Lantana camara* is also planted as a hedge and the leaves have a strong and characteristic odour; the flowers occur in a head and they produce succulent black-purple fruits in a large mass. *Gmelina arborea* produce olive-coloured drupes from yellow zygomorphic flowers. *Tectona grandis* (teak) produces white flowers in large panicles; the fruits have a characteristic inflated calyx. Both *G. arborea* and *T. grandis* are important species in afforestation programmes especially in the savanna zones of Nigeria.

The flowers are hermaphroditic and zygomorphic. They are arranged in heads, spikes or cymes. The calyx is four- to five-lobed, toothed and usually persistent. The corolla is gamopetalous and tubular, with four or five segments; corolla lobes are imbricate. The stamens are epipetalous with four (rarely two or five) two-celled anthers that open lengthwise. The ovary is superior and two- to eight-celled, but generally four-celled. The style is terminal and simple. Species with four corolla lobes usually have a cross-shaped appearance.

The fruit is a drupe or a berry and the calyx is persistent in fruit. When the fruit is drupaceous, it may have between one and four stones in one, two or four cells.

Clerodendrum is a large genus of erect or scandent shrubs. Their flowers are usually brightly coloured and arranged in panicle-like or umbelliform cymes. The calyx is usually petaloid and the stamens and stigma are often conspicuously exserted. The tubular corolla is pentamerous and the fruits are four-celled. *C. speciosum* is a common regrowth species of dry forest and derived savanna. The cymose inflorescence of bright red flowers is characteristic (Fig. 4.4.1). *C. volubile* is another common regrowth species with characteristic flowers

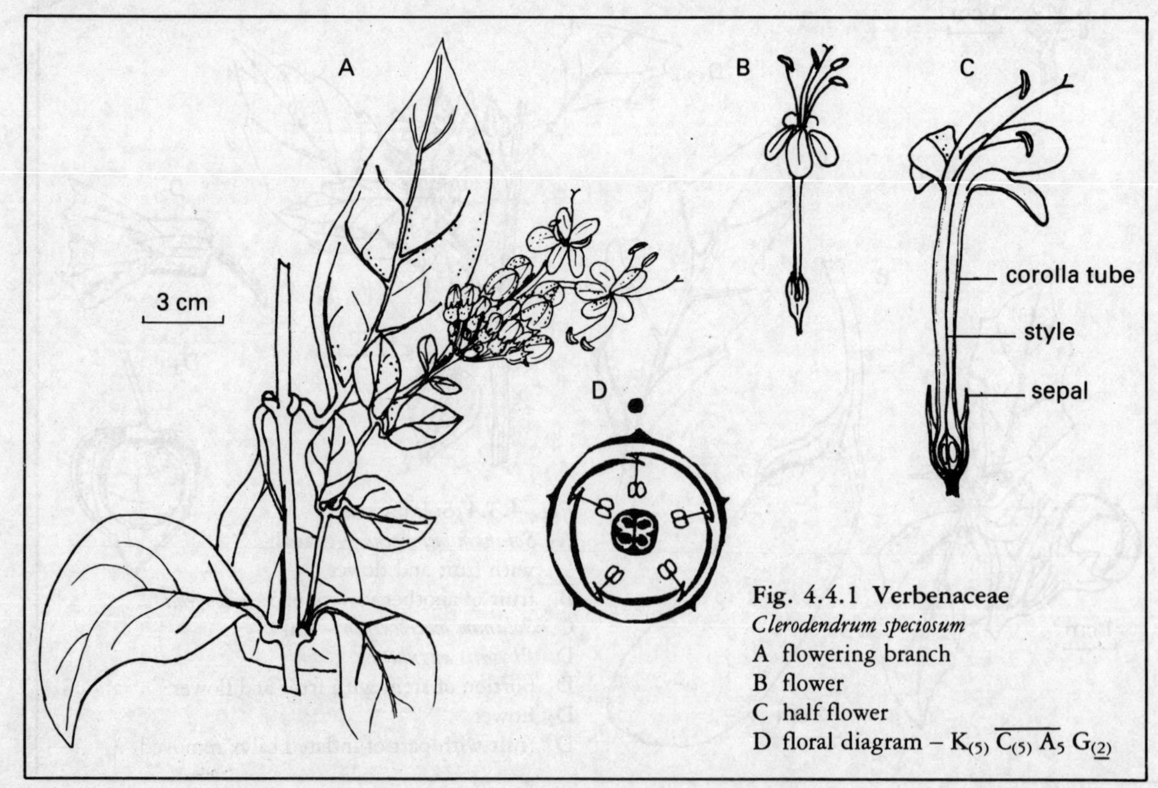

Fig. 4.4.1 Verbenaceae
Clerodendrum speciosum
A flowering branch
B flower
C half flower
D floral diagram – $K_{(5)}\ \overline{C_{(5)}\ A_5}\ G_{\underline{(2)}}$

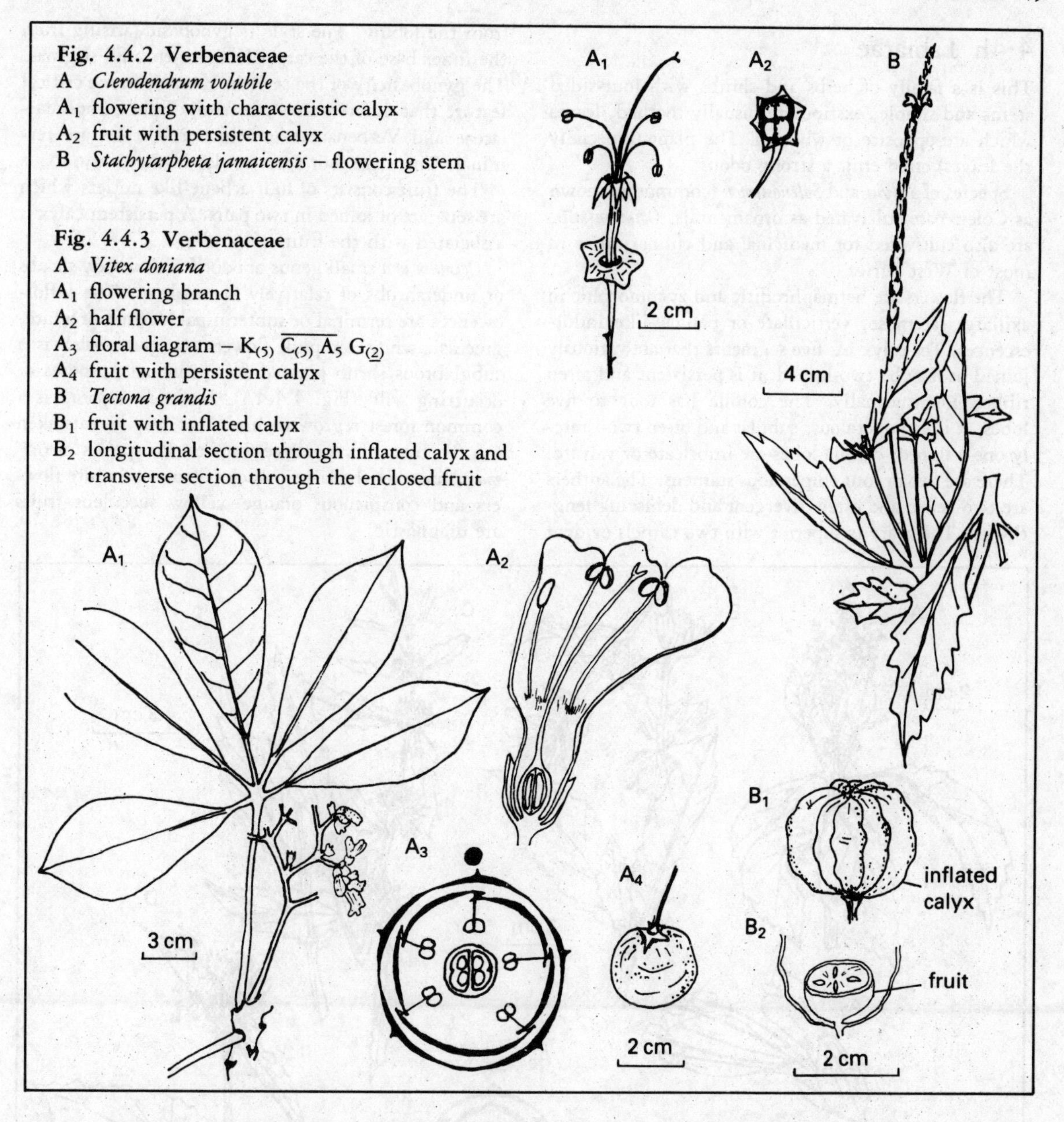

Fig. 4.4.2 Verbenaceae
A *Clerodendrum volubile*
A_1 flowering with characteristic calyx
A_2 fruit with persistent calyx
B *Stachytarpheta jamaicensis* – flowering stem

Fig. 4.4.3 Verbenaceae
A *Vitex doniana*
A_1 flowering branch
A_2 half flower
A_3 floral diagram – $K_{(5)}\ \overline{C_{(5)}\ A_5}\ G_{(\underline{2})}$
A_4 fruit with persistent calyx
B *Tectona grandis*
B_1 fruit with inflated calyx
B_2 longitudinal section through inflated calyx and transverse section through the enclosed fruit

whose corolla-tubes fall off to leave a green, four-lobed fruit which is subtended by a flat, pale green calyx (Fig. 4.4.2A).

Stachytarpheta is a genus of herbaceous or more or less woody plants with opposite leaves and blue flowers in characteristic spicate terminal or subterminal inflorescences. Each flower is subtended by a persistent bract. Only two stamens are present in this genus. *S. indica* (*S. jamaicensis*) is an erect herbaceous plant of waste places with glabrous stems (Fig. 4.4.2B) while *S. cayennensis* is a much-branched undershrub of waste places with narrow spikes.

Vitex is a genus of trees with digitately compound leaves and small tomentose tetramerous flowers in axillary corymbose inflorescences. *V. doniana* is a common savanna treee which produces green fruits which ripen into black and are edible. The young leaves are also collected and used as a green vegetable (Fig. 4.4.3A).

The fruit characteristics of *Tectona grandis* are illustrated in Figure 4.4.3B.

4·4h Labiatae

This is a family of herbs and shrubs with four-sided stems and simple, exstipulate, usually toothed, leaves which are opposite or whorled. The plant (especially the leaves) often emit a strong odour.

Species of *Salvia* and *Solenostemon* (commonly known as Coleus) are cultivated as ornamentals. *Ocimum* spp. are also cultivated for medicinal and culinary uses in most of West Africa.

The flowers are hermaphroditic and zygomorphic in axillary, racemose, verticillate or panicle-like inflorescences. The calyx has five segments that are variously joined or may be two-lipped; it is persistent and often ribbed longitudinally. The corolla has four to five lobes, it is gamopetalous, tubular and often two- (rarely one-) lipped; corolla lobes are imbricate or valvate. There are two or four epipetalous stamens. The anthers are two-celled and often divergent and dehiscing lengthwise. The ovary is superior with two carpels evident from the lobing. The style is gynobasic (arising from the inner base of the carpel lobes) with bifid stigmas. The gynobasicity of the style in Labiatae is one critical feature that differentiates the family from Scrophulariaceae and Verbenaceae in which the styles are terminal. There are four erect ovules in the ovary.

The fruit consists of four achene-like nutlets which are separate or joined in two pairs. A persistent calyx is associated with the fruit.

Ocimum is a small genus of odoriferous woody shrubs or undershrubs of relatively open places. The inflorescences are terminal or subterminal racemes of white, greenish white, or pink flowers. *O. gratissimum* is a subglabrous shrub planted for medicinal purposes or occurring wild (Fig. 4.4.4A). *Hoslundia opposita* is a common forest regrowth shrub that may be mistaken (on the basis of vegetative features) for species of *Ocimum*; however the large panicles of cream–white flowers and conspicuous orange–yellow succulent fruits are diagnostic.

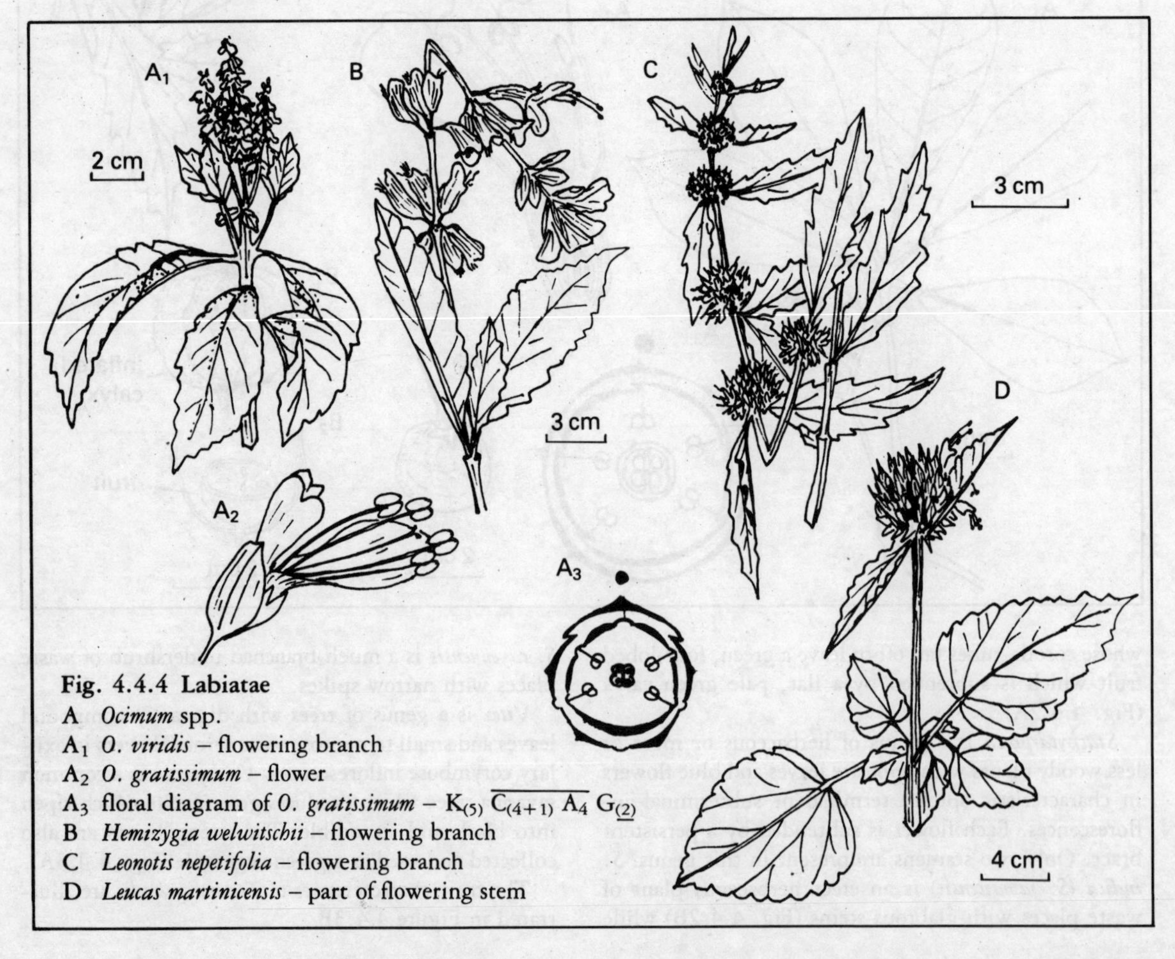

Fig. 4.4.4 Labiatae
A *Ocimum* spp.
A_1 *O. viridis* – flowering branch
A_2 *O. gratissimum* – flower
A_3 floral diagram of *O. gratissimum* – $K_{(5)}\ \overline{C_{(4+1)}\ A_4}\ G_{(2)}$
B *Hemizygia welwitschii* – flowering branch
C *Leonotis nepetifolia* – flowering branch
D *Leucas martinicensis* – part of flowering stem

Solenostemon spp. (Coleus) are also very common weeds of open and/or wet places. Like the ornamental varieties, the terminal spikes of zygomorphic flowers are characteristic.

Hemizygia welwitschii is an odoriferous perennial in dry stony places in the savanna. The white flowers occur in clusters of pink bracts (Fig. 4.4.4B). *H. bracteosa* is a taller plant (about 1 m) in swampy savanna locations.

Leonotis nepetifolia is a robust herb ranging in height from about 1 m to 2 m (Fig. 4.4.4C). The plants are usually encountered around human habitations.

Leucas martinicensis is an odoriferous, erect and usually branched annual with dense whorls of small flowers at the nodes (Fig. 4.4.4D). The plant is about 0·5 m tall and is found mostly in the savanna.

4·5 Scrophulariales

4·5i Scrophulariaceae

This is a family of erect or prostrate herbs or (rarely) small trees. The leaves are mostly opposite or verticillate or (less often) alternate; they are exstipulate. The stems may be square in cross-section.

One common ornamental species in our area of study is *Russelia equisetiformis* which is a herb with verticillate leaves and scarlet tubular flowers. Some species of *Striga* parasitise the roots of Guinea corn, millet and even wild savanna grasses.

The flowers are hermaphroditic and zygomorphic. The calyx is imbricate or valvate and usually joined. The corolla is usually tubular with imbricate segments. There are two or four (rarely five) epipetalous stamens which alternate with corolla segments, and a fifth stamen is usually reduced to a staminode. The one- or two-celled anthers dehisce lengthwise. The ovary is superior, and usually two-celled with a terminal style. The ovules are numerous on two axile placentae. The flowers are solitary in axillary fascicles or the inflorescence is a spike.

The fruit is a capsule or a berry with numerous seeds which may be flattish with more or less rugose seed coats.

Scoparia dulcis is a widespread tropical, usually shrubby, weed up to about 1 m tall; the stems are ribbed and glabrous with crenulate and opposite leaves; the whitish or bluish flowers with subactinomorphic corolla, occur singly or in pairs in the leaf axils (Fig. 4.5.1A). *Capraria biflora* is an erect and branched undershrub about 1 m tall; the alternate leaves look very much like those of *S. dulcis* and carry a pair of white flowers in their axils. The flowers produce globose capsules having persistent style and calyx.

Buchnera capitata is an erect herb up to 1 m tall. Flowers

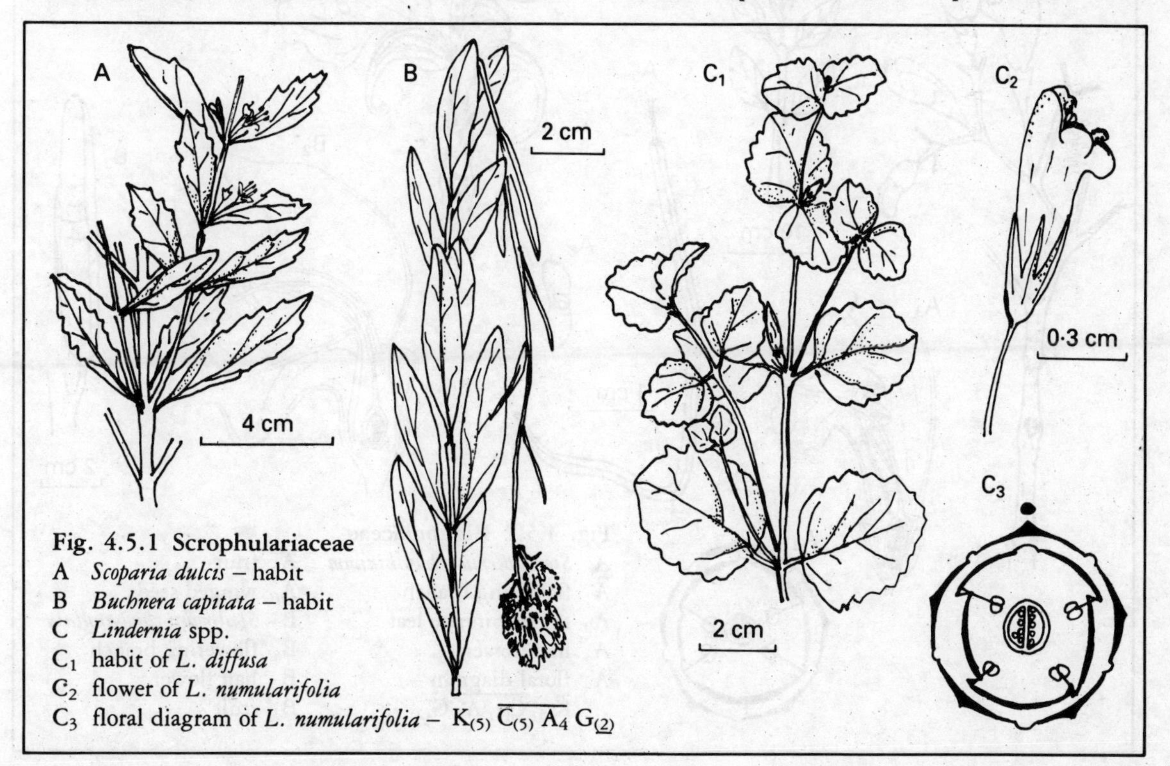

Fig. 4.5.1 Scrophulariaceae
A *Scoparia dulcis* – habit
B *Buchnera capitata* – habit
C *Lindernia* spp.
C_1 habit of *L. diffusa*
C_2 flower of *L. numularifolia*
C_3 floral diagram of *L. numularifolia* – $K_{(5)}\ \overline{C_{(5)}\ A_4}\ G_{(\underline{2})}$

are white or mauve in headlike spikes up to about 2cm long. The plant is black when dry (Fig. 4.5.1B).

Striga is a genus of herbs with terminal spikes of white, yellow, pink or purple flowers. Their leaves are usually lanceolate and rough or scabrous. *S. primuloides* is an erect leafless parasite with yellowish flowers, while *S. hermonthica* is an erect herb which is known to parasitise the roots of *Sorghum* (Guinea corn).

Lindernia is a genus of small prostrate or ascending herbs with opposite leaves and ribbed stems. The zygomorphic flowers are yellow, white, pink or purple, terminal or axillary. *L. diffusa* and *L. numulariifolia* (Fig. 4.5.1C) are widespread throughout our study area in mesic and damp, but open, locations such as lawns.

4·5j Bignoniaceae

Members of this family in our area of study are mainly trees or shrubs in forest and savanna. Their leaves are usually large, compound, imparipinnate, opposite and exstipulate. Axillary buds may develop foliaceous growths which resemble stipules. The stems are usually characterised by prominent leaf scars and lenticels.

Many introduced species are ornamental plants; these include *Jacaranda mimosifolia* and *Tecoma stans* (yellow tecoma). Many wild species are planted around dwellings and are probably important medicinal plants; *Newbouldia laevis* is planted as a fence.

The flowers are usually large, showy, hermaphroditic, mostly zygomorphic and pentamerous. The calyx is campanulate, closed or open in bud, truncate or five-toothed or spathaceous. The corolla is usually of five joined and imbricate segments two of which form an upper lip and three of which form the lower, or abaxial, lip. The stamens are four or two in number and alternate with the corolla segments. Staminodes may be present. Stamens are epipetalous with the anthers connivent in pairs or (rarely) free. They are two celled and they open lengthwise; the anther lobes are joined to the connective by their apex. A disc is usually present around the ovary. The pistil is of two joined superior carpels and one terminal, two-lipped style. The ovary is usually two celled with axile placentae or one-celled with parietal bifid placentae. The ovules are numerous.

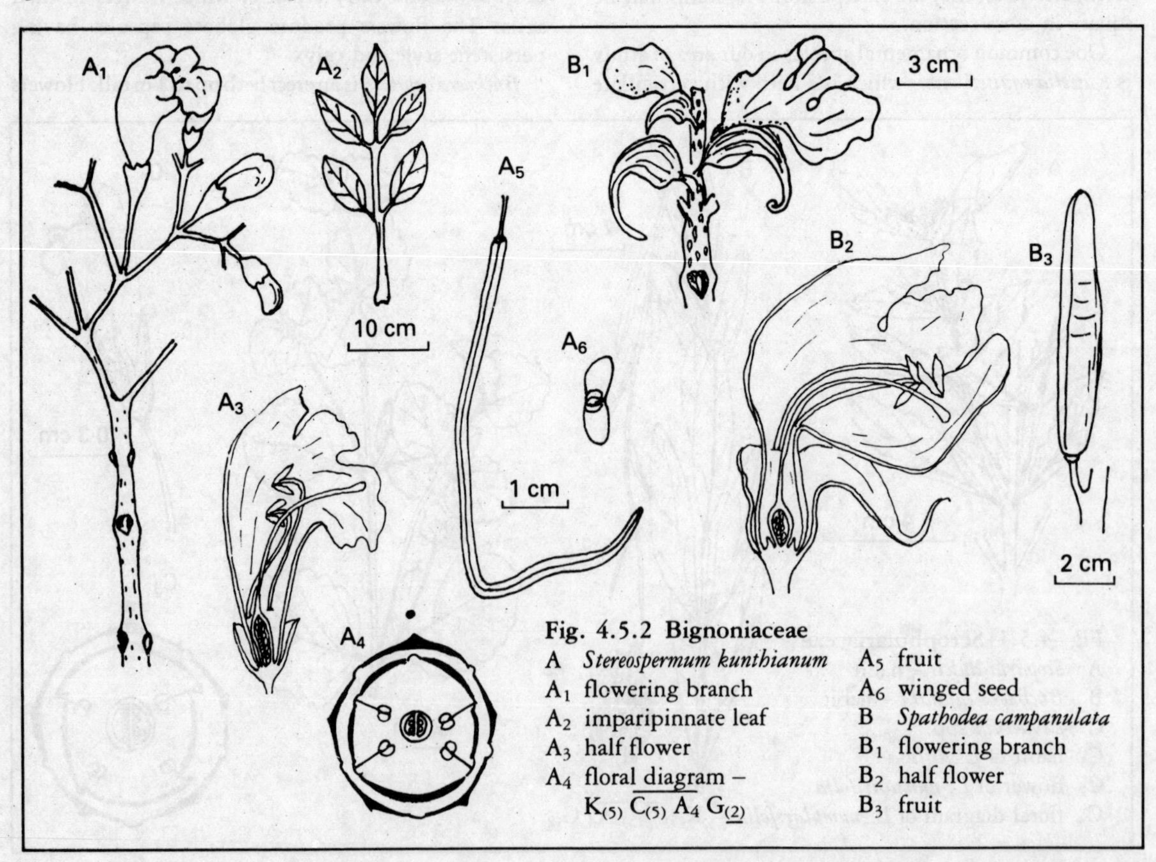

Fig. 4.5.2 Bignoniaceae
A *Stereospermum kunthianum*
A_1 flowering branch
A_2 imparipinnate leaf
A_3 half flower
A_4 floral diagram – $K_{(5)}\ \overline{C_{(5)}\ A_4}\ G_{\underline{(2)}}$
A_5 fruit
A_6 winged seed
B *Spathodea campanulata*
B_1 flowering branch
B_2 half flower
B_3 fruit

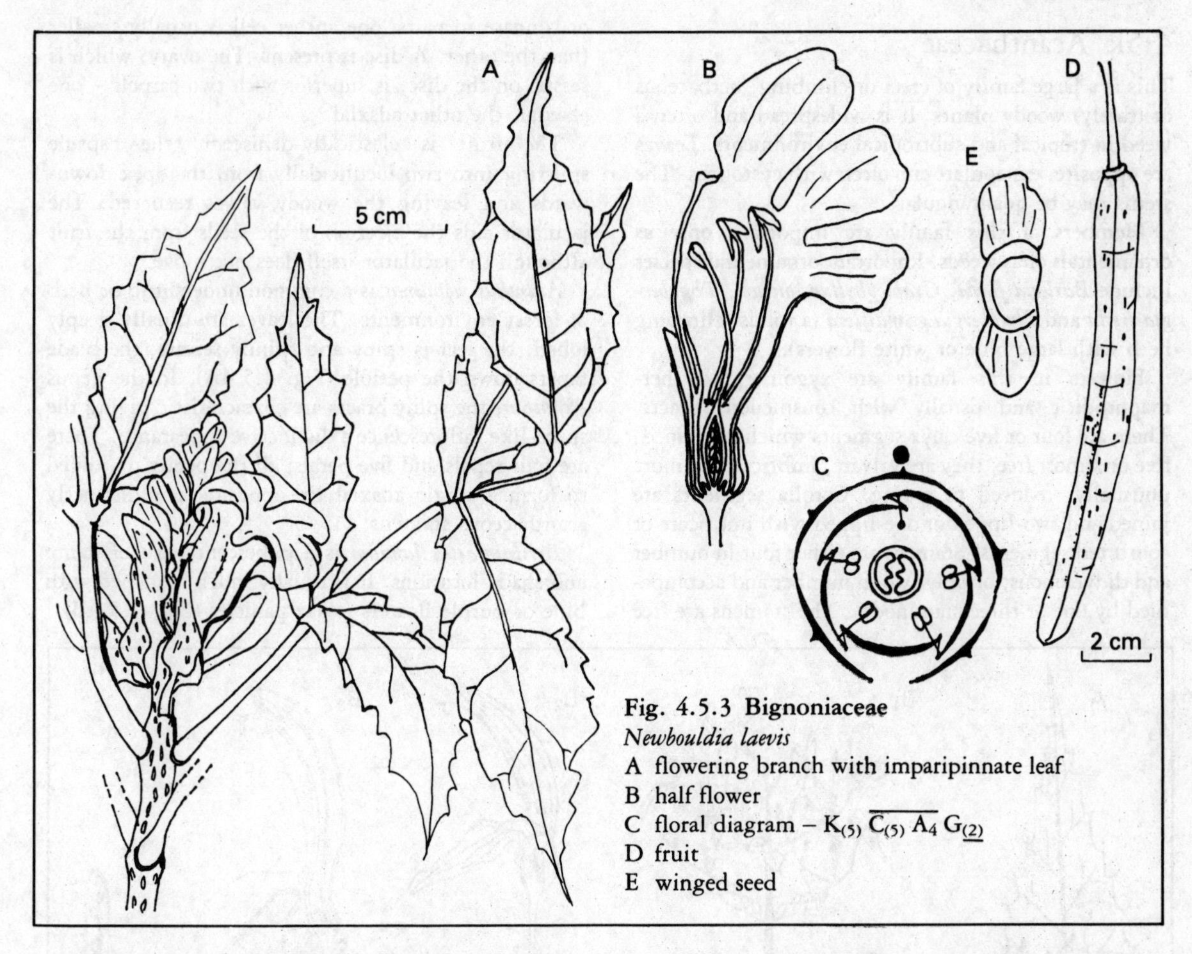

Fig. 4.5.3 Bignoniaceae
Newbouldia laevis
A flowering branch with imparipinnate leaf
B half flower
C floral diagram – $K_{(5)}\ \overline{C_{(5)}\ A_4}\ G_{\underline{(2)}}$
D fruit
E winged seed

The fruit is a capsule or fleshy and indehiscent. It is usually elongate (up to 0·5 m long and 10 cm in diameter in some species). The seeds usually possess membranous wings.

Stereospermum and *Kigelia* are woody genera which differ from other genera in that their calyces are not spathaceous. *Stereospermum kunthianum* is a small tree or shrub in savanna environments. It has a grey bark with brownish young branches. The pink or purplish flowers are produced in large drooping panicles during the dry season when the plant is leafless (Fig. 4.5.2A). *S. acuminatissimum* is a medium-sized forest tree with a straight grey bole. The flowers occur in conspicuous corymbose inflorescences. In both species of *Stereospermum*, the fruits are slender and they are about 0·5 m long. *Kigelia africana* is a small much-branched tree with lax, hanging racemes of large flowers which have an obnoxious odour and are of variable colours (pink, purple, red or yellow). The fruit is large and cylindrical (up to 0·6 m long) with reddish dots on the surface.

Spathodea, *Markhamia*, and *Newbouldia* have spathaceous calyces. *Spathodea campanulata* is a conspicuous tree of fringing forest. The flame-like flowers are bright red with yellow margins. The open end of the spathaceous calyx is adaxial (Fig. 4.5.2B). The fruit is an erect capsule about 15 cm long. *Markhamia tomentosa* is a small tree of fringing forest with yellowish tomentose branches. The flowers are yellow with purple streaks and they are borne in terminal tomentose racemes. The fruit is a tomentose, elongate–linear capsule about 1·5 cm broad and 50 cm long. *Newbouldia laevis* may grow into a fair-sized forest tree. The leaves may have up to four pairs of leaflets apart from the terminal one. The flowers range from light pink to mauve in colour and they are produced in dense and erect terminal or axillary spike-like panicles. The open side of the spathaceous calyx is abaxial. The purplish glandular spots on the calyx attract certain black ants. The fruit is elongate (about 30 cm long) and cylindrical and the seeds are winged (Fig. 4.5.3).

4·5k Acanthaceae

This is a large family of erect or climbing, herbaceous or (rarely) woody plants. It is widespread and often a weed in tropical and subtropical environments. Leaves are opposite, exstipulate and often with cystoliths. The stems may be quadrangular.

Members of this family are important only as ornamentals or as weeds. Important ornamental species include *Barleria flava*, *Graptophyllum pictum*, *Thunbergia erecta* and *Thunbergia grandiflora* (a robust climbing herb with large blue or white flowers).

Flowers in this family are zygomorphic, hermaphroditic and usually with conspicuous bracts. There are four or five calyx segments which are joined, free or almost free; they are valvate, imbricate or, more unusually, reduced to a ring. Corolla segments are joined and two-lipped or one-lipped with imbricate or contorted segments. Stamens are either four in number and didynamous, or only two in number and accompanied by one or three staminodes. The stamens are free or connate in pairs; one anther cell is usually smaller than the other. A disc is present. The ovary, which is sessile on the disc, is superior with two carpels – one abaxial, the other adaxial.

The fruit is elastically-dehiscent, the capsule splitting into two loculicidally from the apex downwards and leaving the woody valves recurved. The jaculator aids the ejection of the seeds from the fruit although the jaculator itself does not move.

Acanthus montanus is a common undershrub or herb of forest environments. The leaves are usually deeply lobed, the leaf is spiny and spinily serrate; the blade tapers down the petiole (Fig. 4.5.4A). In the genus *Acanthus*, the spiny bracts are characteristic giving the spike-like inflorescence a distinctive appearance. There are four sepals and five petals; all the petals are joined to form a single abaxial lip opposite four distinctly acanthaceous stamens.

Brillaintaisia lamium is a pubescent herb of damp and shady locations. It is usually much branched with blue or purple flowers in lax panicles (Fig. 4.5.4B).

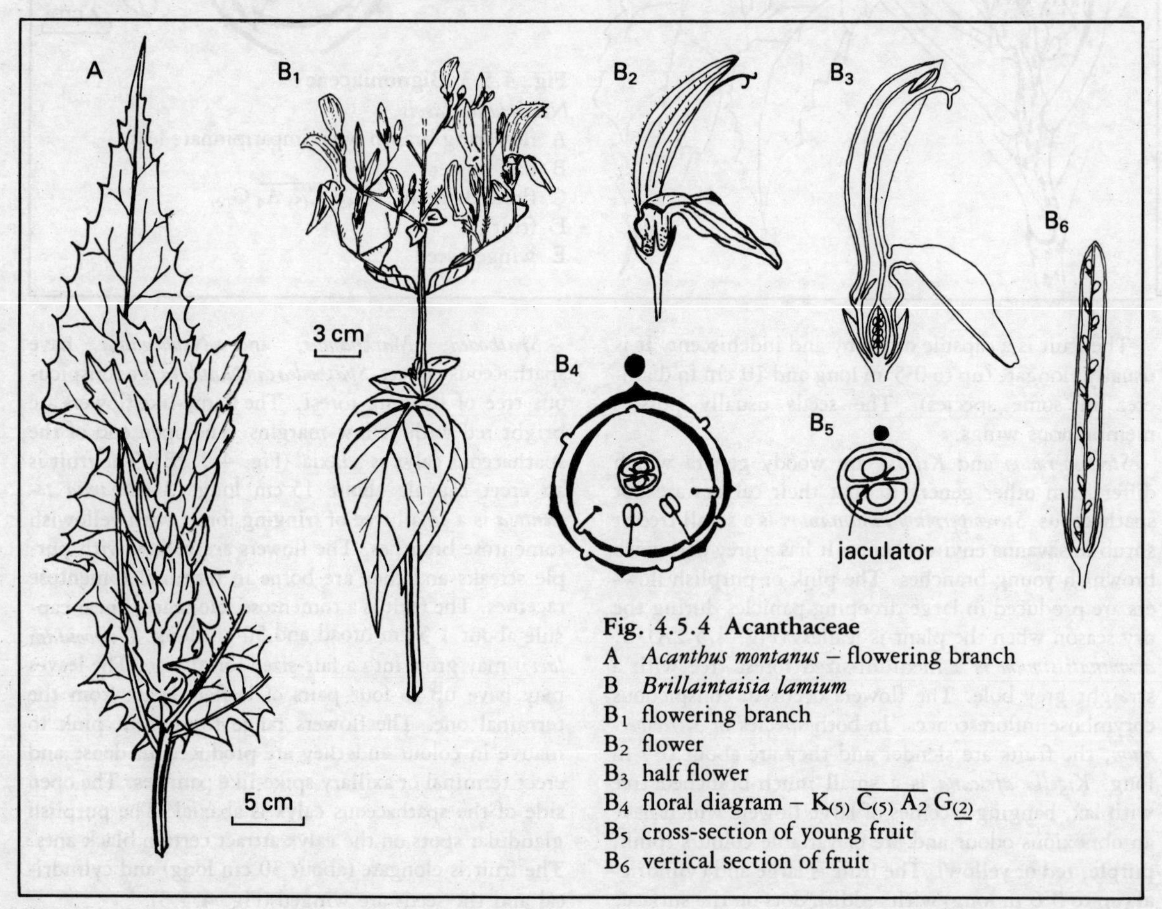

Fig. 4.5.4 Acanthaceae
A *Acanthus montanus* – flowering branch
B *Brillaintaisia lamium*
B_1 flowering branch
B_2 flower
B_3 half flower
B_4 floral diagram – $K_{(5)}\ \overline{C_{(5)}\ A_2}\ G_{\underline{(2)}}$
B_5 cross-section of young fruit
B_6 vertical section of fruit

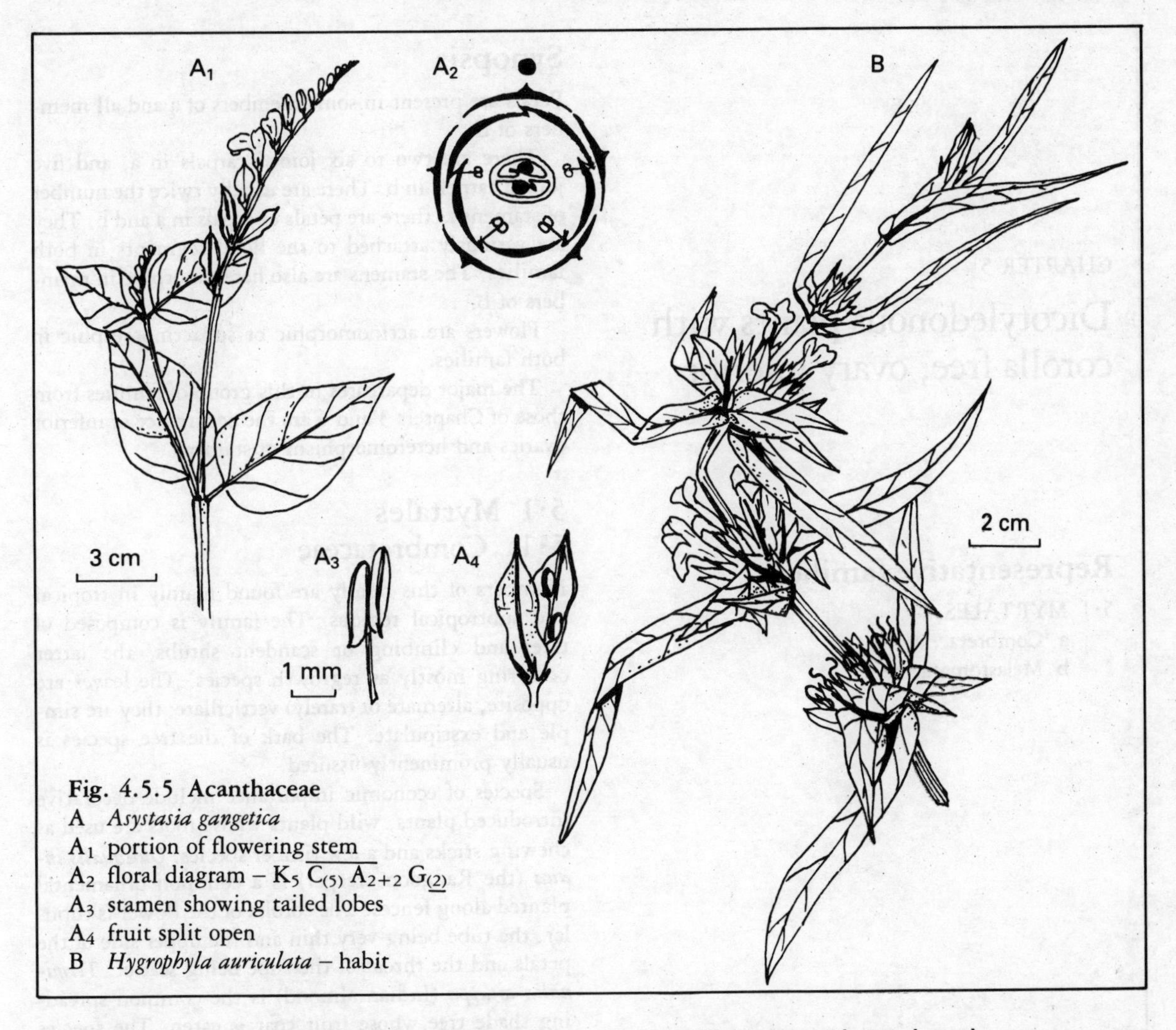

Fig. 4.5.5 Acanthaceae
A *Asystasia gangetica*
A_1 portion of flowering stem
A_2 floral diagram – $K_5\ \underline{C_{(5)}}\ A_{2+2}\ \underline{G_{(2)}}$
A_3 stamen showing tailed lobes
A_4 fruit split open
B *Hygrophyla auriculata* – habit

Asystasia gangetica is a common straggling weed of open locations. The one-sided raceme of white flowers with a purple throat is diagnostic (Fig. 4.5.5A). There are four two-celled anthers which are tailed. *A. calycina* is another plant found in relatively more shaded places in similar geographical locations to *A. gangetica*, but the flowers of the former are white, and they are larger in more open racemes.

Justicia is a genus of *Asystasia*-like plants. However, they have only two stamens per flower and their flowers are either in cylindrical, typically acanthaceous, spikes or in dense nodal clusters. *J. insularis* is a straggling weed with nodal clusters of pink, purple or (rarely) white flowers. *J. flava* is an erect herb with a square and pubescent stem and with yellow flowers in pubescent spikes. It is a widespread weed.

Hygrophyla auriculata is a stout erect herb of wet places up to 2 m tall. The square stem carries blue or purple flowers in dense axillary clusters with several strong spines among the bracts (Fig. 4.5.5B).

Phaulopsis falcisepala is a scandent herb of shaded locations in disturbed forest environments. The square stems are reddish purple and much-branched. The flowers are white and small in spikes with the prominent bracts typical of the family.

Nelsonia canescens is a small prostrate herb of wet open locations. The entire plant is pubescent and dull grey and it produces a grey and erect spike of flowers. The spike is about 3 cm long and the flowers are purple or pink.

CHAPTER 5

Dicotyledonous plants with corolla free, ovary inferior

Representative families

5·1 MYRTALES
- a Combretaceae
- b Melastomataceae

Synopsis

Petals are present in some members of a and all members of b.

There are two to six joined carpels in a, and five joined carpels in b. There are usually twice the number of stamens as there are petals or sepals in a and b. They are variously attached to the floral segments in both families. The stamens are also heteromorphic in members of b.

Flowers are actinomorphic or subactinomorphic in both families.

The major departures in this group of families from those of Chapters 3 and 4 are the occurrence of inferior ovaries and heteromorphism of stamens.

5·1 Myrtales

5·1a Combretaceae

Members of this family are found mainly in tropical and subtropical regions. The family is composed of trees and climbing or scandent shrubs, the latter occurring mostly as regrowth species. The leaves are opposite, alternate or (rarely) verticillate; they are simple and exstipulate. The bark of the tree species is usually prominently fissured.

Species of economic importance include decorative introduced plants, wild plants whose roots are used as chewing sticks and a few timber species. *Quisqualis indica* (the Rangoon creeper) is a common ornamental planted along fences. The corolla of the flower is tubular, the tube being very thin and the upper side of the petals and the throat of the tube being scarlet. *Terminalia catappa* (Indian almond) is the common spreading shade tree whose fruit coat is eaten. The species whose roots are used for chewing sticks in southwestern Nigeria are *Anogeissus leiocarpus* and *Terminalia glaucescens*. Important timber species include *Terminalia superba* and *T. ivorensis*; their commercial names are Idigbo and Afara respectively. *Terminalia* spp. are of considerable importance in afforestation trials in the forest regions of Nigeria.

The flowers are actinomorphic, more or less regular and tetramerous or pentamerous. The calyx tube is adnate to the ovary with four to eight segments or lobes; the calyx lobes are valvate. There are four or five petals or petals are absent. Stamens are four, eight or ten in number, rarely more. There are usually two whorls of stamens: the lower ones are opposite the sepals while the upper ones alternate with them; both whorls are inserted on the receptacle tube or the hypanthium. The anthers are versatile and they open

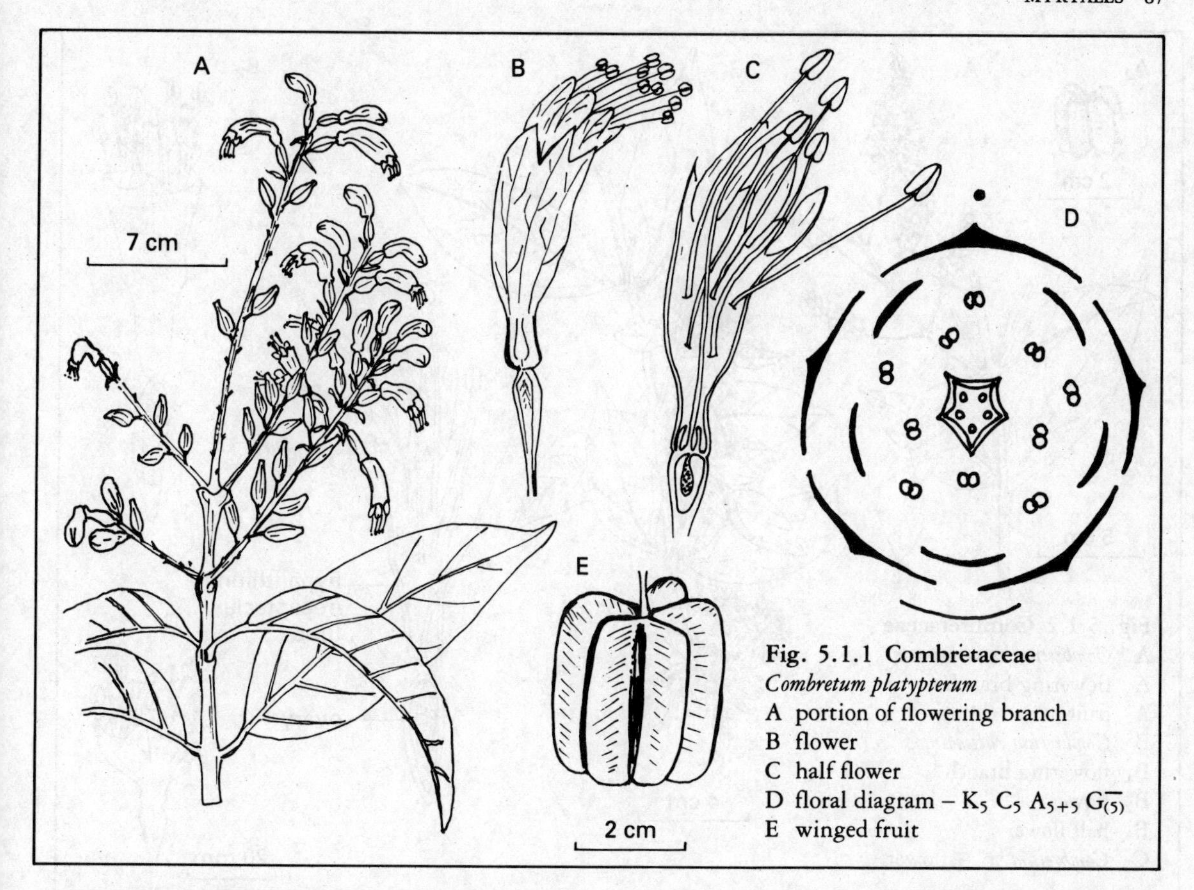

Fig. 5.1.1 Combretaceae
Combretum platypterum
A portion of flowering branch
B flower
C half flower
D floral diagram – $K_5\ C_5\ A_{5+5}\ \overline{G}_{(5)}$
E winged fruit

by longitudinal slits. An epigynous disc is present. The ovary is inferior and one-celled. There are two to six joined carpels with two to six ovules hanging from the top of the cell. Externally the ovary is ridged or winged.

The fruit usually has two or more wings; it is mostly indehiscent. The seeds are pendulous with a small radicle and a convolute or contorted cotyledon.

Combretum is a genus of trees, shrubs and woody, climbing lianes widespread in forest and savanna environments. Their flowers are easily distinguished from those of *Terminalia* in that petals are present. The flowers are usually quite small and they have white, yellow or red petals usually in large racemose inflorescences. The fruits of *Combretum* spp. are also diagnostic because of the wings (four or five of them) on their fruits; the wings are papery and may be brightly coloured in some species. *C. platypterum* is a scandent shrub or forest liane, usually climbing very high and producing large panicles of greyish red flowers. The fruits are red at first, turning brown. This is a prominent species especially in regrowth forest (Fig. 5.1.1).

C. smeathmannii is also a scandent forest shrub becoming a liane; the large axillary panicles carry numerous flowers which are olive green when young and mature into pale white. The fruits are pale pink or pale purple when young (Fig. 5.1.2A). *C. racemosum* is a scandent forest regrowth species with red flowers in congested inflorescences which are subtended by white bracteate leaves with conspicuous green veins (Fig. 5.1.2B). *C. paniculatum* is a prominent climbing or scandent shrub in the savanna. The large panicles of orange–red flowers are produced when the plant is leafless.

Other species of *Combretum* are small trees in savanna or forest fringe regions. *C. glutinosum* is a savanna tree about 12 m tall. The flowers have cream or yellow petals and filaments. The leaves are glutinous above and white, hairy and reticulate on the lower surface. *C. molle* is also a savanna tree about 10 m tall with straight bole and grey, deeply fissured bark; the petals and filaments are cream coloured. Many other tree species of *Combretum* are known in our study area; apart from their characteristic fruits, the deep fissures on their barks and size of their flowers (Fig. 5.1.2C) are

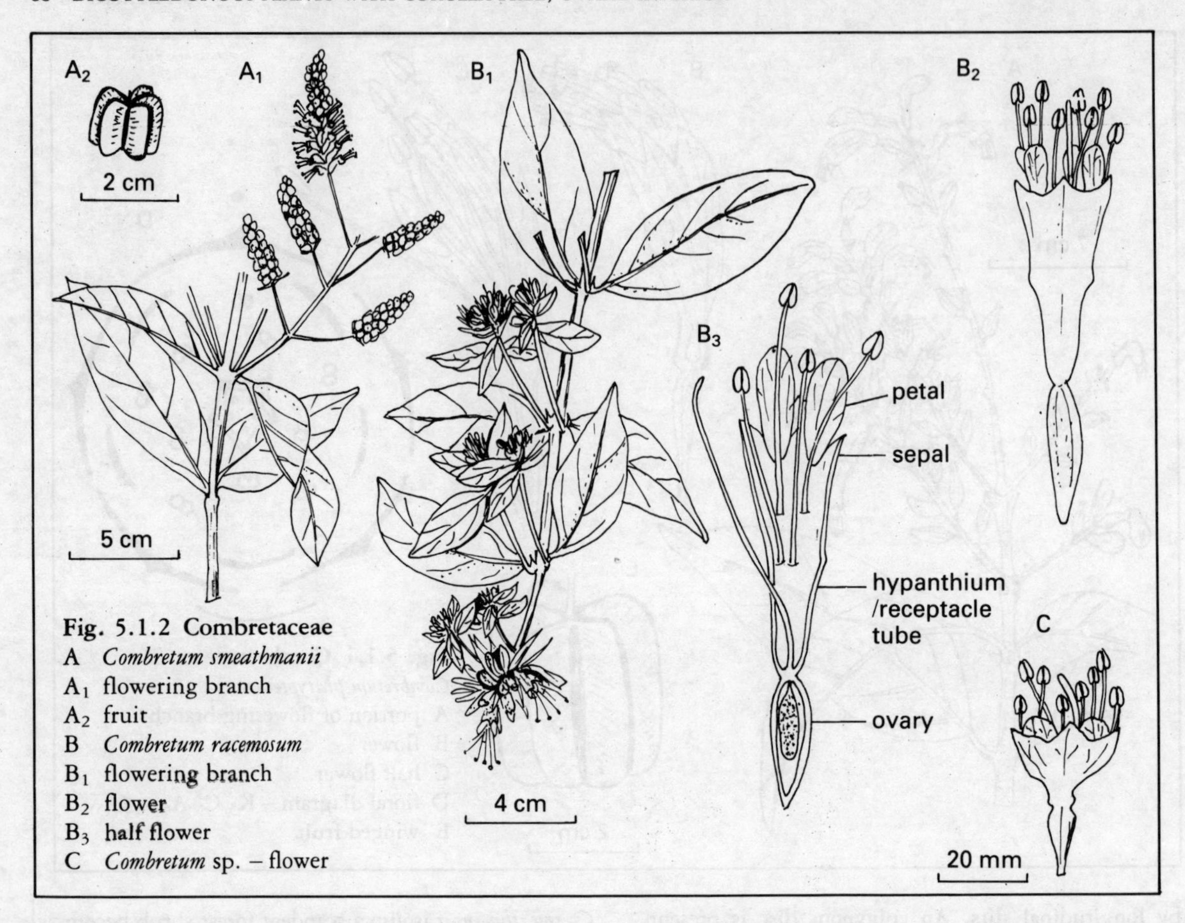

Fig. 5.1.2 Combretaceae
A *Combretum smeathmanii*
A_1 flowering branch
A_2 fruit
B *Combretum racemosum*
B_1 flowering branch
B_2 flower
B_3 half flower
C *Combretum* sp. – flower

important in identifying members of the genus.

Terminalia is a medium-sized genus of trees and shrubs with leaves densely clustered at the end of the branches, winged fruits produced from axillary spikes of whitish flowers and grey or dark barks with wavy and deep fissures. *T. catappa* is the common shade tree planted in villages and towns (Fig. 5.1.3). There is a whole belt of savanna vegetation in which different species of *Terminalia* are dominant. The greyish velutinous leaves and pinkish grey winged fruits are characteristic. *T. glaucescens* is a common savanna tree (Fig. 5.1.4A). The fruits of two important forest species – *T. superba* and *T. ivorensis* – are shown in Fig. 5.1.4B and C; *T. superba* may be distinguished from *T. ivorensis* by means of the regular whorl of branches in the former. *T. superba* is used extensively in afforestation programmes.

Guiera senegalensis is a common shrub in sandy waste places in savanna and semi-desert zones. The velutinous leaves are conspicuously grey and the inflorescence is capitate or umbelliform, producing characteristic hairy fruits with remains of the calyx sitting persistently on them (Fig. 5.1.4D).

Anogeissus leiocarpus is a tall tree of savanna with grey velvety leaves and grey bark (dark in older trees). The drooping branches have green inflorescences (Fig. 5.1.4E) which are diagnostic. The flowers, which are apetalous, appear in the dry season.

Conocarpus erectus is a mangrove shrub up to 3 m tall. The branchlets are slightly winged and the lanceolate leaves taper into a biglandular petiole; the inflorescence, flowers and fruits are very much like those of *Anogeissus leiocarpus*.

5·1b Melastomataceae

This is a family of herbs, shrubs and trees with opposite branches and usually square stems (especially the herbaceous species). The leaves are opposite or verticillate with three to nine parallel main veins and prominent horizontal veins joining the main veins (Figs 5.1.5 – 5.1.7). The leaves are exstipulate. The herbaceous

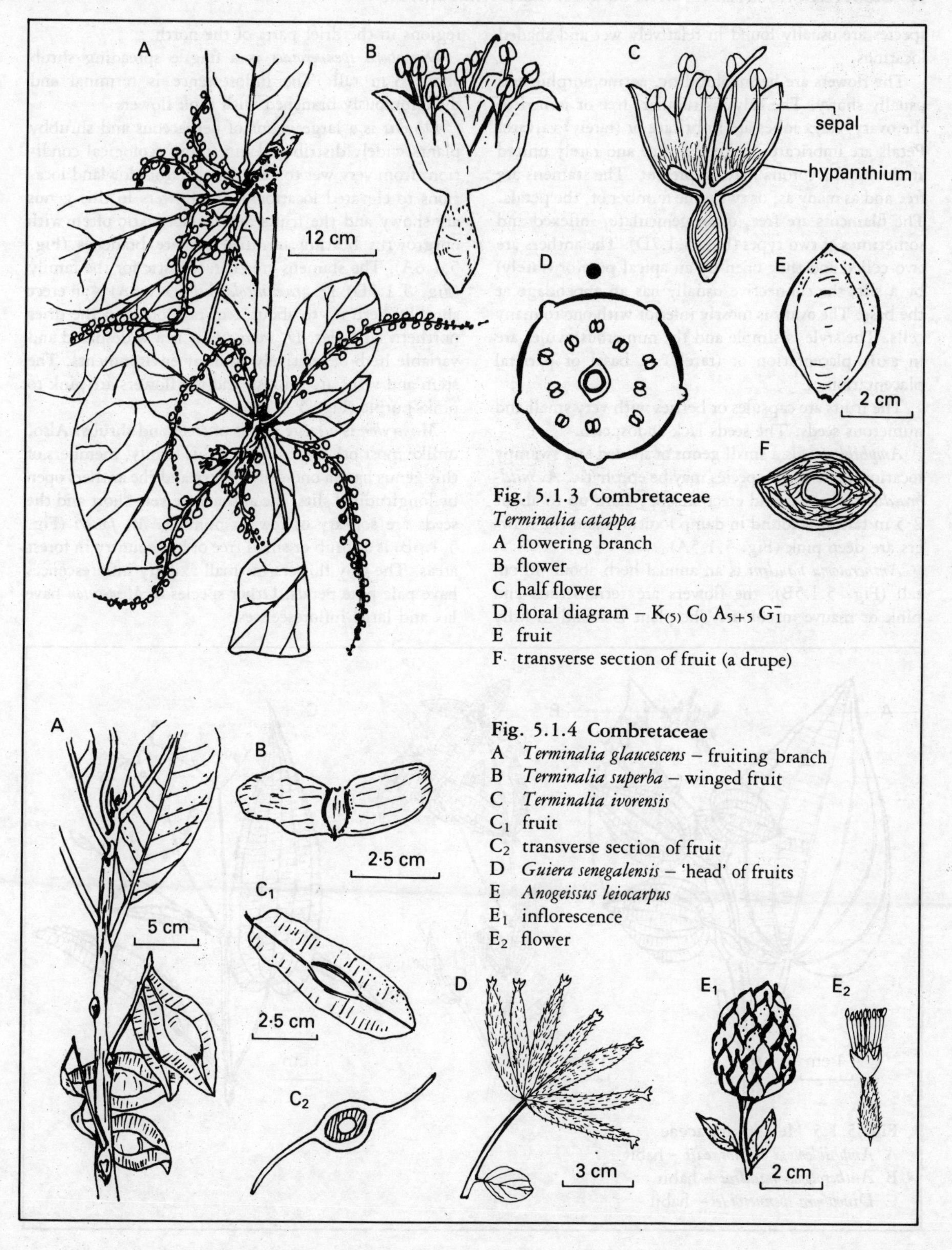

Fig. 5.1.3 Combretaceae
Terminalia catappa
A flowering branch
B flower
C half flower
D floral diagram – $K_{(5)}\ C_0\ A_{5+5}\ \bar{G}_1$
E fruit
F transverse section of fruit (a drupe)

Fig. 5.1.4 Combretaceae
A *Terminalia glaucescens* – fruiting branch
B *Terminalia superba* – winged fruit
C *Terminalia ivorensis*
C_1 fruit
C_2 transverse section of fruit
D *Guiera senegalensis* – 'head' of fruits
E *Anogeissus leiocarpus*
E_1 inflorescence
E_2 flower

species are usually found in relatively wet and shaded locations.

The flowers are hermaphroditic, actinomorphic and usually showy. The calyx is tubular, free or joined to the ovary; calyx lobes are imbricate or (rarely) valvate. Petals are imbricate, contorted, free and rarely united at the base; a corona may be present. The stamens are free and as many as, or twice the number of, the petals. The filaments are free, often geniculate, inflexed and sometimes of two types (Fig. 5.1.7D). The anthers are two-celled and they open by an apical pore or (rarely) by a slit; the connective usually has an appendage at the base. The ovary is mostly inferior with one to many cells. The style is simple and the numerous ovules are in axile placentation or (rarely) in basal or parietal placentation.

The fruits are capsules or berries with very small and numerous seeds. The seeds lack endosperm.

Amphiblemma is a small genus of shaded and swampy locations; one of the species may be epiphytic. *A. mildbraedii* is a robust and erect shrubby herb up to about 2·5 m tall; it is found in damp locations and the flowers are deep pink (Fig. 5.1.5A).

Antherotoma naudini is an annual herb about 30 cm tall (Fig. 5.1.5B); the flowers are tetramerous and pink or mauve in colour. The plant is found in hilly regions in the drier parts of the north.

Dinophora spenneroides is a fragile spreading shrub about 3 m tall. The inflorescence is terminal and dichotomously branched with pink flowers.

Dissotis is a large genus of herbaceous and shrubby plants widely distributed in varying ecological conditions from very wet to very dry and from lowland locations to elevated locations. The flowers in this genus are showy and the fruits are characteristic often with outgrowths bearing terminal stellate branches (Fig. 5.1.6A). The stamens are characteristic for the family (Fig. 5.1.7D). *D. graminicola* (Fig. 5.1.6A) is an erect shrubby herb up to about 2 m tall found in the drier northern latitudes. *D. rotundifolia* is a widespread and variable herb of relatively swampy environments. The stem and veins are pinkish and the flowers are pink to pink–purple (Fig. 5.1.7A).

Memecylon is a large genus of trees and shrubs. Also, unlike most other members of this family, members of this genus have a one-celled ovary and the anthers open by longitudinal slits; the anthers are very short and the seeds are solitary or few in number. *M. fosteri* (Fig. 5.1.6B) is a shrub or small tree of hill country in forest areas. The tiny flowers in small axillary inflorescences have pale blue petals. Other species of *Memecylon* have lax and large inflorescences.

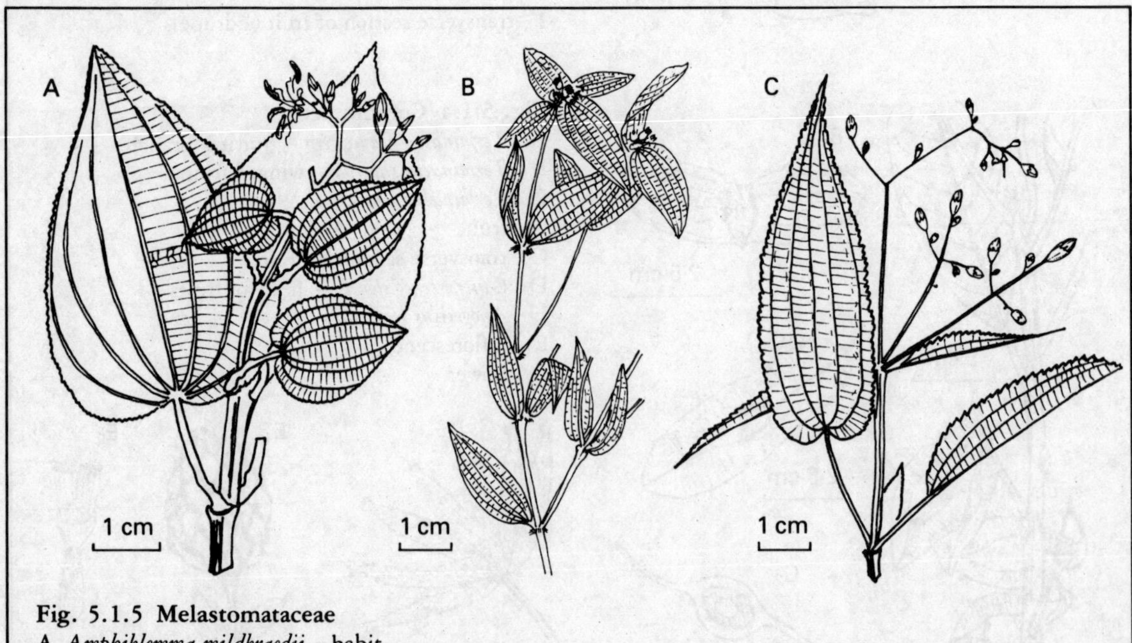

Fig. 5.1.5 Melastomataceae
A *Amphiblemma mildbraedii* – habit
B *Antherotoma naudini* – habit
C *Dinophora spenneroides* – habit

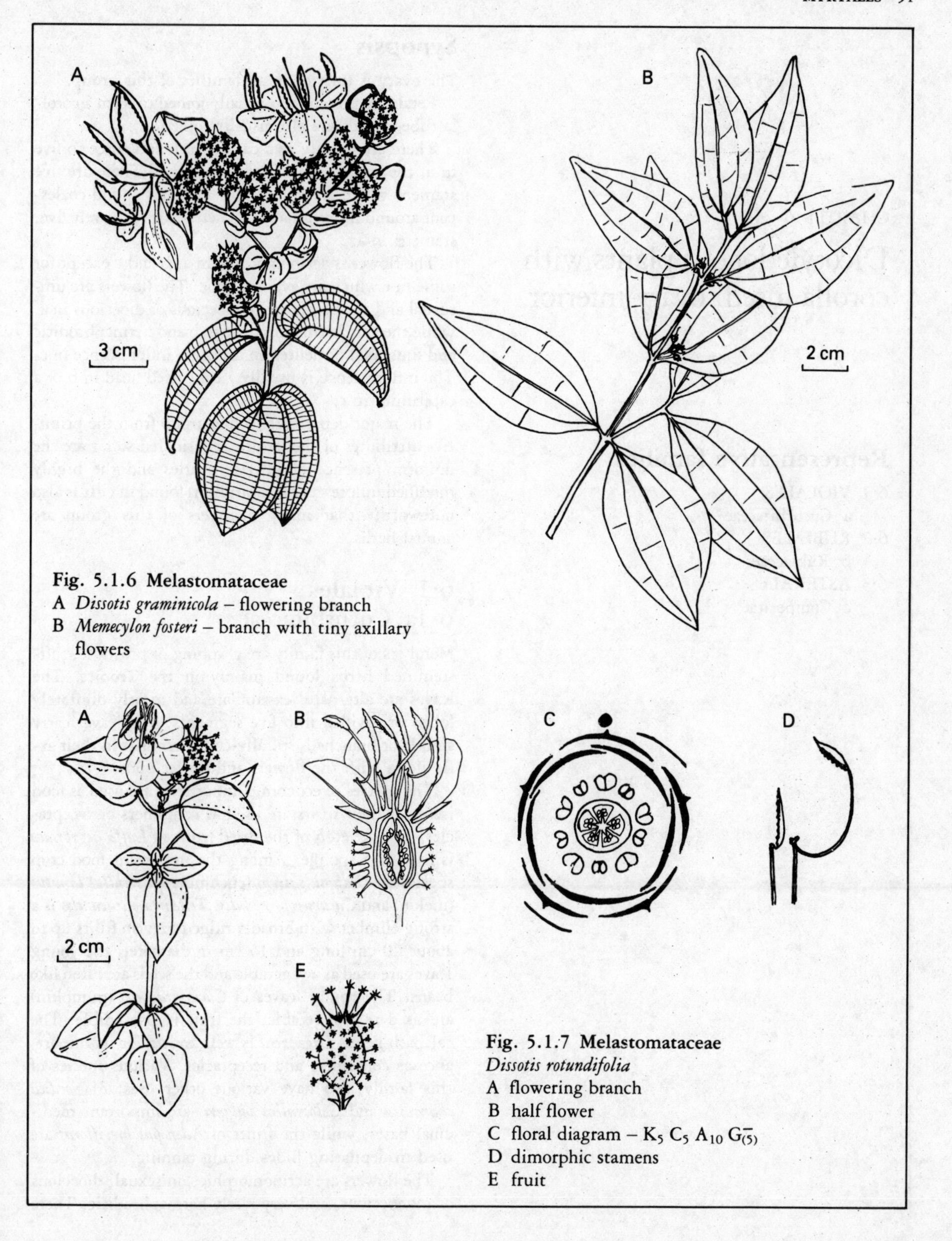

Fig. 5.1.6 Melastomataceae
A *Dissotis graminicola* – flowering branch
B *Memecylon fosteri* – branch with tiny axillary flowers

Fig. 5.1.7 Melastomataceae
Dissotis rotundifolia
A flowering branch
B half flower
C floral diagram – $K_5\ C_5\ A_{10}\ \overline{G}_{(5)}$
D dimorphic stamens
E fruit

CHAPTER 6

Dicotyledonous plants with corolla fused, ovary inferior

Representative families

6·1 VIOLALES
    a Cucurbitaceae
6·2 RUBIALES
    b Rubiaceae
6·3 ASTERALES
    c Compositae

Synopsis

The ovary is inferior in all families of this group.

Petals are present and usually joined to form a corolla tube. Sepals are also usually joined.

There are usually two carpels in b and three to five in a; the carpels are joined in a and b. There are five stamens which are epipetalous in b and c, and coalescent around the stigma in c. There are three, rarely five, stamens in a.

The flower is actinomorphic in a, b and c except for some in c which are zygomorphic. The flowers are unisexual and the plants are monoecious or dioecious in a, while they are hermaphroditic in b and hermaphroditic and unisexual or neuter on the same inflorescence in c. The inflorescence is usually a congested head in b or a capitulum in c.

The major departures in this group from the primitive attributes of the families considered so far are the uniform presence of inferior ovaries and the highly modified inflorescence (capitulum) found in c. It is also noteworthy that many members of this group are annual herbs.

6·1 Violales
6·1a Cucurbitaceae

Members of this family are climbing or prostrate, soft-stemmed herbs found mainly in the Tropics. The leaves are alternate, exstipulate and mostly digitately lobed or divided into five segments. The leaves carry simple or branched, spirally-coiled tendrils in their axils along with the flowers and flower buds.

The species of economic importance are used as food items or their fruits are used as containers or receptacles. The skeleton of the dried fruits of *Luffa aegyptiaca* is used as a sponge. Among the important food crop species are *Cucumis sativus* (cucumber), *Citrullus lanatus* (melon) and *Cucumeropsis edulis*. *Telfairia occidentalis* is a strong climber with broadly ridged greyish fruits up to about 40 cm long and 15 cm in diameter; the young leaves are used as a vegetable and the seeds are eaten like beans. The young leaves of *Cucurbita pepo* (pumpkin) are used as a vegetable; the fruit is also edible. The calabash genus *Lagenaria* is well known for its importance as containers and receptacles. Various species of this family also have various other uses; *Momordica charantia* and *Colocynthis vulgaris* are important medicinal bases, while the fruits of *Adenopus breviflorus* are used in depilating hides during tanning.

The flowers are actinomorphic, unisexual, dioecious or monoecious, and very rarely hermaphroditic. There are five petals and five sepals which are sometimes free to the base. Petal lobes are often imbricate or induplicate valvate. Except in *Luffa*, where five stamens are found, the male flower usually has three stamens with curved, folded or fused anthers; one of the three anthers is two-celled while the remainder are four-celled. In the female flower, the ovary is inferior with three to five joined carpels at the top of which occurs a style bearing the same number of stigmas; staminodes are often present in the female flowers.

The fruit is a berry with a hard rind sometimes variously patterned in deep green and light green. There are numerous flat non-endospermous seeds. The seed coats may be entirely woody or the seeds may be variously woody only at the margins.

The genera *Adenopus*, *Lagenaria* and *Cayaponia* are distinguished from the others by the presence of a pair of glands at the top of the leaf petioles or on each side of a decurrent leaf base. *Adenopus breviflorus* is a perennial high climber in dry, secondary forests. It has white flowers and green ellipsoid fruits with cream blotches of various sizes; it is used along with ashes for removing hairs from hides. *Cayaponia africana* is a strong climber in regrowth forest. It has fruits about

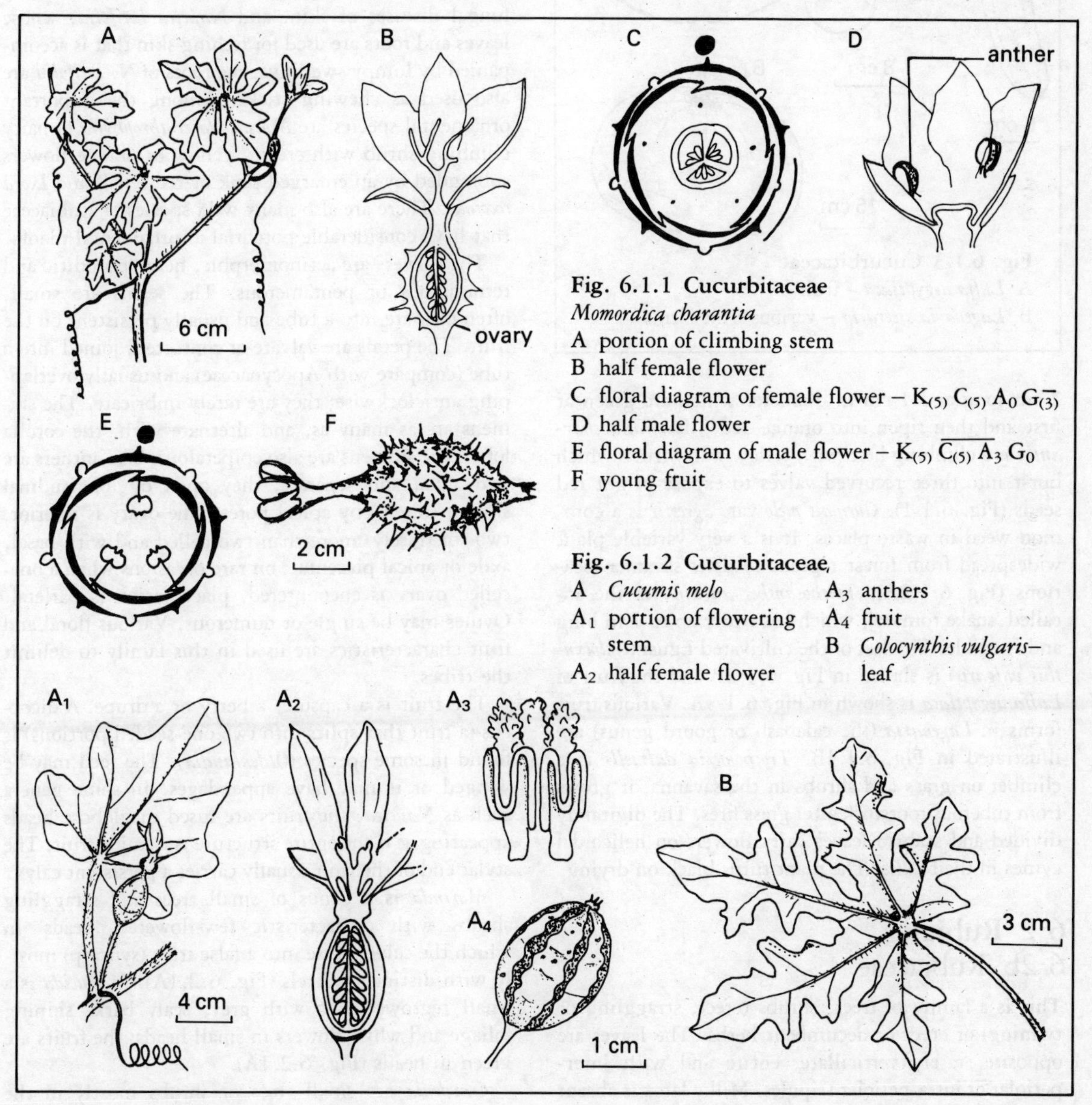

Fig. 6.1.1 Cucurbitaceae
Momordica charantia
A portion of climbing stem
B half female flower
C floral diagram of female flower – $K_{(5)}\ C_{(5)}\ A_0\ G_{\overline{(3)}}$
D half male flower
E floral diagram of male flower – $K_{(5)}\ \overline{C_{(5)}\ A_3}\ G_0$
F young fruit

Fig. 6.1.2 Cucurbitaceae,
A *Cucumis melo*
A_1 portion of flowering stem
A_2 half female flower
A_3 anthers
A_4 fruit
B *Colocynthis vulgaris*– leaf

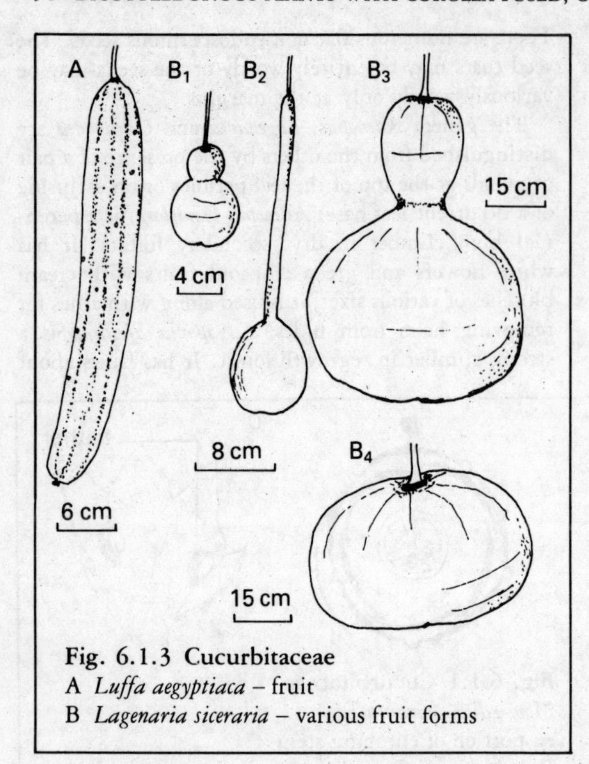

Fig. 6.1.3 Cucurbitaceae
A *Luffa aegyptiaca* – fruit
B *Lagenaria siceraria* – various fruit forms

2 cm long and 1 cm in diameter which are green at first and then ripen into orange–red. *Momordica charantia* is a climbing herb with yellow warty fruits which burst into three recurved valves to expose bright red seeds (Fig. 6.1.1). *Cucumis melo* var. *agrestis* is a common weed in waste places. It is a very variable plant widespread from forest regions to drier savanna locations (Fig. 6.1.2A). *Trichosanthes anguina* is the so-called 'snake tomato', which has fruits up to 1 m long and is edible. The leaf of the cultivated Egusi (*Colocynthis vulgaris*) is shown in Fig. 6.1.2B and the fruit of *Luffa aegyptiaca* is shown in Fig. 6.1.3A. Various fruit forms in *Lagenaria* (the calabash or gourd genus) are illustrated in Fig. 6.1.3B. *Trochomeria dalzielli* is a climber on grass and shrubs in the savanna; it grows from tuberous rootstock after grass fires. The digitately divided and scabrid leaves carry flowers on helicoidal cymes in their axils. The plant turns black on drying.

6·2 Rubiales

6·2b Rubiaceae

This is a family of trees, shrubs (erect, straggling or twining) or erect or decumbent herbs. The leaves are opposite, rarely verticillate, entire and with interpetiolar or intra-petiolar stipules. Milky latex is absent in this family. The absence of milky latex and the presence of stipules and an inferior ovary are major attributes that separate this family from Apocynaceae.

Some members yield timber and the coffee plant is a member of the family too. A number of herbaceous species are important in indigenous medicine, while a few species are introduced as ornamentals. The important timber species are *Nauclea diderrichii* and *Mitragyna ciliata*. Three species of the coffee plant are cultivated in West Africa; they are *Coffea liberica*, *C. arabica* and *C. canephora*. The species of medicinal importance include *Mitracarpum scabrum*, which is used for curing fungal diseases of skin, and *Nauclea latifolia*, whose leaves and roots are used for itching skin that is accompanied by lumpy swellings; the roots of *N. latifolia* are also used as chewing sticks. Among the important ornamental species are *Mussaenda erythrophylla* (a hairy climbing shrub with cream, yellow or orange flowers subtended by an enlarged pink or red sepal) and *Ixora coccinea*. There are also many wild species of Rubiaceae that have considerable potential as ornamental plants.

The flowers are actinomorphic, hermaphroditic and tetramerous or pentamerous. The sepals are small, often connate into a tube and usually persistent on the fruit. The petals are valvate or contorted, joined into a tube (compare with Apocynaceae) and usually overlapping anticlockwise; they are rarely imbricate. The stamens are as many as, and alternate with, the corolla lobes; the stamens are also epipetalous. The anthers are two-celled and separate; they open by longitudinal slits or (rarely) by apical pores. The ovary is inferior, two- or (rarely) more-than-two-celled and with basal, axile or apical placentae; on rare occasions when a one-celled ovary is encountered, placentation is parietal. Ovules may be single or numerous. Various floral and fruit characteristics are used in this family to delimit the tribes.

The fruit is a capsule, a berry or a drupe. A dicoccus (a fruit that splits into two one-seeded portions) is found in some species (*Oldenlandia*). The seed may be winged or it may have appendages. In some genera such as *Nauclea*, the fruits are fused in globose heads appearing as if the entire structure is a single fruit. The stylar end of the fruit usually carries a persistent calyx.

Morinda is a genus of small trees and straggling shrubs with characteristic few-flowered 'heads' in which the calyces fuse into a false fruit (syncarp) mostly with distinct pedicels (Fig. 6.2.1A). *M. lucida* is a small regrowth tree with grey, scaly bark, shining foliage and white flowers in small heads; the fruits are green in heads (Fig. 6.2.1A).

Gardenia are small trees or shrubs mostly in the

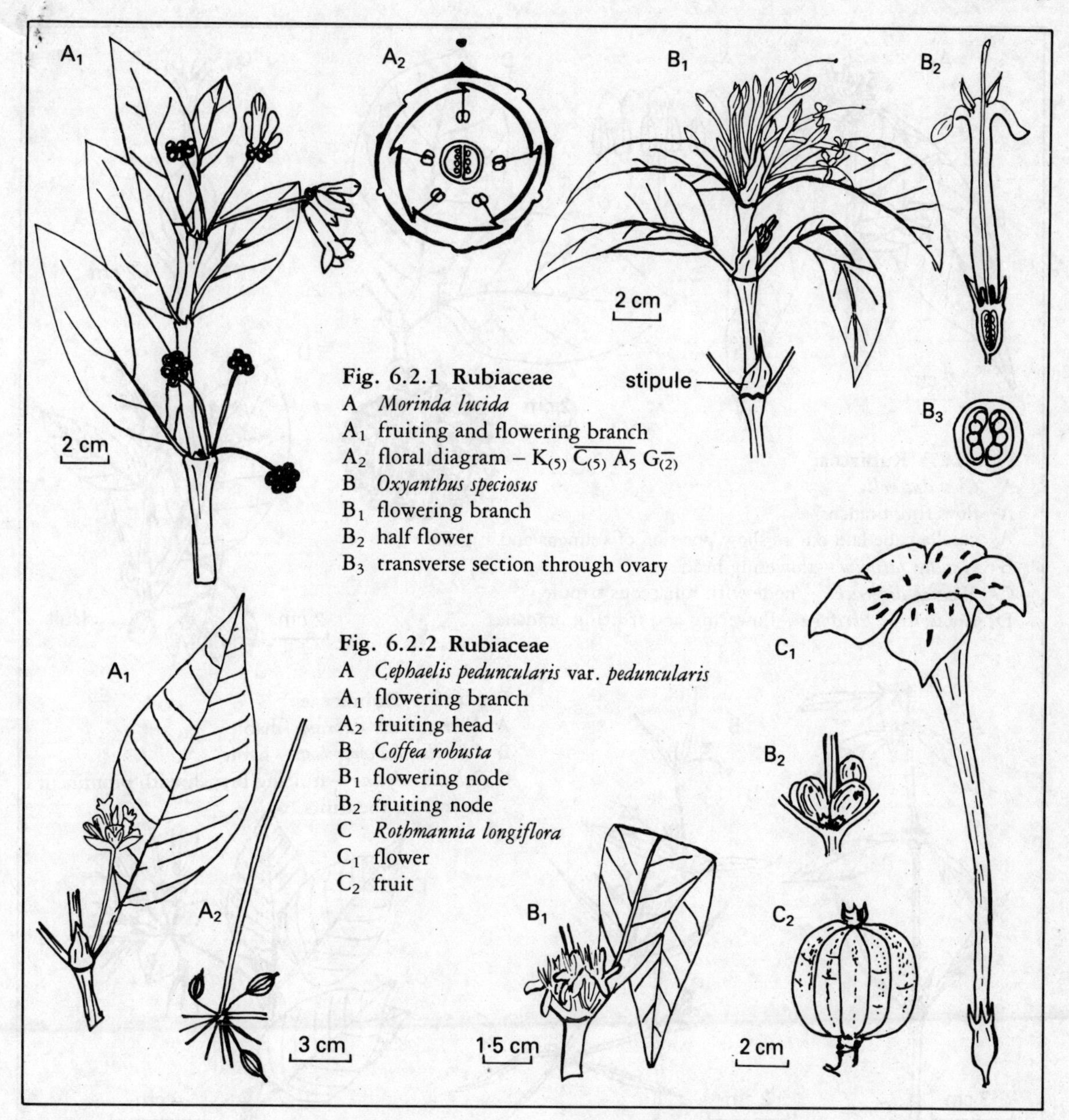

Fig. 6.2.1 Rubiaceae
A *Morinda lucida*
A_1 fruiting and flowering branch
A_2 floral diagram – $K_{(5)}\ \overline{C_{(5)}\ A_5}\ G_{\overline{(2)}}$
B *Oxyanthus speciosus*
B_1 flowering branch
B_2 half flower
B_3 transverse section through ovary

Fig. 6.2.2 Rubiaceae
A *Cephaelis peduncularis* var. *peduncularis*
A_1 flowering branch
A_2 fruiting head
B *Coffea robusta*
B_1 flowering node
B_2 fruiting node
C *Rothmannia longiflora*
C_1 flower
C_2 fruit

savanna and dry forest zones. The stems are grey and twisted and produce white fragrant flowers with a strongly contorted tubular corolla in the dry season. The bark is usually scaly. The fruits are rough grey and may be about 5 cm long.

Oxyanthus speciosus is a small forest tree with deep green and shiny foliage. The flowers are fragrant; they are produced in axillary fascicles (Fig. 6.2.1B).

Cephaelis peduncularis is a very variable forest shrub about 1 m tall. The flowers are white followed by numerous distinctly pedunclate ridged fruits in regular heads (Fig. 6.2.2A).

The genus *Coffea* is characterised by prominent stipules, usually numerous white axillary flowers followed by sessile one- or two-stoned drupes in axillary clusters (Fig. 6.2.2B).

Rothmannia is a genus of forest shrubs and small trees with prominent trumpet-shaped solitary flowers in terminal or axillary positions. The fruits are large, indehiscent, two-celled capsules with numerous seeds

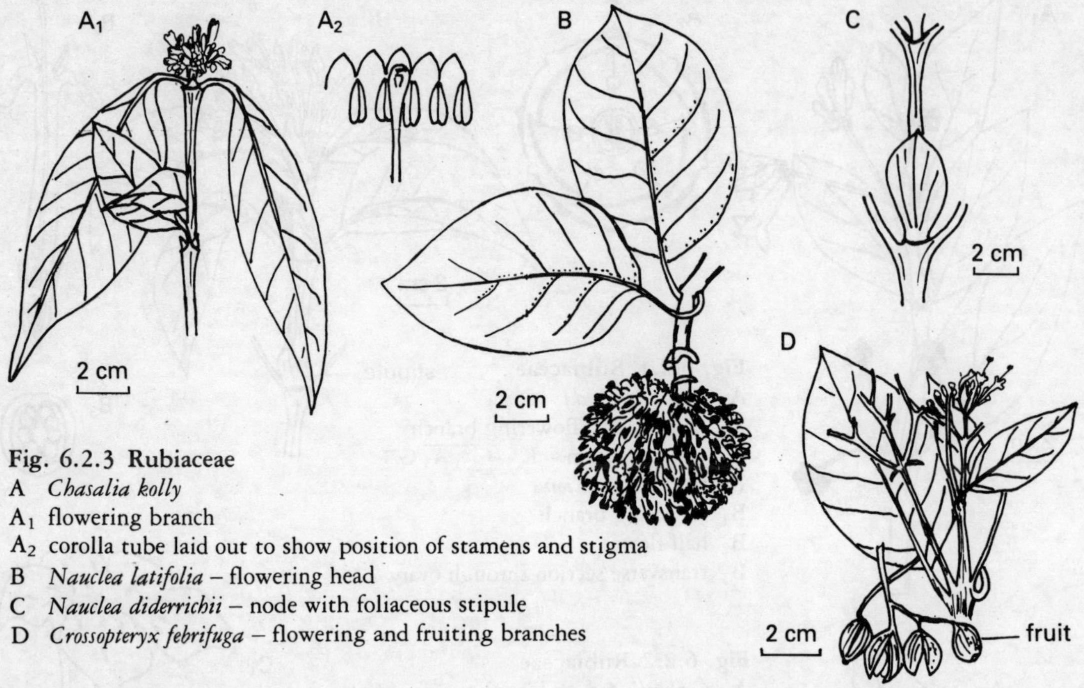

Fig. 6.2.3 Rubiaceae
A *Chasalia kolly*
A_1 flowering branch
A_2 corolla tube laid out to show position of stamens and stigma
B *Nauclea latifolia* – flowering head
C *Nauclea diderrichii* – node with foliaceous stipule
D *Crossopteryx febrifuga* – flowering and fruiting branches

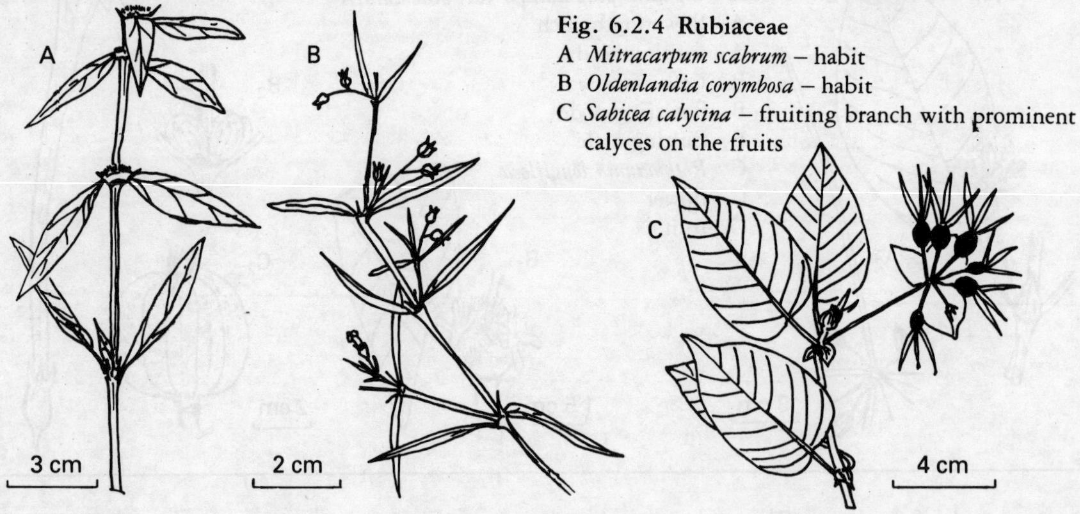

Fig. 6.2.4 Rubiaceae
A *Mitracarpum scabrum* – habit
B *Oldenlandia corymbosa* – habit
C *Sabicea calycina* – fruiting branch with prominent calyces on the fruits

in axile placentation. The stipules fall early so that they are observed in young shoots only. The flower and fruit of *R. longiflora* are illustrated in Fig. 6.2.2C.

Chasalia kolly is a soft-stemmed shrub about 1–3 m tall in forest regions. The flowers are white or pink; the calyx and peduncle are purple. (Fig. 6.2.3A).

Nauclea is a genus of shrubs and trees found in savanna and forest locations. The occurrence of flowers in globose heads, the prominent stipules and the large shining leaves are characteristic. *N. latifolia* is a shrub or small tree of dry forest and savanna; the inflorescence consists of white sweet-scented flowers in spherical heads appearing in the dry season; the greyish syncarp is about the size of a golf ball (Fig. 6.2.3B). *N. diderrichii* is a tall forest tree with a distinct bole and low buttress; the stipules are prominent (Fig. 6.2.3C).

Crossopteryx febrifuga is widespread in West Africa. It is a savanna tree or shrub up to about 10 m tall with scaly bark. The flowers are numerous in dense terminal corymbs; they are creamy white and have an unpleasant odour. The fruits partially split into two valves to expose their irregularly-winged seeds (Fig. 6.2.3D).

Mitracarpum scabrum is an annual weed around 30 cm tall or much smaller. The flowers are white and crowded at the nodes within the divided stipules. The leaves are elliptic to obovate with whitish veins.

Oldenlandia is a genus of small scandent or erect herbs with linear leaves and usually white flowers in axillary or terminal heads. The flowers are either many and subsessile, or few or solitary and long-pedicelled. The fruit in the genus is a dicoccus. *Oldenlandia corymbosa* is a common annual weed with small pedunculate white flowers. It is widespread in open places in forest and savanna (Figure 6.2.4B).

The woody climbing habit is found in the genera *Mussaenda*, *Uncaria*, *Sabicea* and *Sherbournia*. *Uncaria africana* is a shrub of secondary forest up to about 20 m high with paired and recurved axillary hooks. The flowers occur in dense spherical heads at terminal positions. *Sherbournia bignoniiflora* is a straggling or climbing shrub. The green calyx has red markings and the corolla tube, which is about 3 cm in diameter and about 5 cm long, is pale pink. The plant occurs in forest regions. *Sabicea calycina* is a slender climbing plant of secondary forest especially in regrowth conditions. The flowers are white and produce red fruits, while the calyx and stipules are pinkish purple (Fig. 6.2.4C).

Borreria is a large genus of herbaceous and woody plants commonly found in dry grasslands. The flowers are white and aggregated in large numbers at the nodes, and at the end of branches. In this genus, the stipules are often divided into many linear segments and the fruit is a small capsule which splits into two. *B. octodon*, *B. filiformis*, *B. scabra* and *B. filifolia* are shown in Fig. 6.2.5A, B, C and D

Diodia is another genus very much like *Borreria*; however the fruits in the latter separate into two halves, but the seeds are not released from these halves.

Geophila (Fig. 6.2.5E) is a genus of creeping herbs on the forest floor; they root at the nodes. The flowers are white and star-like with four petals; they are unmistakable on forest floors. Their fruits are shining and scarlet or blue.

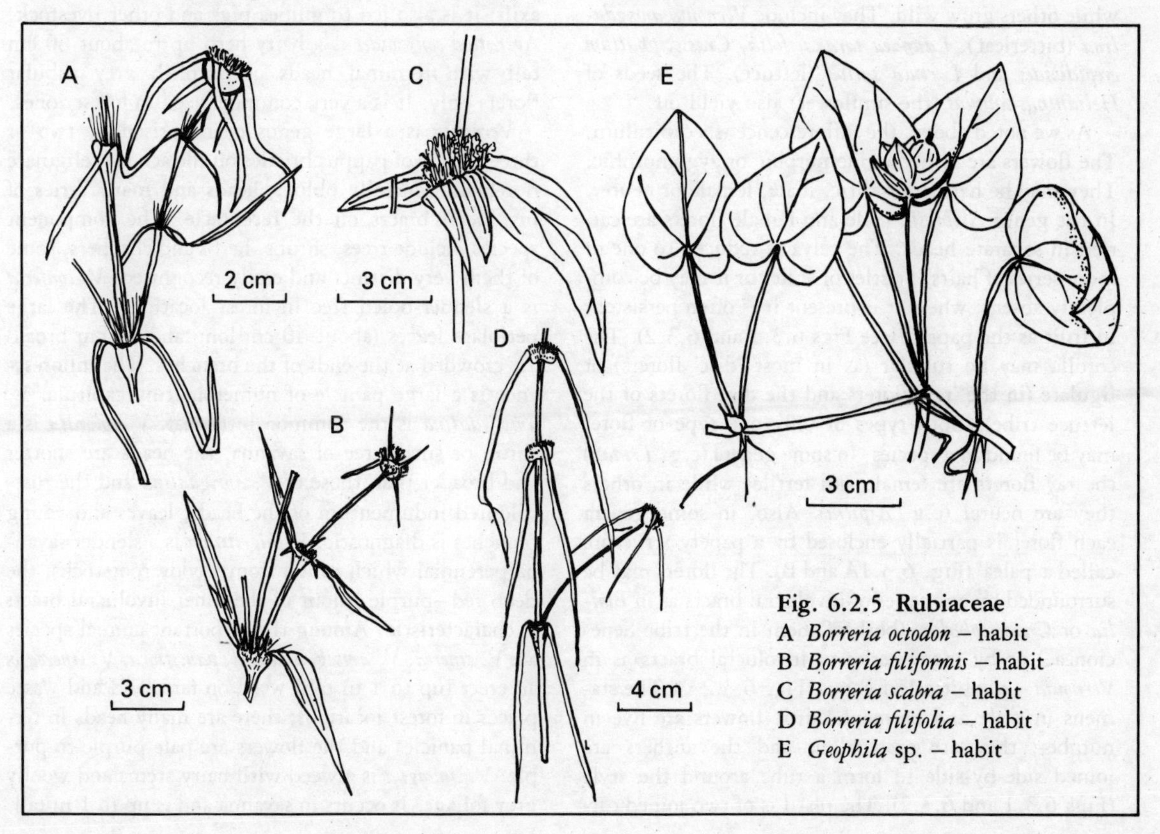

Fig. 6.2.5 Rubiaceae
A *Borreria octodon* – habit
B *Borreria filiformis* – habit
C *Borreria scabra* – habit
D *Borreria filifolia* – habit
E *Geophila* sp. – habit

6·3 Asterales

6·3c Compositae

Compositae is a very large cosmopolitan family. A few members are trees, a few are climbing herbs but most of them are weedy herbs or shrubs. The inflorescence is a flat head (capitulum) with many flowers and the head is subtended by an involucre of bracts in one or more series. The leaves are simple (pinnately lobed in some species), alternate or opposite and prickly in some genera. Winged stems are found in some genera and milky latex is diagnostic for the lettuce tribe – Cichorieae.

Members of this family are important as weeds, as ornamentals and as green vegetables. Important weed species in this family include *Eupatorium odoratum* which is widespread in the disturbed areas of the forest, *Tridax procumbens* (Fig. 6.3.1A), *Ageratum conyzoides* which grows in large populations, *Aspilia* spp. (Fig. 6.3.1B) which are widely distributed and *Melanthera* spp. which are morphologically very much like *Aspilia* spp. The ornamental species include *Zinnia*, *Tagetes* (marigold) and *Helianthus annuus* (sunflower). Some of the important vegetable species are cultivated while others grow wild. They include *Vernonia amygdalina* (bitterleaf), *Launaea taraxacifolia, Crassocephallum crepidioides* and *Lactuca sativa* (lettuce). The seeds of *Helianthus annuus* (the sunflower) also yield oil.

As we noted above, the inflorescence is a capitulum. The flowers are either actinomorphic or zygomorphic. They may be hermaphroditic, male, female or neuter. In the genus *Ambrosia* male and female florets are carried in separate heads. The calyx is reduced to one or more series of hairs, bristles or scales or it may be completely absent; where it is present it is often persistent in fruit as the pappus (see Figs 6.3.1 and 6.3.2). The corolla may be tubular (as in most 'disc' florets) or ligulate (in the 'ray' florets and the disc florets of the lettuce tribe). Both types or only one type of floret may be found in a species. In some genera (e.g. *Tridax*) the ray florets are female and fertile, while in others they are neuter (e.g. *Aspilia*). Also, in some genera each floret is partially enclosed by a papery structure called a 'palea' (Fig. 6.3.1A and B). The florets may be surrounded by one series of involucral bracts as in *Emilia* or *Crassocephalum* (both of them in the tribe Senecioneae) or by many series of involucral bracts as in *Vernonia* – the bitterleaf genus (Fig. 6.3.2A). The stamens in male and hermaphroditic flowers are five in number; they are epipetalous and the anthers are joined side-by-side to form a tube around the style (Figs 6.3.1 and 6.3.2). The pistil is of two joined carpels with a single basal ovule in an inferior ovary.

The fruit is a cypsela (a form of achene carrying a pappus at the stylar end), or a pseudonut in the absence of a pappus. The pappus may have various appendages to aid dispersal. In many genera, the disc achenes and the ray achenes are distinctly different in size and morphology.

Tridax procumbens is a common weed, widespread from the coast to the drier north. A flowering portion of the stem and details of flower are illustrated in Fig. 6.3.1A.

Aspilia is a fairly large and variable genus with opposite leaves and capitula with yellow, white, purple or cream ray florets. *A. africana* is a variable species (Fig. 6.3.1B) with golden yellow ray florets and is often fed to livestock especially rabbits and guinea pigs. *A. africana* may be easily confused with members of the genus *Melanthera*; however, in the latter, the pappus is made up of separate bristles or teeth instead of being united into a cup as in the former. *Synedrella nodiflora* and *Ageratum conyzoides* are commonly found growing in the same localities as *Aspilia africana*. *Synedrella nodiflora* is a small herb with tiny yellow flowers in few-flowered capitula at the nodes in leaf axils; it is also fed to guinea pigs and other livestock. *Ageratum conyzoides* is a hairy herb up to about 60 cm tall with terminal heads of purplish grey tubular florets only. It is a very common weed in forest zones.

Vernonia is a large genus characterised by two or three whorls of pappus bristles on the achene, eligulate florets in generally oblong heads and many series of involucral bracts on the receptacle. The component species include trees, shrubs, herbs and climbers, some of them very distinct and easily recognised. *V. conferta* is a slender-boled tree in forest locations. The large petiolate leaves (about 40 cm long and 15 cm broad) are crowded at the ends of the branches. The inflorescence is a large panicle of numerous tiny capitula. *V. amygydalina* is the common bitterleaf. *V. colorata* is a shrub or small tree of savanna; the heads are shorter and broader than those of *V. amygdalina* and the rust-coloured indumentum of the heads, leaves and young branches is diagnostic. *V. nigritiana* is a slender savanna perennial which grows from woody rootstocks; the deep red–purple colour of the inner involucral bracts is characteristic. Among the important annual species are *V. cinerea*, *V. ambigua* and *V. pauciflora*. *V. cinerea* is an erect (up to 1 m tall) weed on farmland and waste places in forest locations; there are many heads in terminal panicles and the flowers are pale purple to purple. *V. ambigua* is a weed with hairy stems and woolly grey foliage. It occurs in savanna and is up to 1 m tall.

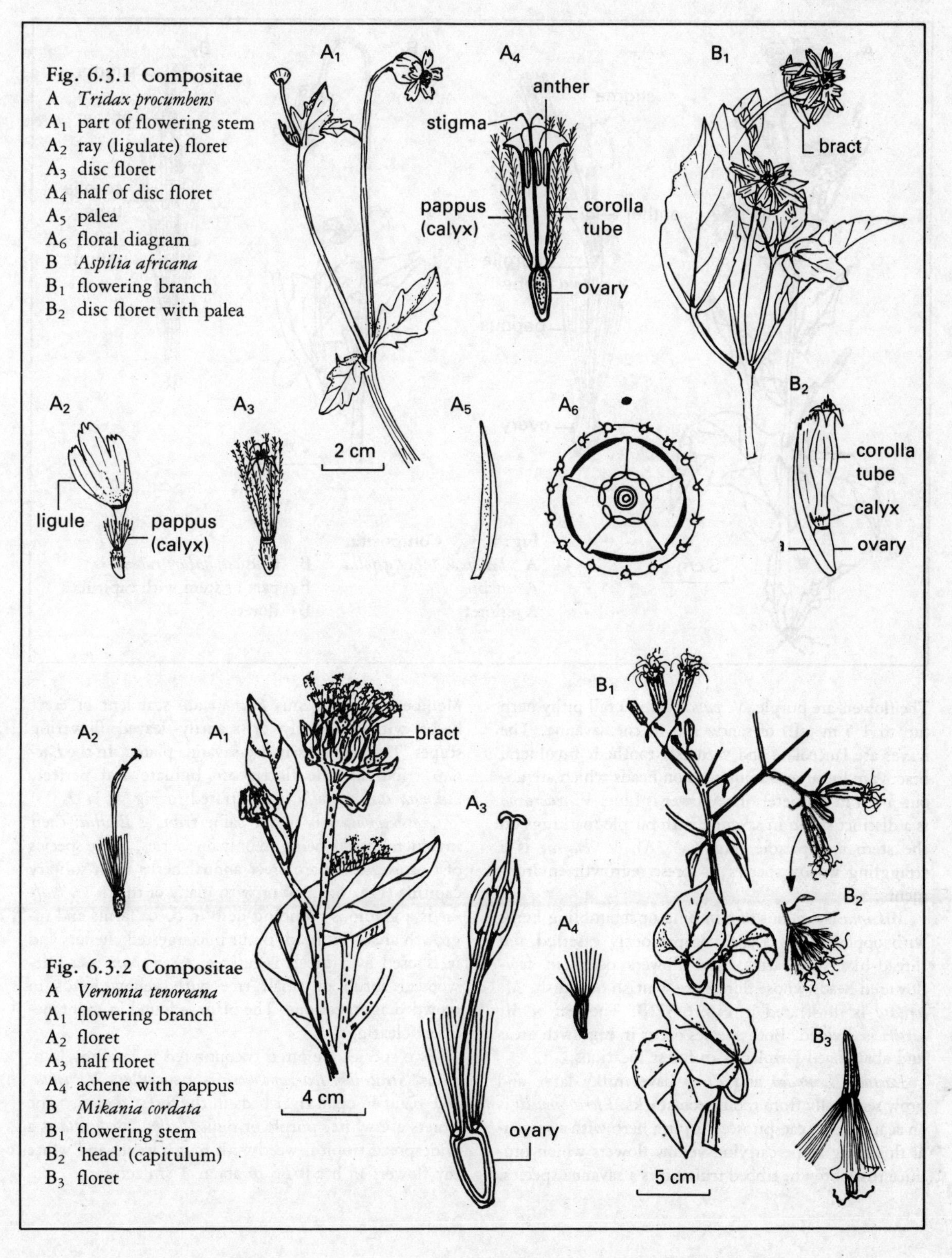

Fig. 6.3.1 Compositae
A *Tridax procumbens*
A_1 part of flowering stem
A_2 ray (ligulate) floret
A_3 disc floret
A_4 half of disc floret
A_5 palea
A_6 floral diagram
B *Aspilia africana*
B_1 flowering branch
B_2 disc floret with palea

Fig. 6.3.2 Compositae
A *Vernonia tenoreana*
A_1 flowering branch
A_2 floret
A_3 half floret
A_4 achene with pappus
B *Mikania cordata*
B_1 flowering stem
B_2 'head' (capitulum)
B_3 floret

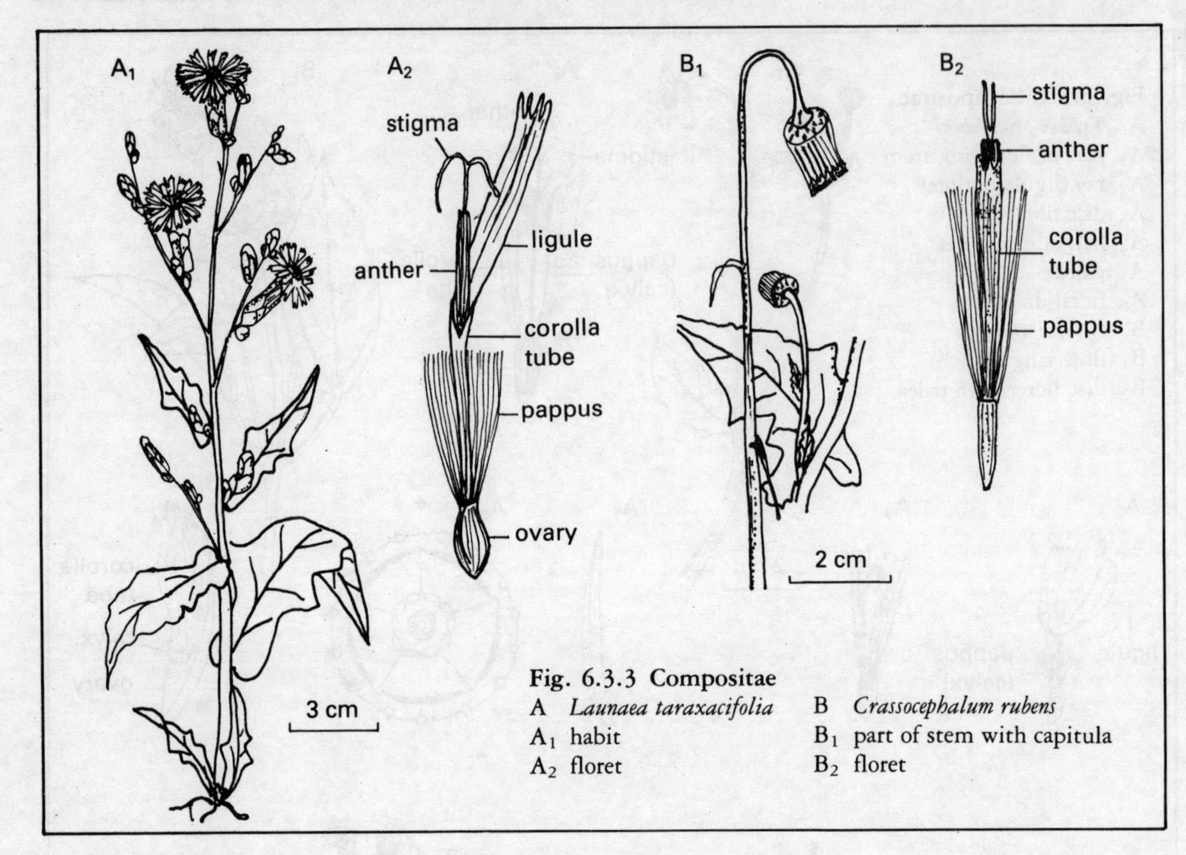

Fig. 6.3.3 Compositae
A *Launaea taraxacifolia*
A_1 habit
A_2 floret
B *Crassocephalum rubens*
B_1 part of stem with capitula
B_2 floret

The flowers are purple. *V. pauciflora* is a tall pithy herb (up to 1·5 m tall) of sandy soil in the savanna. The leaves are lanceolate and variously toothed. Involucral bracts are linear and numerous on heads which are about 3 cm in diameter. The flower is blue. *V. tenoreana* is a distinct shrub in savanna with purple markings on the stem and petiole (Fig. 6.3.2A). *V. biafrae* is a straggling woody species in forest regrowth environments.

Mikania is a genus of climbing or scrambling herbs with opposite leaves and conspicuously exserted and thread-like style arms. The flowers occur in few-flowered heads whose florets are whitish or bluish. *M. cordata* is illustrated in Fig. 6.3.2B. The leaf of *M. carteri* is divided. Both species occur in regrowth areas and abandoned farmlands in forest locations.

Lactuca, Launaea and *Picris* have milky latex and grow seasonally from robust rootstocks. *Picris humilis* is an acaulescent, caespitose, purplish herb with a seasonal flowering scape carrying yellow flowers which produce rusty brown, ribbed fruits; it is a savanna species. Members of the genus *Lactuca* are scandent or erect herbs with decidedly or scantily leaved flowering scapes. They are mostly dry-savanna plants. In the *Lactuca* tribe, all the florets are ligulate and perfect. *Launaea taraxacifolia* is illustrated in Fig. 6.3.3A.

Crassocephalum is in the same tribe as *Emilia*; their involucral bracts being in only one series. Some species of *Crassocephalum* are erect annual herbs with solitary capitula (Fig. 6.3.3B) or with many of them. *C. biafrae* is a glabrous climbing herb in cocoa farms and regrowth areas. Its heads occur in congested clusters and it is used as a green vegetable. *C. mannii* is a soft-wooded shrub or small tree with yellow heads in crowded aggregations. The plant is found in montane-forest clearings.

Two species are often encountered in swampy locations. *Struchium sparganophora* is a succulent *Vernonia*-like plant in open river beds in the forest regions. The florets are white, purple or pink. *Eclipta prosptrata* is a widespread tropical weed with rough leaves and white ray flowers in heads up to about 1 cm across.

CHAPTER 7

Monocotyledonous plants

Representative families

Synopsis

The ovary is superior in most of j, in some of l and in a, b, c, h and i. The ovary is inferior in k, some of l and in d, e, f, g, m and n. There are usually three joined carpels except in b, c, h and i.

Stamens range in number from half to eight. There are usually six stamens in j, k and l. Staminodes are present in a, f, and m with three, half and three functional anthers respectively in members of these families. There are five stamens in g. In d and e there are one and half an anther respectively. In n the single anther is highly specialised. There are three versatile anthers in b, c and h, while the two to eight anthers in i are usually joined to form a synandrium.

The flower is actinomorphic in h, i, j, k, l and m; subactinomorphic in f and zygomorphic in a, b, c, d, e, g and n. Hermaphroditism is the rule in a, d, e, f, g, j, k, l, n and most of b and c; there are rare exceptions in j. Monoecism is found in some of k and l and in most of h and i. Dioecism is common in m.

The inner perianth whorl at least is petaloid in a, d, e, f, g, j, k, l and n, while the perianth is sepaloid in all the other families, but absent or highly reduced in i. The perianth is chaffy, scale-life or lacking in b and c, and it is stiff in h.

Some important vegetative traits are also noticeable in members of this group of families. We find undergound bulbs in some members of j and k, and rhizomes in some members of b, c, d, e, f, g, i and l; the rhizomes are tuberous in i. Corms or corm-like underground stems are also encountered in members of c and n. Most members of this group have parallel-veined leaves, the exceptions being found in m and some members of j. The presence of jointed stems in a, b, c, d, e, f, i and n is a significant departure from the unjointed stems of the dicotyledonous families. The strong tendency to trimerous floral parts is not only significant, but resembles the condition in some of the primitive dicotyledonous plants.

7·1 Commelinales

7·1a Commelinaceae

This is a family of terrestial or aquatic, usually succulent annual or perennial herbs. They usually root at the nodes; the leaves are alternate with conspicuous closed sheaths. A sickle-shaped spathe subtending the inflorescence is found in most genera of this family.

Members of this family are weeds that are very difficult to eradicate; these include species of *Commelina, Cyanotis* and *Aneilema*. Some members of the family are

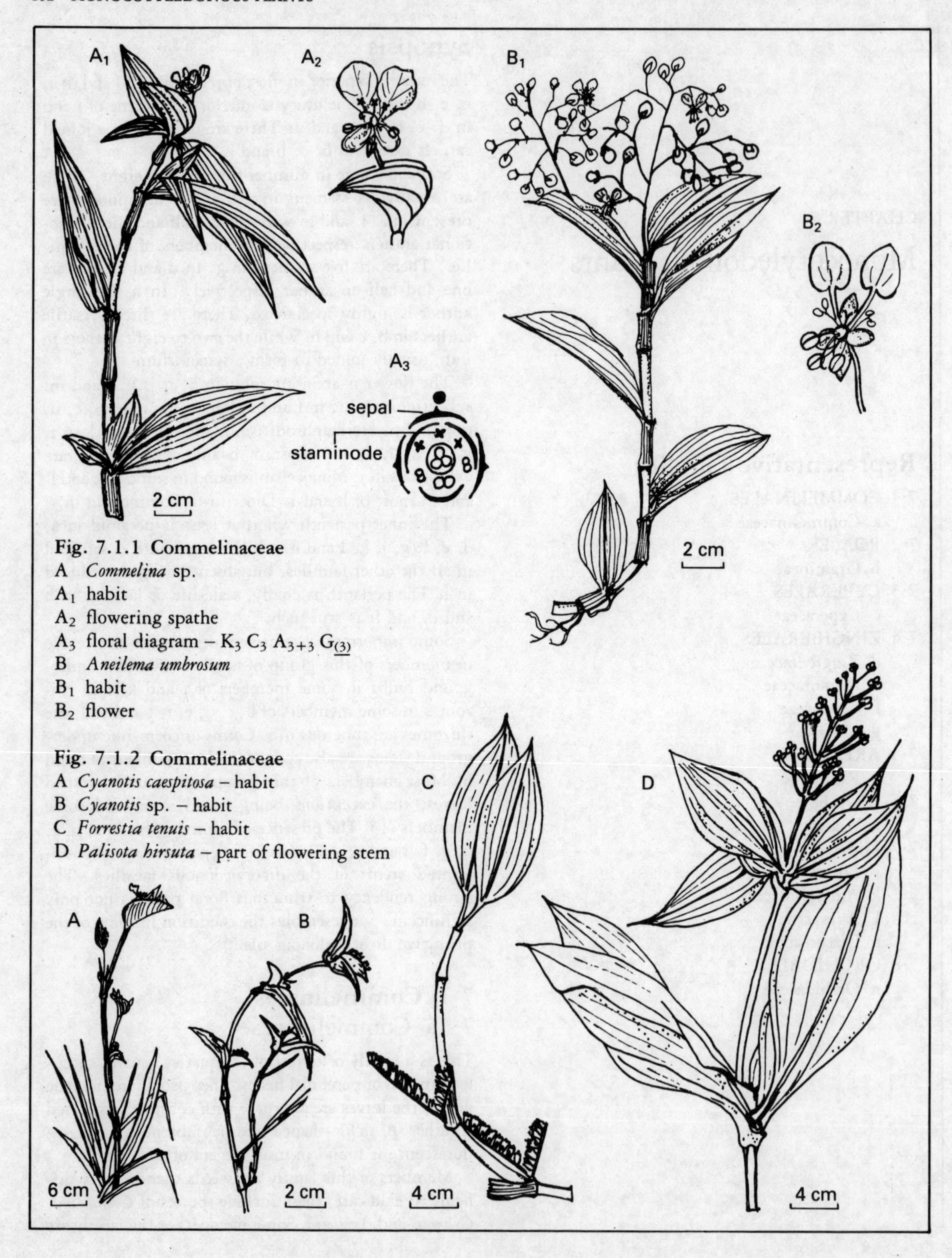

Fig. 7.1.1 Commelinaceae
A *Commelina* sp.
A_1 habit
A_2 flowering spathe
A_3 floral diagram – $K_3\ C_3\ A_{3+3}\ G_{(3)}$
B *Aneilema umbrosum*
B_1 habit
B_2 flower

Fig. 7.1.2 Commelinaceae
A *Cyanotis caespitosa* – habit
B *Cyanotis* sp. – habit
C *Forrestia tenuis* – habit
D *Palisota hirsuta* – part of flowering stem

introduced ornamental plants; these are *Tradescantia virginiana* (with white flowers and tufted purple and green lanceolate leaves), *Zebrina pendula* (a creeping commelina-like plant with leaves which have parallel whitish, zebra-like stripes) and *Rhoeo discolor*.

The flower is actinomorphic or zygomorphic, hermaphroditic and trimerous. The sepals and petals are distinct (three of each) although the latter may be unequal in size. The number of stamens may be six, five with one staminode, three with three staminodes or two with four staminodes. The staminodes have three or four lobes and the anther opens by longitudinal slits. The ovary consists of three superior carpels with one or a few ovules in each cell. We should emphasise the diagnostic value of the spathe of the inflorescence in some genera. The inflorescence is terminal or lateral.

The fruit is a two- or three-valved capsule with one or a few muricate, reticulate or ridged seeds. The fruit is loculicidal or (rarely) indehiscent. *Commelina* and *Aneilema* have striking vegetative similarity but the characteristic spathe of *Commelina* is absent in *Aneilema*. It is also noteworthy that in *Aneilema* the group of three staminodes occur on the adaxial side of the flower opposite the three stamens, while *Commelina* has three adaxial staminodes, two lateral stamens and one large and sagittate abaxial stamen (Fig. 7.1.1). *Commelina* sp. and *Aneilema umbrosum* subsp. *umbrosum* are shown in Fig. 7.1.1A and B respectively.

Cyanotis is a genus of erect herbs with tuberous or bulbous rootstock. Members of *Cyanotis*, like those of *Commelina*, have a spathe, but in *Cynanotis* several smaller ordinary bracts are enclosed in the large spathe (Fig. 7.1.2 A and B). *Cyanotis caespitosa* is a grassland species with bulbous rootstock and blue flowers (Fig. 7.1.2A).

Forrestia tenuis is an erect or ascending herb usually of swampy places in rainforest. The leaves are petiolate and the characteristic bifurcated inflorescences of blue flowers burst through the leaf sheaths to be exposed. The inflorescences are therefore lateral (Fig. 7.1.2C).

Palisota is a genus of erect forest herbs. Their leaf margins are densely hairy and the leaves usually occur in false whorls. The inflorescence lacks the commelina-type spathe. *Palisota hirsuta* is illustrated in Fig. 7.1.2D.

7·2 Poales

7·2b Gramineae

This is the grass family – a very large cosmopolitan family which may be annual or perennial, growing in tufts or tussocks, creeping and rooting at the nodes or rhizomatous. The family resembles the sedge family in many respects but it differs in that the stems of grasses are cylindrical in cross-section, the sheath is split on the opposite side of the blade and the leaf is ligulate. Some members of the family – the bamboos – are woody and attain three-like proportions.

The economic importance of the grass family cannot be over-emphasized. Many cereal crops belong in the family – *Zea mays* (maize, corn), *Sorghum bicolor* (Guinea corn), *Pennisetum americanum* (millet), *Oryza sativa* (rice) and *Eleusine coracana* (finger millet). *Saccharum officinarum* (sugar cane) is also a member of this family. Various wild grasses such as *Hyparrhenia* spp., *Pennisetum* spp. and *Jardinea congoensis* are used for mat-making and fencing, while other species, especially *Imperata cylindrica* and *Schizachyrium* spp., are used for roofing. Many species are important range and fodder species for cattle and yet others are important weeds.

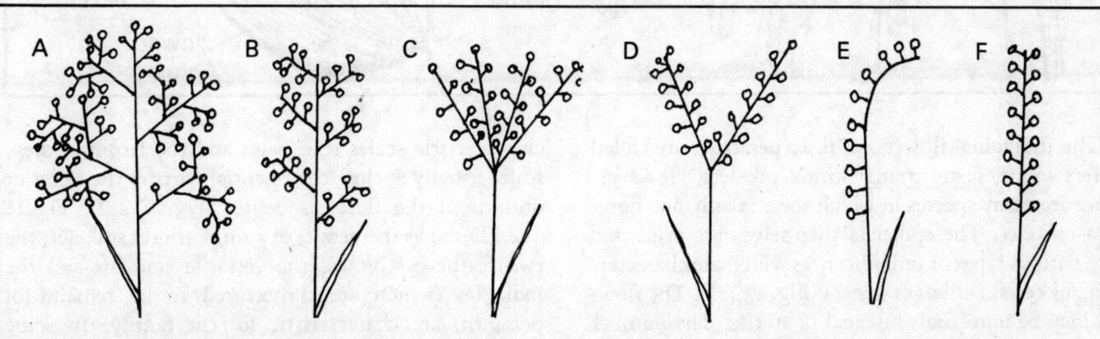

Fig. 7.2.1 Inflorescence types in Gramineae
A panicle in *Sorghum* and *Panicum*
B compound raceme in *Leptochloa*
C digitate inflorescence in *Digitaria*
D digitate inflorescence in *Paspalum conjugatum* and *Andropogon*
E unbranched inflorescence in *Ctenium*
F unbranched inflorescence in *Pennisetum*

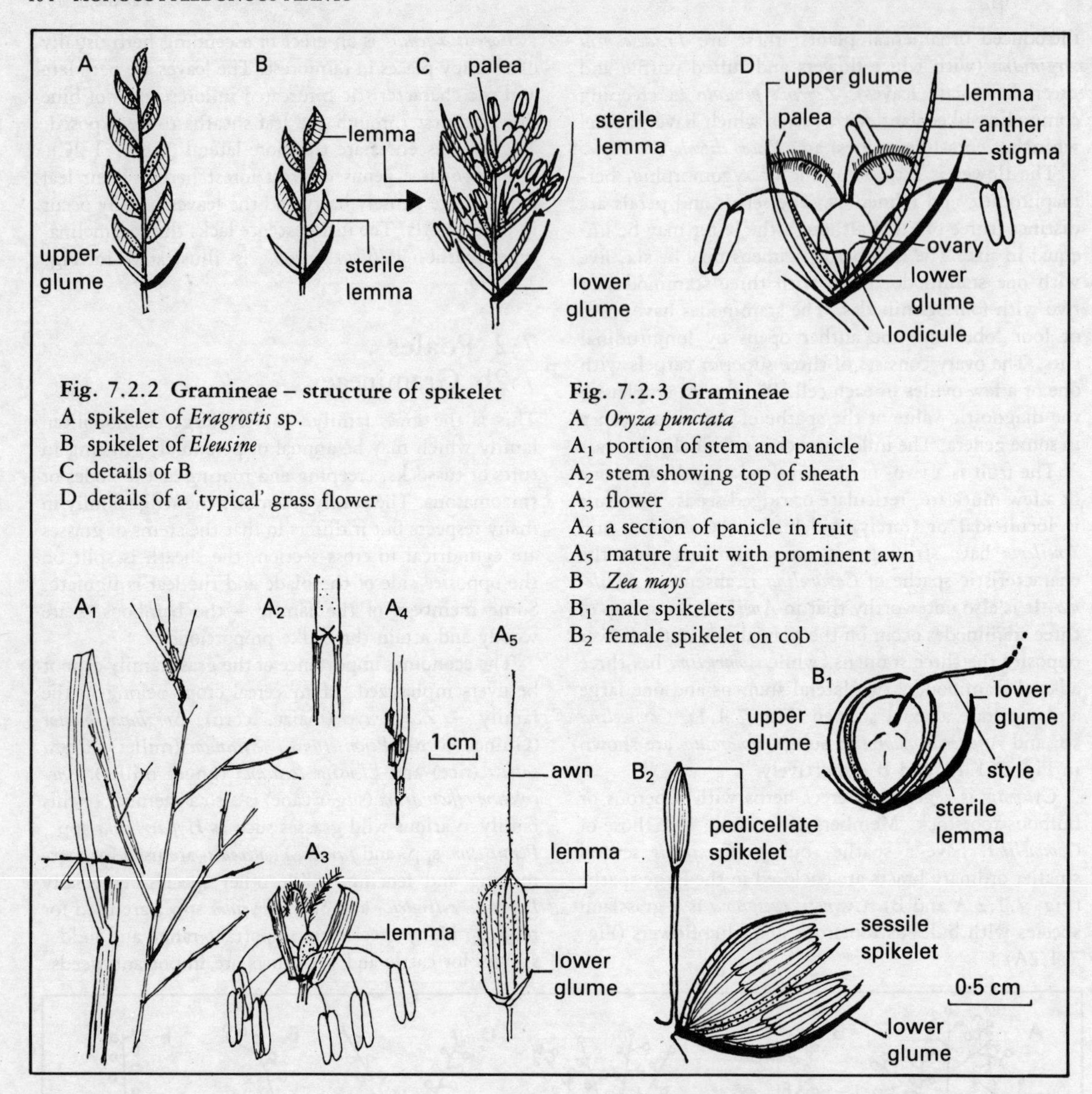

Fig. 7.2.2 Gramineae – structure of spikelet
A spikelet of *Eragrostis* sp.
B spikelet of *Eleusine*
C details of B
D details of a 'typical' grass flower

Fig. 7.2.3 Gramineae
A *Oryza punctata*
A_1 portion of stem and panicle
A_2 stem showing top of sheath
A_3 flower
A_4 a section of panicle in fruit
A_5 mature fruit with prominent awn
B *Zea mays*
B_1 male spikelets
B_2 female spikelet on cob

The individual flowers, as in Cyperaceae, are called florets and these are arranged into spikelets. However, there are many species in which there is only one floret in a spikelet. The spikelets themselves are organised into various types of inflorescences which are characteristic for certain tribes or genera (Fig. 7.2.1). The flowers may be unisexual, bisexual or sterile. The number of florets per spikelet is also supposed to indicate evolutionary trends and taxonomic relationships. The structure of the spikelet is such that two scales subtend the floret(s); these are the lower glume and the upper glume. The glumes enclose the florets, which have two characteristic scales (the palea and the lemma); these scales usually enclose the essential parts of the floret or nothing if the floret is neuter (Fig. 7.2.2). Figure 7.2.2D shows the details of a single-floret spikelet; the two feathery stigmas, the versatile stamens and the lodicules (which are conjectured to be remains of perianth) are characteristic for the family. In some genera, a long bristle (the awn) develops from the lemma.

The fruit in grasses is a caryopsis (characterised as indehiscent with fused testa and pericarp) which develops from a single carpel; it is also called a grain.

The genus *Oryza* includes the cultivated rice and belongs in a tribe of grasses with six stamens, much-reduced glumes and an awned lemma. The ligules are also quite prominent, forming a collar around the stem. This genus, mostly of swamp plants, is represented in Nigeria by *O. punctata* (Fig. 7.2.3A) which is about 1 m tall with paniculate inflorescences.

Zea mays is one of the monoecious grasses, the cob being the female inflorescence and the tassel being the male inflorescence. There are two kinds of male spikelet – sessile and pedicellate (Fig. 7.2.3B); the female spikelet is shown in Fig. 7.2.3B.

Sorghum has both wild and cultivated species. Like other genera of the tribe Andropogoneae, *Sorghum* spp. have two kinds of spikelet – the male pedicellate ones and the hermaphroditic sessile ones (Fig. 7.2.4A, B). *S. arundinaceum* is a forest regrowth grass with loose drooping panicles (Fig. 7.2.4A).

Andropogon and *Hyparrhenia* are genera of the tribe Andropogoneae and are widespread in the Nigerian savanna. Their inflorescences are raceme pairs which are borne as leafy 'panicles'; the raceme pairs emerge from papery spathes of different lengths. The two genera differ in many ways. In *Andropogon* the leaf blades are broader, the awns are thinner and shorter, the raceme axes are thicker and they do not bend backwards at maturity. There are also many more spikelets per raceme pair than in *Hyparrhenia*. *Andropogon gayanus* and *A. tectorum* are the northern gamba grass and the southern gamba grass respectively; there are eight or so other species of *Andropogon* in Nigeria. *Hyparrhenia* is also a fairly large genus of tall grasses

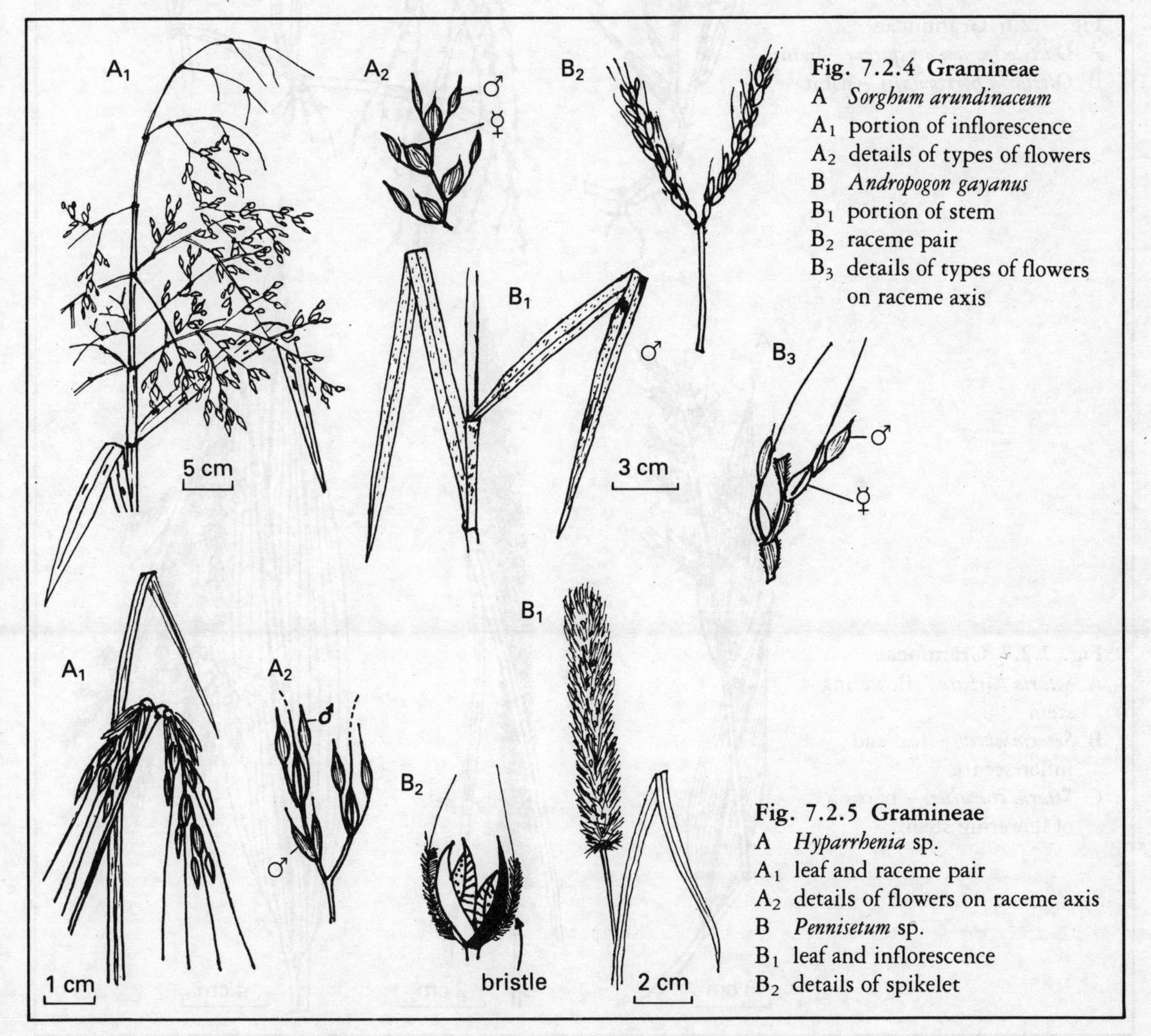

Fig. 7.2.4 Gramineae
A *Sorghum arundinaceum*
A_1 portion of inflorescence
A_2 details of types of flowers
B *Andropogon gayanus*
B_1 portion of stem
B_2 raceme pair
B_3 details of types of flowers on raceme axis

Fig. 7.2.5 Gramineae
A *Hyparrhenia* sp.
A_1 leaf and raceme pair
A_2 details of flowers on raceme axis
B *Pennisetum* sp.
B_1 leaf and inflorescence
B_2 details of spikelet

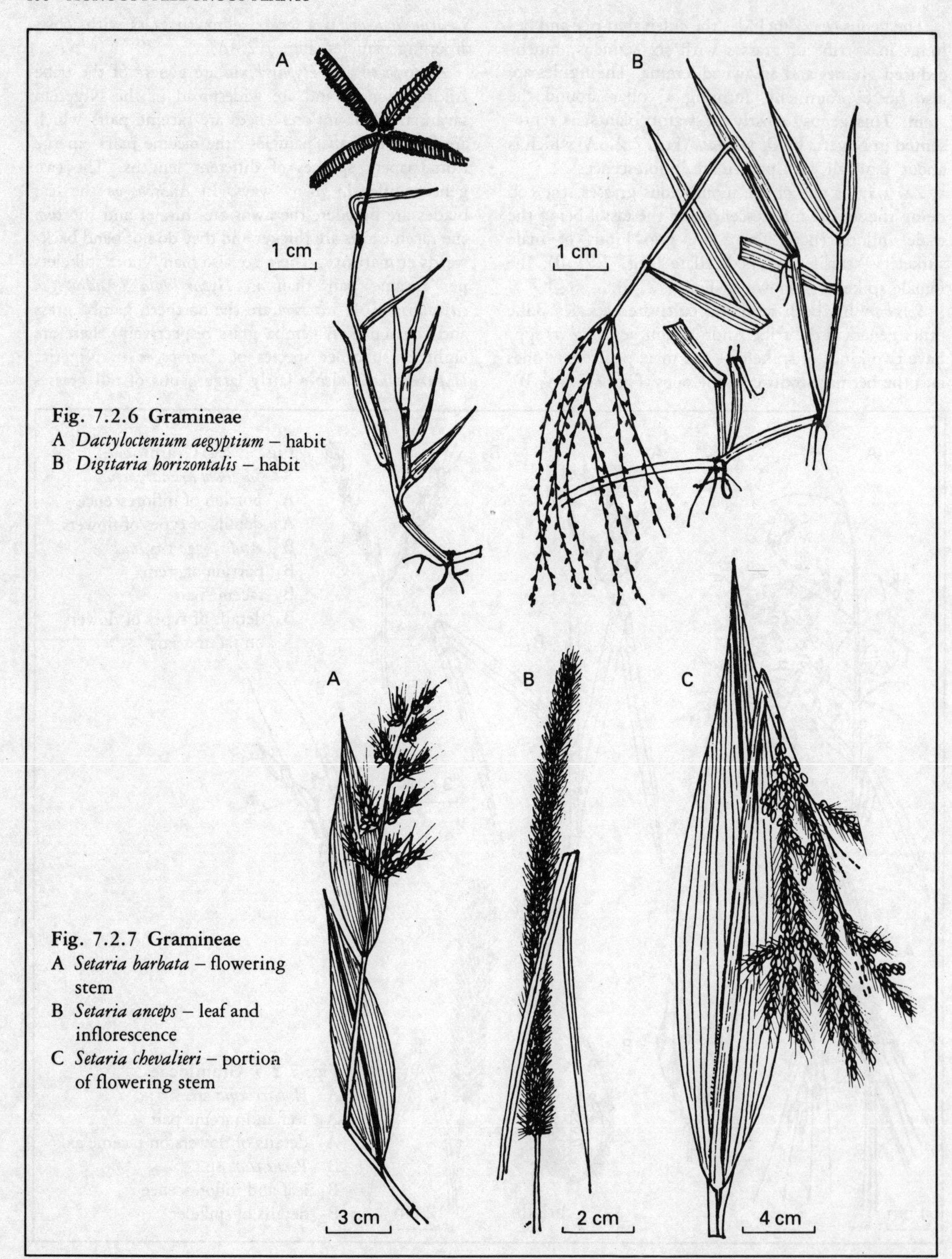

Fig. 7.2.6 Gramineae
A *Dactyloctenium aegyptium* – habit
B *Digitaria horizontalis* – habit

Fig. 7.2.7 Gramineae
A *Setaria barbata* – flowering stem
B *Setaria anceps* – leaf and inflorescence
C *Setaria chevalieri* – portion of flowering stem

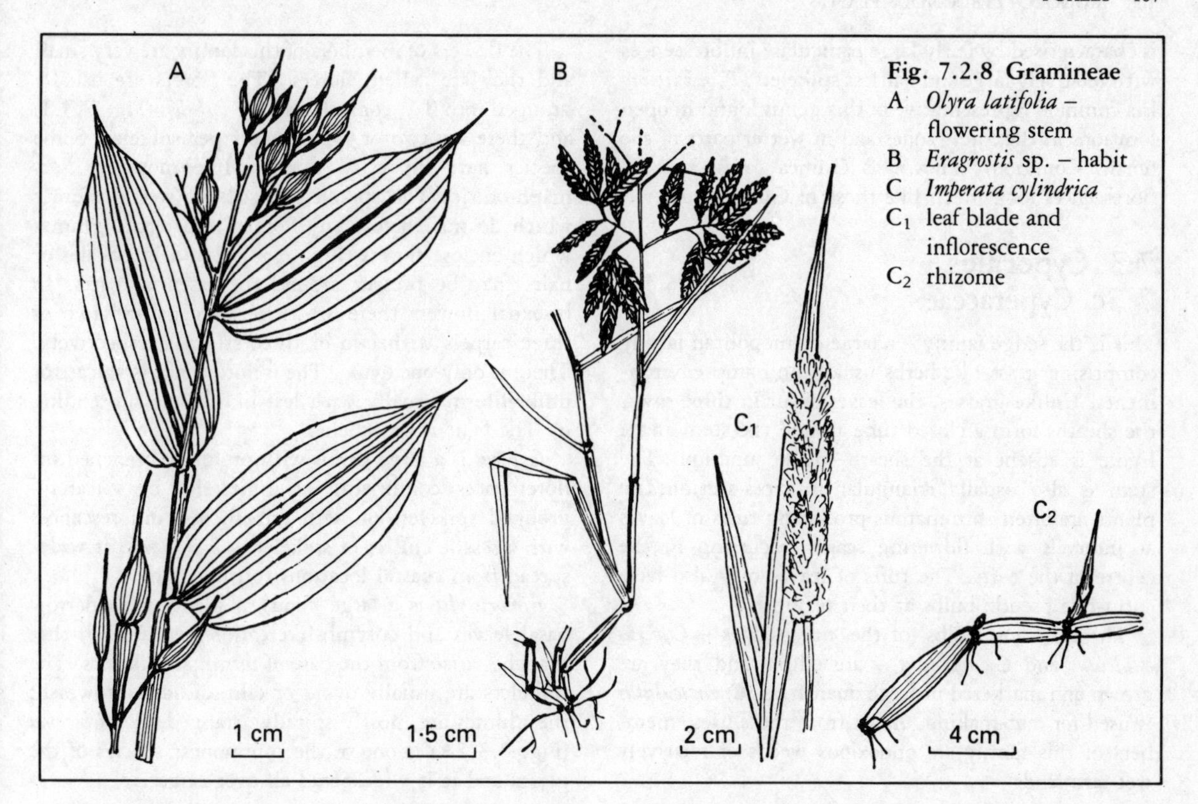

Fig. 7.2.8 Gramineae
A *Olyra latifolia* – flowering stem
B *Eragrostis* sp. – habit
C *Imperata cylindrica*
C_1 leaf blade and inflorescence
C_2 rhizome

with prominent awns and raceme pairs that are strongly bent backwards at maturity (Fig. 7.2.5A); they occupy relatively poor soils and they grow in very large numbers.

Pennisetum is the elephant grass and millet genus. The genus is characterised by unbranched inflorescences which look very much like a bottle brush. The spikelets are subtended by bristles which may or may not be hairy (Fig. 7.2.5B). They resemble some species of *Setaria*, but in the latter the bristles do not fall with the spikelets. The important species in this genus are *P. purpureum* (elephant grass). *P. subangustum* and *P. pedicellatum*.

Various other species of Gramineae are shown in Figs 7.2.6, 7.2.7 and 7.2.8. *Dactyloctenium aegyptium* and *Digitaria horizontalis* are widespread weed species throughout Nigeria.

Setaria barbata (Fig. 7.2.7A) is a common weed in open forest locations. *Setaria chevalieri* is a large plant with lax, drooping inflorescences (Fig. 7.2.7C), while *S. megaphylla* is another large (up to about 1·5 m tall) plant found along rivers. The latter has rigid and compact inflorescences about 1 m long; both of them are species of forest areas. *Setaria anceps* is a species of grassland swamps up to 2 m tall (Fig. 7.2.7B), while *S. pallidefusca* is a small plant of weedy habit in open relatively wet grassland.

Olyra latifolia is a bamboo-like straggling forest species with two distinct kinds of spikelet: one pair is small and male, and a solitary one is perfect and awned. The seed is hard, large and white (Fig. 7.2.8A).

Eragrostis is a large genus of plants found mostly in waste places. They have panicles containing numerous flat spikelets (Fig. 7.2.8B) each with many florets. Members of this genus have open panicles; however *E. ciliaris* is a small distinctive plant of dry grassland with a congested narrow and hairy brownish purple inflorescence. Another distinctive species of open and wet forest location is *E. tenella* – small plant (*c.* 15 cm tall), with tiny pale green spikelets about 1 mm long.

Imperata cylindrica is the 'spear grass', so-called because of the very sharp suckers which it produces from white underground rhizomes (Fig. 7.2.8C). The leaves all arise from the soil level and flowering axes are produced in the dry season from the middle of the leaf tufts. The inflorescence is cylindrical, white and woolly (Fig. 7.2.8C). This plant is an obnoxious weed with very stubborn rhizomes in impoverished savanna soils.

Panicum is a large genus in West Africa. The genus

is characterised by fairly large paniculate inflorescences with relatively large and hairless spikelets. *P. maximum* is a common representative of this genus found in open locations in the forest zones and in wetter parts of the north. Commonly known as Guinea grass, the inflorescences look much like those of Guinea corn.

7·3 Cyperales
7·3c Cyperaceae

This is the sedge family – a large cosmopolitan family comprising grass-like herbs usually in damp environments. Unlike grasses, the leaves occur in three rows, the sheaths form a closed tube around the stem and a ligule is absent at the sheath–blade junction. The stem is also usually triangular in cross-section. The plants are often rhizomatous producing tufts of leaves at intervals with flowering scapes occurring in the centre of the tufts. The tufts of leaves may also have corm-like pseudo-bulbs at their origin.

The tuberous 'bulbs' of the 'nut' sedges – *Cyperus esculentus* and *C. rotundus* – are edible and they are grown and marketed in large quantities. *C. articulatus* is used for mat-making. Apart from these uses, members of this family are obnoxious weeds of relatively wet farmlands.

The flowers of members of this family are very small and they are called 'florets'. The florets are usually grouped into flat, conical or terete *spikelets* (Fig. 7.3.1) and there are two or more florets per spikelet. Some species have unisexual florets while most are hermaphroditic. The spikelet consists of sterile glumes which do not enclose any florets, and fertile glumes which enclose florets (Fig. 7.3.1 C and D). Scales or hairs may be present around the essential parts. In bisexual flowers there are three stamens and two or three carpels with two or three stigmas respectively. There is only one ovule. The inflorescence is spicate or umbelliform usually with leaf-like bracts subtending it. The fruit is an achene.

Cyperus is a large genus with profusely branched inflorescences comprising of digitately- or spicately-grouped spikelets or with unbranched inflorescences with a sessile cluster of spikelets. *C. dilatatus* is widespread from coastal locations to the north.

Fimbristylis is a large genus of plants with narrow basal leaves and corymb-like inflorescences such that branches arise from the base of terminal spikelets. The spikelets are usually ovoid or ellipsoid and brownish; the glumes are mostly spirally arranged. *F. dichotoma* (Fig. 7.3.2A) is one of the commonest species of the genus and it is widespread all over Nigeria.

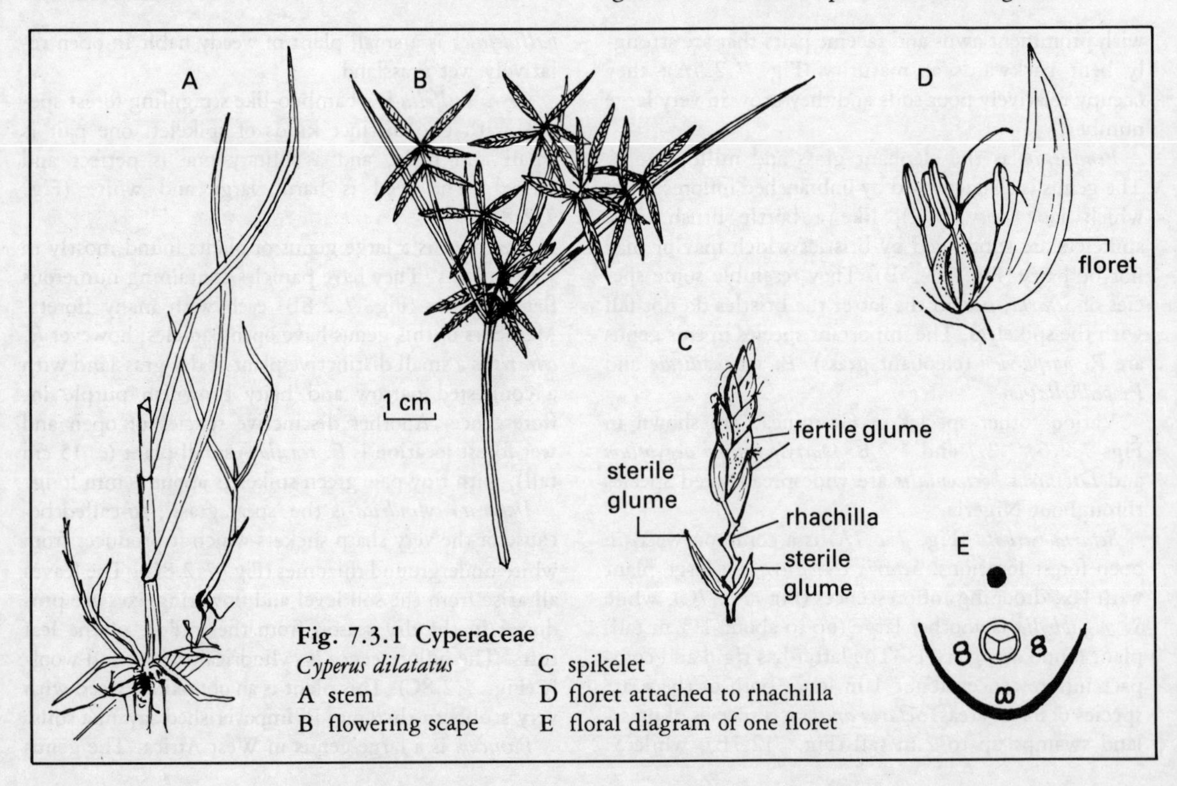

Fig. 7.3.1 Cyperaceae
Cyperus dilatatus
A stem
B flowering scape
C spikelet
D floret attached to rhachilla
E floral diagram of one floret

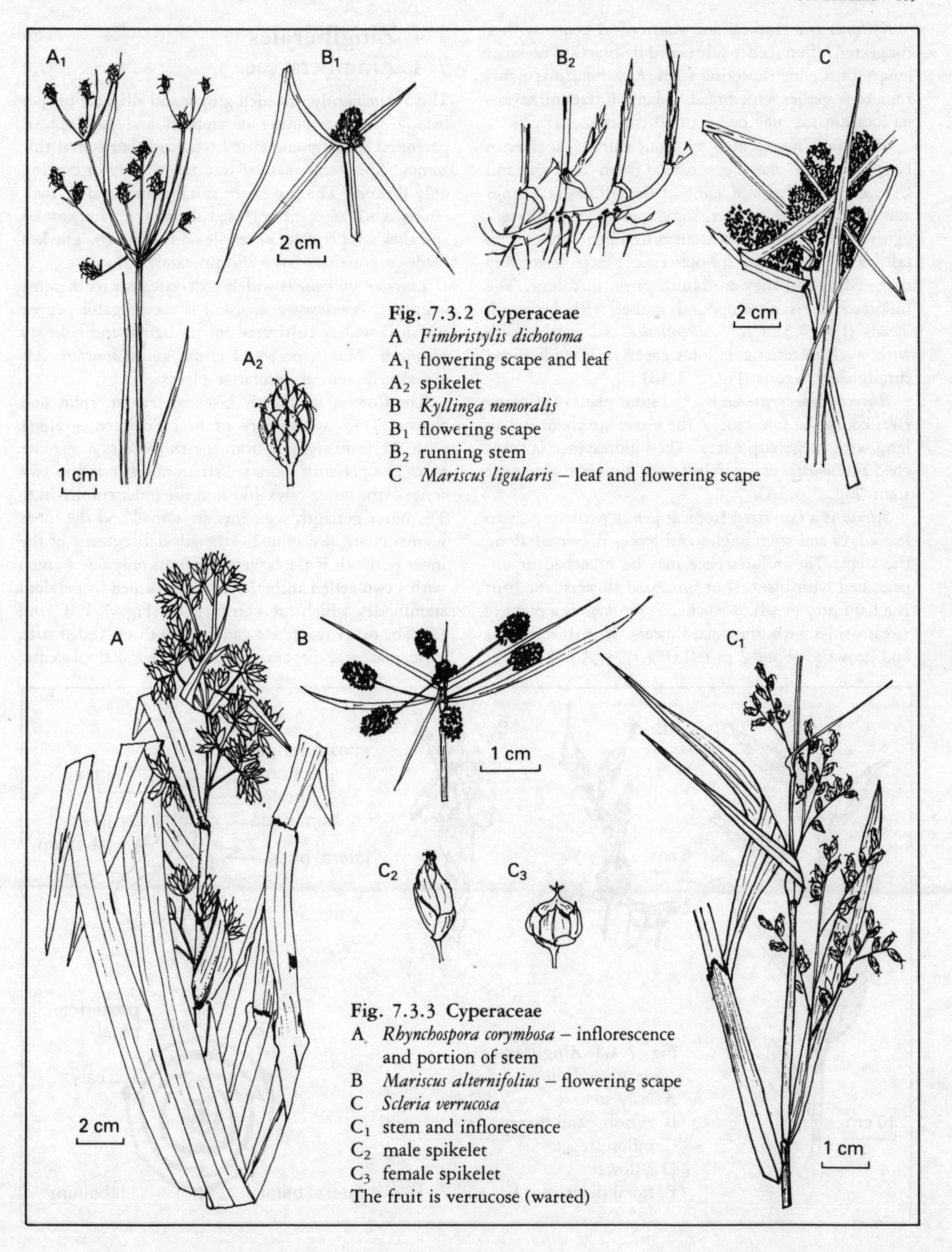

Fig. 7.3.2 Cyperaceae
A *Fimbristylis dichotoma*
A_1 flowering scape and leaf
A_2 spikelet
B *Kyllinga nemoralis*
B_1 flowering scape
B_2 running stem
C *Mariscus ligularis* – leaf and flowering scape

Fig. 7.3.3 Cyperaceae
A *Rhynchospora corymbosa* – inflorescence and portion of stem
B *Mariscus alternifolius* – flowering scape
C *Scleria verrucosa*
C_1 stem and inflorescence
C_2 male spikelet
C_3 female spikelet
The fruit is verrucose (warted)

Kyllinga is a tropical and subtropical genus with a congested inflorescence subtended by bracts of unequal length on a nude flowering stem. *K. nemoralis* is a rhizomatous species widespread in damp forest and savanna locations; it may be up to 30 cm tall.

Mariscus is represented by about thirteen species in Nigeria. They may be confused with *Kyllinga* and *Cyperus* on the basis of some of their floral attributes and their habit. However, some species are easily recognised. *M. ligularis* is a stout tufted plant about 1 m tall. The leaves and inflorescence bracts have very sharp edges and they are bluish green in colour. The inflorescence is a compound umbel with brownish 'heads' (Fig. 7.3.2C). *M. alternifolius* is a variable herb with a characteristic inflorescence; it is widespread throughout Nigeria (Fig. 7.3.3B).

Rhynchospora corymbosa is a tall stout plant of swampy river courses in forest areas; the leaves are about 0·7 m long with very sharp edges. The inflorescence is branched and usually arises in leaf-axils dispersed along the stem (Fig. 7.3.3A).

Scleria is a fair-sized tropical genus with very sharp leaf edges and stem angles and leaves dispersed along the stem. The inflorescence may be branched or unbranched with bisexual or unisexual flowers; the fruit is a hard grey or white achene. *S. verrucosa* is a plant of stream-sides with unisexual flowers. It is rhizomatous and grows to about 2 m tall (Fig. 7.3.3C).

7·4 Zingiberales
7·4 Zingiberaceae

This is the family to which ginger and alligator pepper belong. It is a family of tropical and subtropical, perennial, usually aromatic herbs with horizontal rhizomes. The stems may be very short, leafy or bearing only flowers. The leaves are spirally arranged or two-ranked with an open or closed sheath (cf. Musaceae) – and they are petiolate or sessile on the sheath. The leaf blades are usually large and pinnately veined.

Zingiber officinale is widely cultivated for its rhizome – ginger. *Aframomum melegueta* is the alligator pepper which is widely cultivated for medicinal and culinary purposes. Many species of *Costus* and *Kaempferia* are actual or potential decorative plants.

The flowers are mostly bisexual, zygomorphic and showy. They are solitary or in inflorescences along with the leafy stem or from the rhizome as a separate axis. The perianth lobes are six in number and in two series – the outer calyx-like and the inner corolla-like. The outer perianth segments are joined and the inner segments are also joined – the adaxial segment of the inner perianth is the largest. There is only one stamen with a two-celled anther; it is accompanied by petaloid staminodes which are conspicuous (Fig. 7.4.1C and D). The ovary is inferior and two- to three- celled with axile placentae or one-celled with parietal placenta;

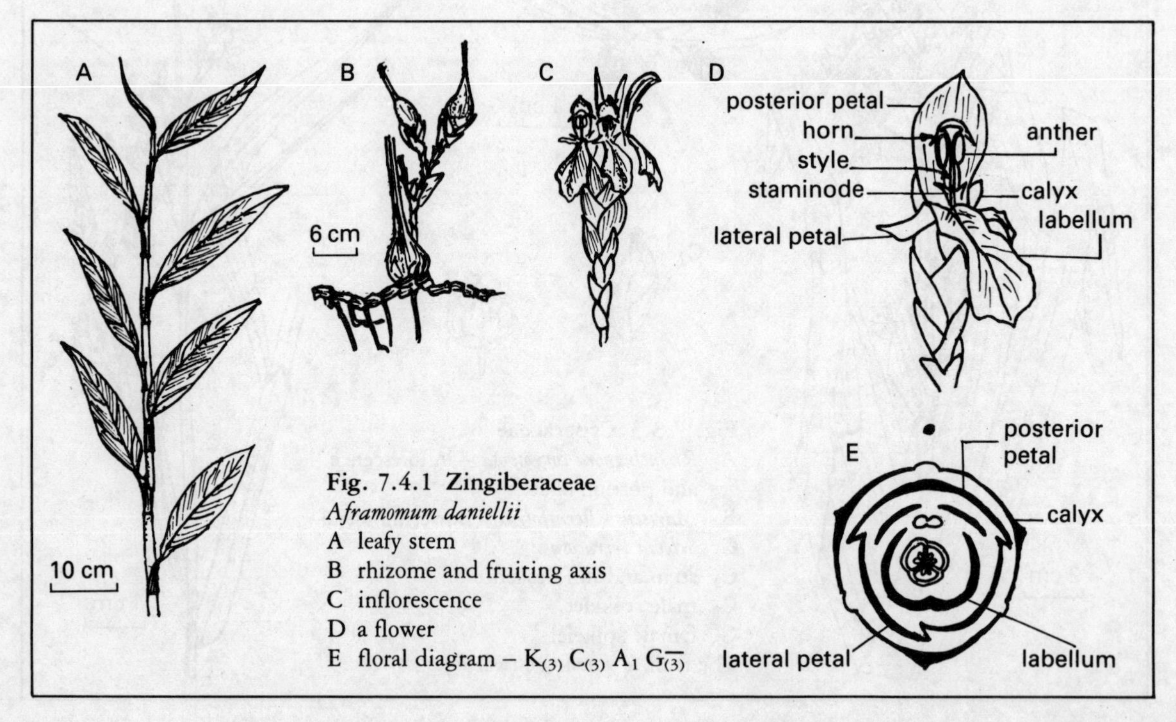

Fig. 7.4.1 Zingiberaceae
Aframomum daniellii
A leafy stem
B rhizome and fruiting axis
C inflorescence
D a flower
E floral diagram – $K_{(3)}\ C_{(3)}\ A_1\ G_{\overline{(3)}}$

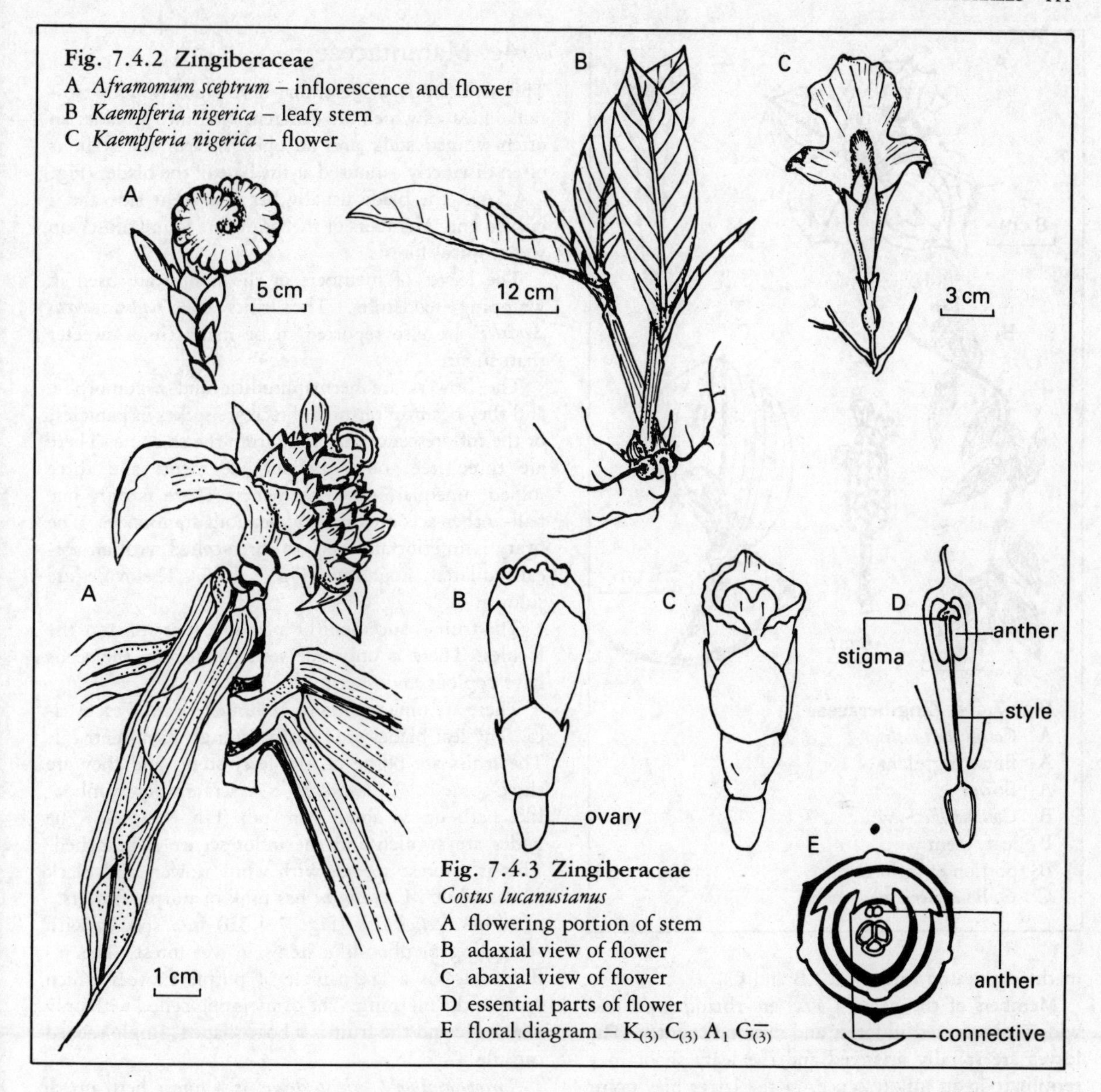

Fig. 7.4.2 Zingiberaceae
A *Aframomum sceptrum* – inflorescence and flower
B *Kaempferia nigerica* – leafy stem
C *Kaempferia nigerica* – flower

Fig. 7.4.3 Zingiberaceae
Costus lucanusianus
A flowering portion of stem
B adaxial view of flower
C abaxial view of flower
D essential parts of flower
E floral diagram – $K_{(3)}\ C_{(3)}\ A_1\ G_{\overline{(3)}}$

ovules are numerous. The style is usually filiform with a funnel-shaped stigma which occupies a groove between the two anther cells (Fig. 7.4.1D).

The fruit is a capsule or it is succulent and indehiscent. The seeds are arillate and they have abundant endosperm.

Aframomum is a genus of forest and savanna *Canna*-like herbs with separate leafy and flowering axes both arising from a horizontal rhizome. The leaves are two-ranked. *A. daniellii* is a large plant (*c.* 5 m tall) of hilly locations in forest edges with relatively high rainfall. The flowers (Fig. 7.4.1C, D) are pinkish orange with an orange labellum. The fruits ripen into red 'berries' (Fig. 7.4.1B). *A. elliotii* is a plant of savanna woodland (*c.* 1·5 m tall) with white flowers and a yellow throat. The plant is abundant in sunny locations and fruits are edible. The flower of *A. sceptrum* is illustrated in Fig. 7.4.2A. This is a species of dense forests; it is up to 2 m tall.

Kaempferia has a non-rhizomatous rootstock and there are relatively few leaves which are two-ranked or solitary. In *K. nigerica* the compact rootstock produces spindle-shaped lateral roots, and the purplish pink flowers usually emerge when the leaves have died back

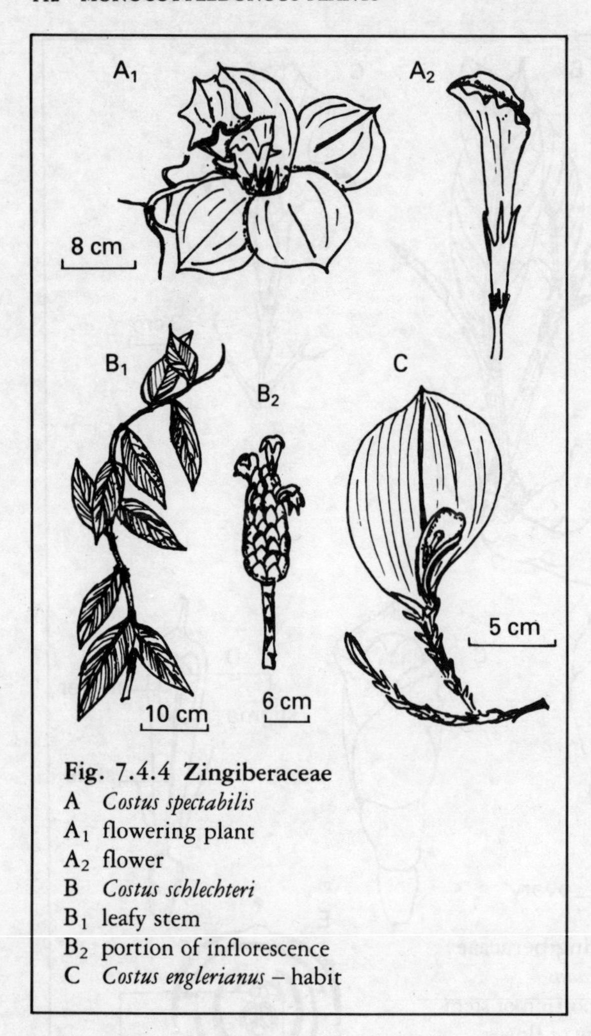

Fig. 7.4.4 Zingiberaceae
A *Costus spectabilis*
A_1 flowering plant
A_2 flower
B *Costus schlechteri*
B_1 leafy stem
B_2 portion of inflorescence
C *Costus englerianus* – habit

in the dry season (Fig. 7.4.2 B and C).

Members of the genus *Costus* are rhizomatous and widely distributed in forest and savanna country. The leaves are spirally arranged and the leafy shoot may terminate in an inflorescence, or the latter may occur on a separate axis as in *Aframomum*. *Costus lucanusianus*. (Fig. 7.4.3) has white flowers tipped with red; it is found in wet locations. It is similar to *C. afer* except that the flowers of the latter are yellow. *C. spectabilis* is a savanna species with a rosette of four suborbicular leaves and yellow flowers (Fig. 7.4.4A). *C. schlechteri* is a species of wet forests with spirally-arranged leaves. The flowers have pink petals and they are borne on a separate flowering axis near the soil surface (Fig. 7.4.4B). *C. englerianus* is yet another peculiar forest species with white flowers and solitary suborbicular leaves arising from a rhizome (Figure 7.4.4C).

7·4e Marantaceae

This is a family of perennial herbs mostly with two-ranked leaves which are differentiated into a blade, an often-winged stalk and an open sheath; the stalk is often distinctly calloused at the base of the blade. (Fig. 7.4.5A). The blade usually has a straight side and a curved one. Members of the family are found mostly in wet tropical forests.

The leaves of members of the family are used in wrapping foodstuffs. The fruits of *Thaumatococcus daniellii* are also reported to be many times sweeter than sugar.

The flowers are hermaphroditic and zygomorphic and they occur in terminal bracteate spikes or panicles, or the inflorescence may arise from the rhizome. There are three free, outer perianth segments and three joined, unequal, inner segments. There is only one half-anther accompanied by petaloid staminodes. The ovary is inferior and one- to three-celled with an apically-dilated, stout style (Fig. 7.4.6C). The ovules are solitary.

The fruit is succulent or a capsule that splits at the locules. There is only one seed per locule; the seeds have copious endosperm.

There are nine species of *Marantochloa* in West Africa. The leaf blades are usually strongly asymmetrical. The fruits are borne on the leafy stems and they are three-seeded. *M. leucantha* is a scrambling bamboo-like herb up to about 4 m tall (Fig. 7.4.5A). The nodes are swollen and the inflorescence is branched. This is a forest species with white flowers and black seeds, while *M. purpurea* has pink or purple flowers.

Thalia welwitschii (Fig. 7.4.5B) is a species with straggling bamboo-like stems in wet forest. The inflorescence is a lax panicle of purple flowers which open in the morning. The ovary is one-celled with only one ovule and the fruit is a boat-shaped, single-seeded capsule.

Sarcophrynium brachystachyum is a forest herb up to 2 m tall with long-petioled leaves and characteristic inflorescences (Fig. 7.4.6A) carrying white flowers which produce red fruits.

Thaumatococcus daniellii (Fig. 7.4.6B) is a herb up to 3 m tall growing in large stands on forest floors. The inflorescence arises from slender horizontal rhizomes. The crimson fruits are produced on or below ground level.

Many species of *Maranta* have variegated markings on their leaves and they are cultivated as ornamentals. *M. arundinacea* (Fig. 7.4.6C) is a common ornamental – the arrowroot.

Fig. 7.4.5 Marantaceae
A *Marantochloa leucantha* – habit
B *Thalia welwitschii* – habit

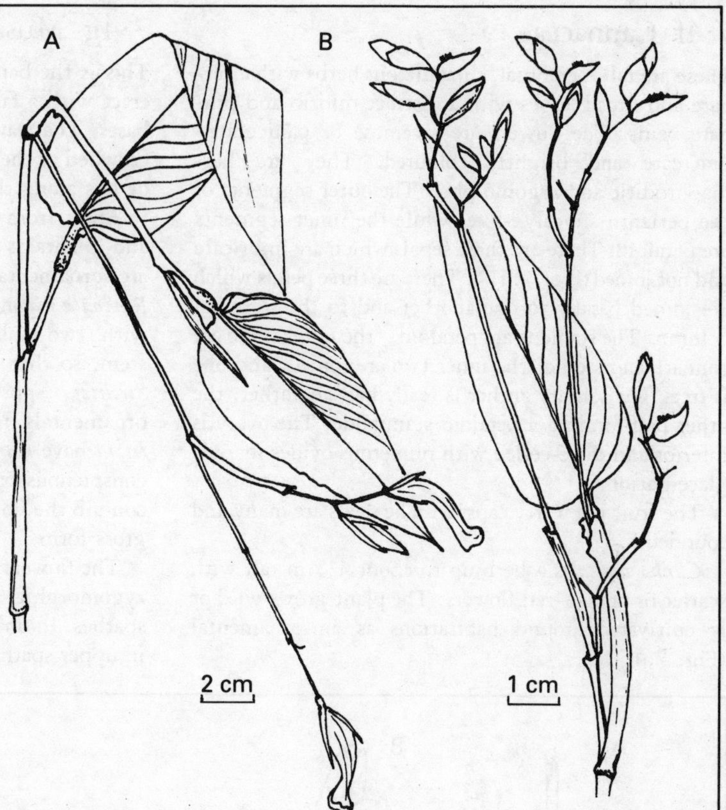

Fig. 7.4.6 Marnantaceae
A *Sarcophrynium brachystachyum* – habit (note fruits)
B *Thaumatococcus daniellii* – habit
C *Maranta arundinacea*
C_1 habit
C_2 flower
C_3 pistil
C_4 floral diagram – $K_3\ C_{(3)}\ A_1\ \overline{G_{(3)}}$

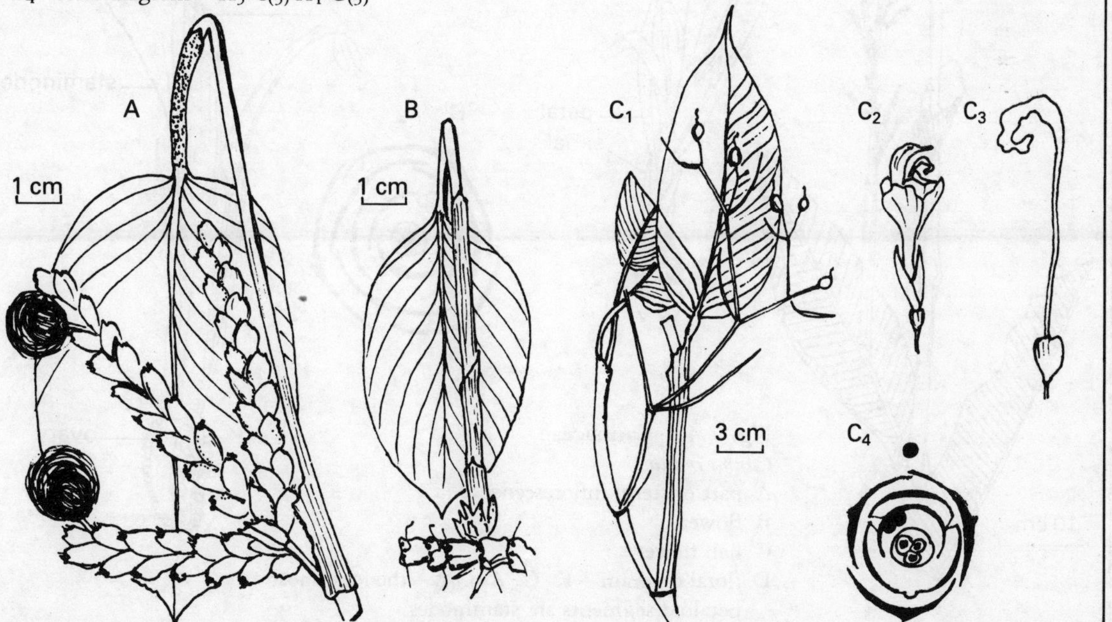

7·4f Cannaceae

These are tall, perennial, rhizomatous herbs with alternate and broad leaves with a distinct midrib and pinnate veins. The flowers are racemose or paniculate–bracteate and brightly coloured. They are hermaphroditic and zygomorphic. The outer segments of the perianth are calyx-like, while the inner segments are petaloid. There are three sepals which are imbricate and not joined (Fig. 7.4.7). There are three petals which are joined basally to one another and to the staminal column. The stamens are petaloid – the outer three are connate and sterile, the inner two are joined, and one is free. The solitary anther is really half an anther, the other half forming a petaloid staminode. The ovary is inferior and three-celled with numerous ovules in axile placentation.

The fruit is a warty capsule. The seeds are many and rounded.

Canna indica is a herb up to about 1·5 m tall with scarlet or orange–red flowers. The plant grows wild or is cultivated around habitations as an ornamental (Fig. 7.4.7).

7·4g Musaceae

This is the banana family. Members of the family are erect with a false stem formed from overlapping leaf-bases. The leaves are spirally arranged, very large and crowded at the end of the false stem or dispersed more or less along the stem.

Apart from *Musa* spp., which are various bananas and 'plantains', other species of economic importance are ornamental plants. These ornamentals include *Ravenala madagascariensis* the so called 'travellers' palm' with two-ranked, banana-like leaves on a distinct stem, so that the entire plant looks like a giant fan. *Strelitzia* spp. is another group of much smaller ornamentals; the commonly-cultivated species of *Strelitzia* have *Canna*-like leaves and orange spathes on a conspicuous axis; the spathes are about 15 cm long and contain the flowers which look like those of bananas in gross form.

The flowers are unisexual or bisexual, trimerous and zygomorphic; they are subtended by conspicuous spathes. In unisexual species, male flowers occur within upper spathes and female flowers occur inside lower

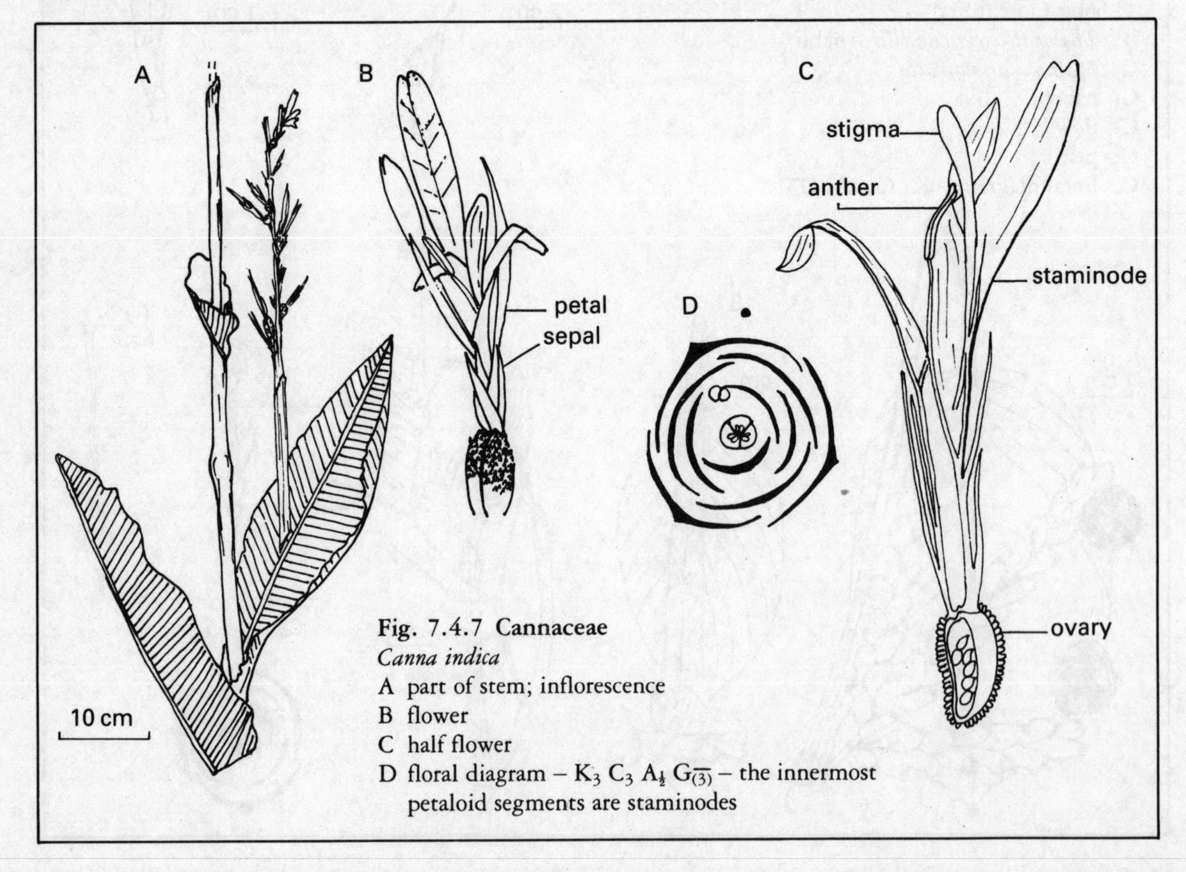

Fig. 7.4.7 Cannaceae
Canna indica
A part of stem; inflorescence
B flower
C half flower
D floral diagram – K_3 C_3 $A_{\frac{1}{2}}$ $G_{\overline{(3)}}$ – the innermost petaloid segments are staminodes

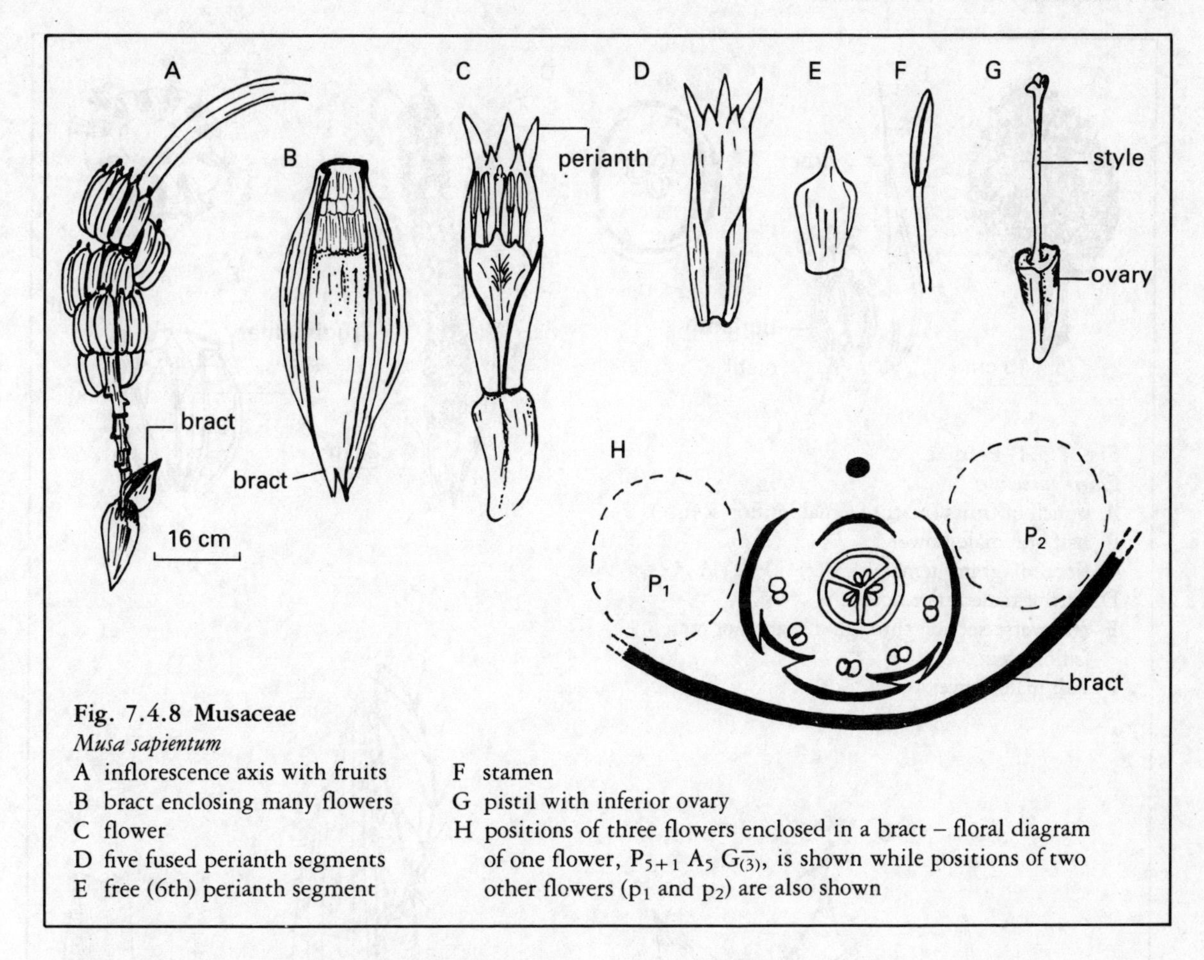

Fig. 7.4.8 Musaceae
Musa sapientum
A inflorescence axis with fruits
B bract enclosing many flowers
C flower
D five fused perianth segments
E free (6th) perianth segment
F stamen
G pistil with inferior ovary
H positions of three flowers enclosed in a bract – floral diagram of one flower, $P_{5+1}\ A_5\ G_{\overline{(3)}}$, is shown while positions of two other flowers (p_1 and p_2) are also shown

spathes. Where a calyx is present, it is tubular, subsequently splitting on one side. The perianth is generally two-lipped. There are five perfect stamens and a rudimentary one, or there are six perfect stamens. The anthers are two-celled on filiform filaments (Fig. 7.4.8). The ovary is inferior and three-celled. The style is filiform and each ovary cell has many ovules.

The fruit is succulent, indehiscent and three-loculed. There are numerous seeds with hard seed coats and copious endosperm.

Details of the inflorescence and floral structure in *Musa sapientum* are shown in Fig. 7.4.8. *M. paradisiaca* is the species with large fruits, while *M. nana* fruits are dwarf. A wild member of this family is *Ensete gilletii*; it is a wild banana of hilly grassland locations. The spirally-arranged leaves are dispersed all over the stem and the terminal inflorescence produces unisexual flowers. The fruits, which are about 6 cm long with orange pulp, split longitudinally to expose the brown seeds.

7·5 Arecales

7·5h Palmae (arecaceae)

This is the family of the palms. Some of them are large trees with leaves crowded at the end of a bole which is rarely branched, or they are slender climbing plants usually with vicious hooks. The leaves are usually spirally arranged on the stem; they are either pinnately compound (in the feather palms) or palmately compound (in the fan palms). They are mostly tropical and subtropical plants.

Economic species include various introduced ornamentals, certain food crops and fibre-producing plants. The Cuban royal palm (*Oreodoxa regia*) is a common decorative palm with a smooth, spindle-shaped bole. The coconut palm (*Cocos nucifera*), the oil palm (*Elaeis guineensis*), and the date palm (*Phoenix dactylifera*) are well-known crop plants. Palm wine is also produced in commercial quantities from oil palm and *Raphia* spp.

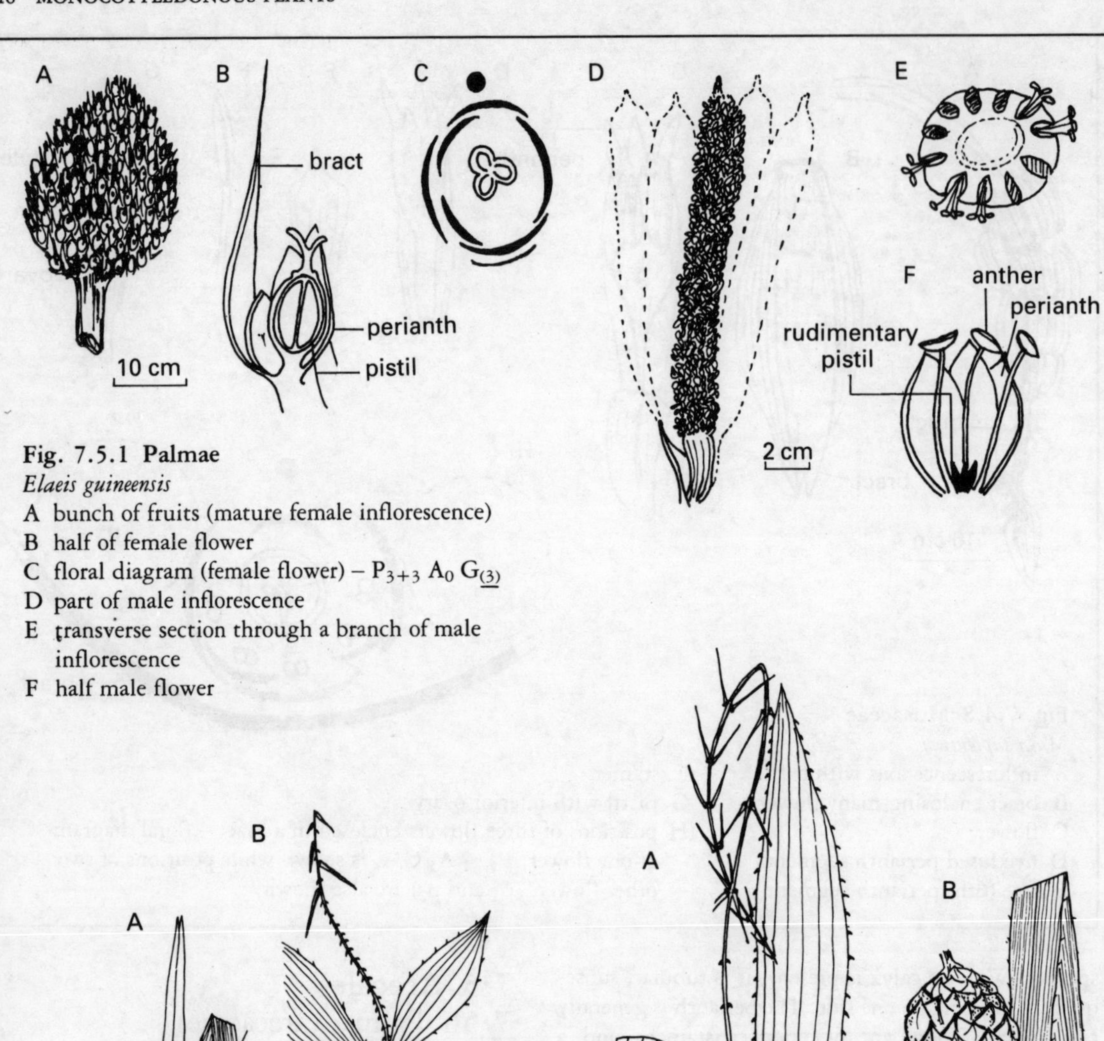

Fig. 7.5.1 Palmae
Elaeis guineensis
A bunch of fruits (mature female inflorescence)
B half of female flower
C floral diagram (female flower) – $P_{3+3}\ A_0\ G_{\underline{(3)}}$
D part of male inflorescence
E transverse section through a branch of male inflorescence
F half male flower

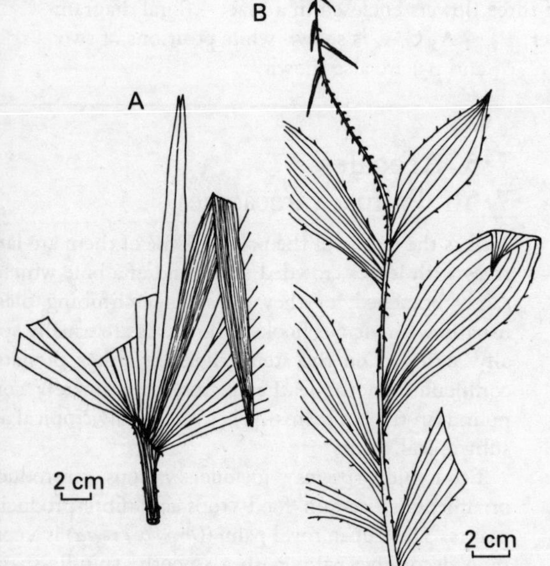

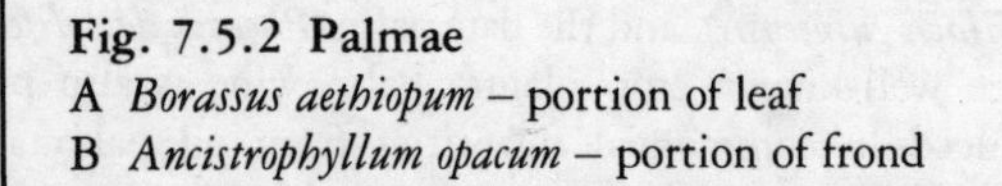

Fig. 7.5.2 Palmae
A *Borassus aethiopum* – portion of leaf
B *Ancistrophyllum opacum* – portion of frond

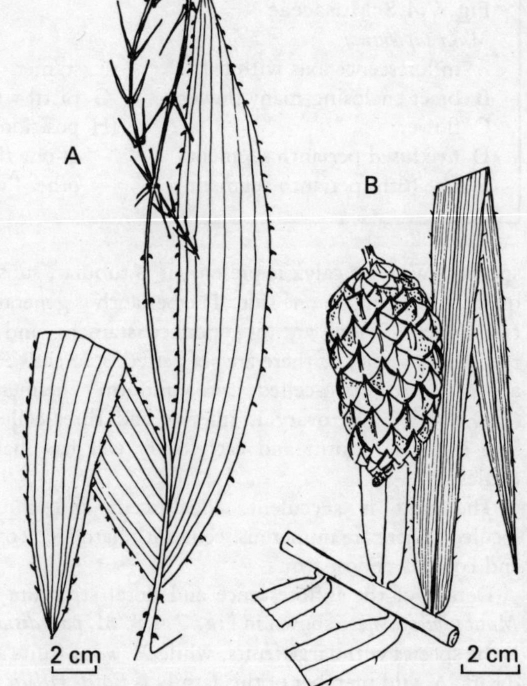

Fig. 7.5.3 Palmae
A *Eremospatha hookeri* – portion of frond
B *Raphia hookeri* – fruit and portion of frond

The flowers are small, actinomorphic and trimerous. The climbing palms (rattans) are monoecious with male and female flowers on separate inflorescences, while the tree palms are either dioecious, or monoecious with male and female flowers on the same or separate inflorescences. The perianth segments are six in number and a prominent bract may be present. The male flowers may have a rudimentary pistil. There are six stamens. The pistil consists of three joined carpels. The inflorescence is a panicle with spicate ultimate branches or a spike. Some of the floral characteristics of *Elaeis guineensis* are shown in Fig. 7.5.1. The ovary is superior with one to three locules. The ovules are solitary.

The fruit is a berry or drupe with one to three cells. The exocarp is usually fibrous or covered with many reflexed scales. The seeds are free or adherent to the endocarp; endosperm may be present but the embryo is usually small.

The tree palms have either pinnate leaves or their leaves are fan-like (Figs. 7.5.2 and 7.5.3). Apart from oil palm, species of *Raphia* and *Phoenix* have pinnate leaves. The species of *Raphia* are found mostly along river courses; their fruits and flowers are diagnostic (Fig. 7.5.3B) and they have very rigid and narrow straps of material wrapped around their stems. The fruit and a portion of the stem of *R. hookeri* are shown in figure 7.5.3B. A common palm of river courses in the savanna is *Phoenix reclinata*, a palm (about 10 m tall) producing a large bunch of yellow drupes and growing in large clusters.

Hyphaene thebaica and *Borassus aethiopum* are grassland species with flabellate (fan-like) leaves (Fig. 7.5.2A). *H. thebaica* is a branched, dioecious palm of hot and dry locations, while *B. aethiopum* is the unbranched, dioecious palm with barrel-shaped stems and is found in the wetter parts of the savanna.

Portions of the leaves of two climbing palms – *Ancistrophyllum opacum* and *Eremospatha hookeri* – are shown in Figs. 7.5.2B and 7.5.3A. Many of the climbing palms are species of wet forest regions.

7·6 Arales

7·6i Araceae

This is a family of succulent perennial herbs – the cocoyam family. Members have tuberous rootstocks or elongate rhizomes; they are mostly of the Cocoyam habit or (rarely) woody or climbing. One species – *Pistia stratiotes* (water lettuce) – is a distinct aquatic species. The peculiar flowering structure known as the spadix is diagnostic. Some species in this family are climbing plants (usually with clasping roots). The leaves are radical, solitary or cauline and distichous or spirally arranged on the stem.

Species of economic significance include ornamental plants (such as *Caladium* spp.) and food crops, which are generally subsumed under the name cocoyams – *Xanthosoma sagittifolia* (*X. mafaffa*) and *Colocasia esculenta*.

The flowers are unisexual without perianths or rarely hermaphroditic with perianths. The male flower has

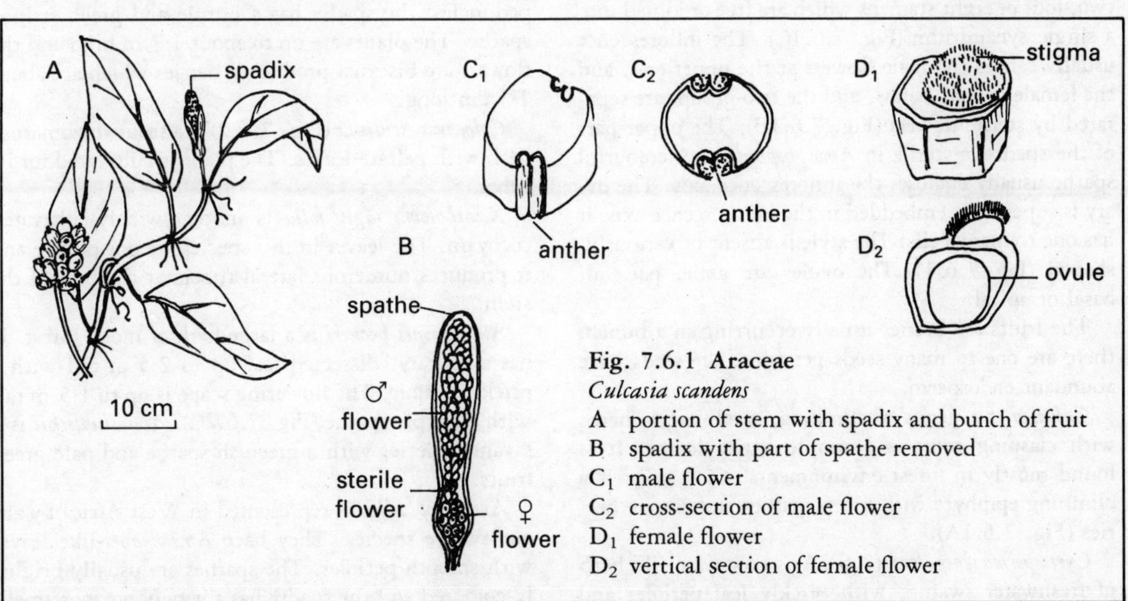

Fig. 7.6.1 Araceae
Culcasia scandens
A portion of stem with spadix and bunch of fruit
B spadix with part of spathe removed
C_1 male flower
C_2 cross-section of male flower
D_1 female flower
D_2 vertical section of female flower

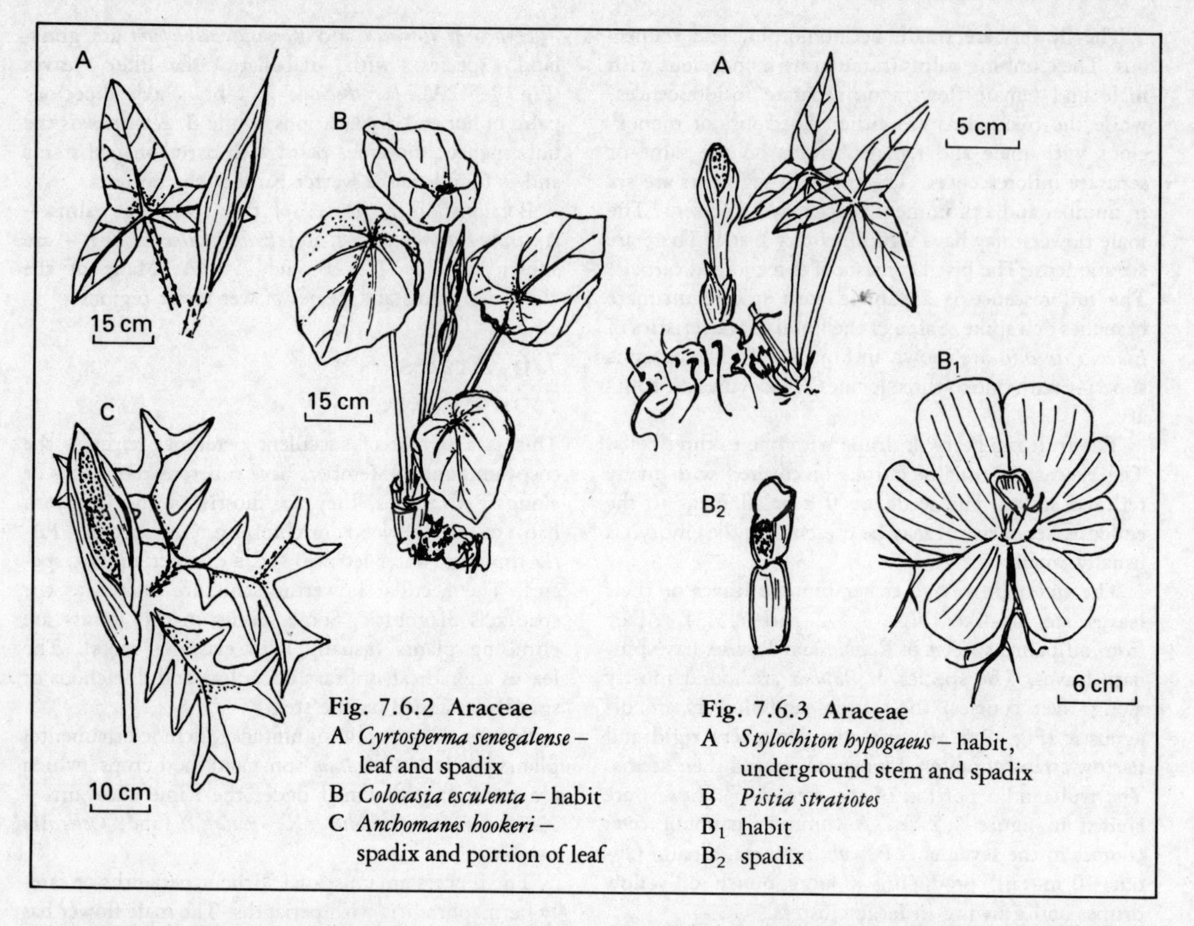

Fig. 7.6.2 Araceae
A *Cyrtosperma senegalense* – leaf and spadix
B *Colocasia esculenta* – habit
C *Anchomanes hookeri* – spadix and portion of leaf

Fig. 7.6.3 Araceae
A *Stylochiton hypogaeus* – habit, underground stem and spadix
B *Pistia stratiotes*
B_1 habit
B_2 spadix

two, four or eight stamens which are free or joined into a single synandrium (Fig. 7.6.1C). The inflorescence usually carries the male flowers at the upper end, and the female flowers below, and the two groups are separated by sterile flowers (Fig. 7.6.1B). The upper part of the spadix is sterile in *Amorphophallus*. A colourful spathe usually encloses the inflorescence axis. The ovary is superior or embedded in the inflorescence axis; it has one to many cells. The style is absent or variously-shaped (Fig. 7.6.1). The ovules are axile, parietal, basal or apical.

The fruits are berries usually occurring in a bunch; there are one to many seeds per fruit; the seeds have abundant endosperm.

Culcasia is a genus of climbing, rarely erect, herbs with clasping roots and simple entire leaves. It is found mostly in forest environments. *C. scandens* is a climbing epiphyte with green spathes and orange berries (Fig. 7.6.1A).

Cyrtosperma senegalense (7.6.2A) is a cocoyam-like herb of freshwater swamps with prickly leaf petioles and peduncles; the spadix has a purple and green striped spathe. The plants are up to about 1·7 m high and the flowers are bisexual producing berries which are about 1·5 cm long.

Colocasia esculenta (Fig. 7.6.2) is a stout rhizomatous herb with peltate leaves. The plant is cultivated for its tuber.

Xanthosoma sagittifolia is another widely-cultivated cocoyam. The leaves in this species are not peltate and it produces numerous lateral tubers at the base of the stem.

Anchomanes hookeri is a large herb of moist forest. It has a solitary, dissected leaf up to 2·5 m tall with a prickly petiole. The flowering scape is up to 1·5 m tall with a purple spathe (Fig. 7.6.2C). *A. welwitschii* is a savanna species with a greenish spathe and pale green fruits.

Amorphophallus is represented in West Africa by about twelve species. They have *Anchomanes*-like leaves with smooth petioles. The spathes are usually brightly-coloured and the spadix has a very obnoxious smell.

They are mostly plants of the savanna.

Stylochiton is a genus of small herbs about 30 cm or so tall with sagittate or hastate leaves and roots that produce a sticky yellow juice. The flowers are unisexual, but remains of perianth are discernible; the flowers appear after the leaves. *S. hypogaeus* (Fig. 7.6.3A) is a common plant of open grassland locations.

Pistia stratiotes (water lettuce) is illustrated in Fig. 7.6.3B.

7·7 Liliales

7·7j Liliaceae

This is a family mostly of perennial herbs, or rarely soft-wooded shrubs, with underground stems (bulbs, rhizomes or tubers). The stem is erect, climbing, leafy or scapose. This family resembles Amaryllidaceae, except that the flowers of Liliaceae are never in an umbel (with the single exception of *Smilax* which is easily distinguishable and is even usually treated as a separate family) and ovaries are superior.

The genera of this family are important mostly as decorative plants. The decorative genera are *Aloe*, *Asparagus*, *Chlorophytum* and *Urginea*.

The flowers are hermaphroditic, rarely unisexual, actinomorphic or rarely zygomorphic; they are trimerous, sometimes showy but never in umbels except in *Smilax*. The perianth is corolla-like with or without a tube. There are usually six perianth segments which are imbricate; bracts are often present. The stamens are six in number, hypogynous or adnate to and opposite the perianth segments. The filaments are usually free and the two-celled anthers open by longitudinal slits. The ovary is superior, three-celled with axile placentae, or one-celled with parietal placentae; ovules are numerous and usually in two series inside each cell.

The fruit is a capsule or berry, globose or three-lobed with numerous seeds having copious endosperm.

Chlorophytum is widespread – a genus usually with branched spikes of dull flowers; the petals are not joined and they bend back to expose three-lobed fruits. The habit and details of floral characteristics of a representative species is shown in Fig. 7.7.1A. Members of this genus grow from a short underground rootstock and not a bulb.

Albuca abyssinica is a herb with a bulb. The inflorescence is up to 2 m tall with large pendulous yellow flowers and leathery strap-like leaves (Fig. 7.7.1B).

Gloriosa superba has a characteristic habit with leaf

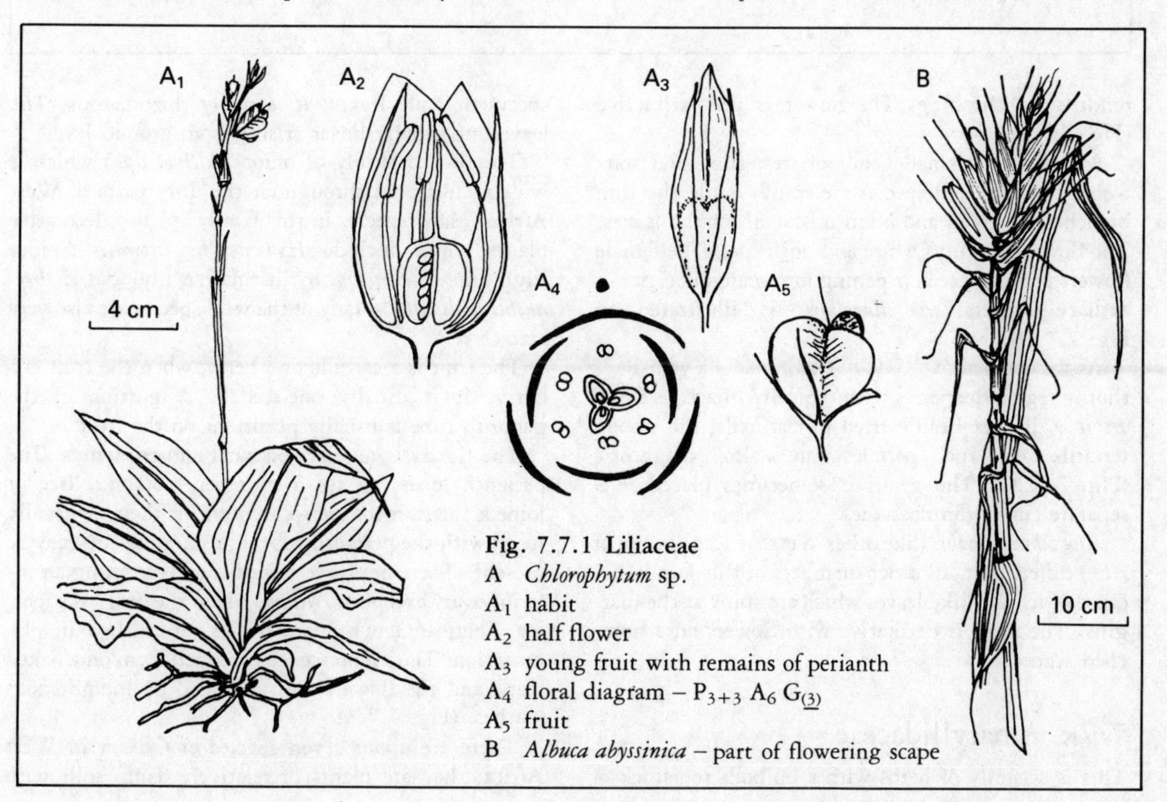

Fig. 7.7.1 Liliaceae
A *Chlorophytum* sp.
A_1 habit
A_2 half flower
A_3 young fruit with remains of perianth
A_4 floral diagram – P_{3+3} A_6 $G_{\underline{(3)}}$
A_5 fruit
B *Albuca abyssinica* – part of flowering scape

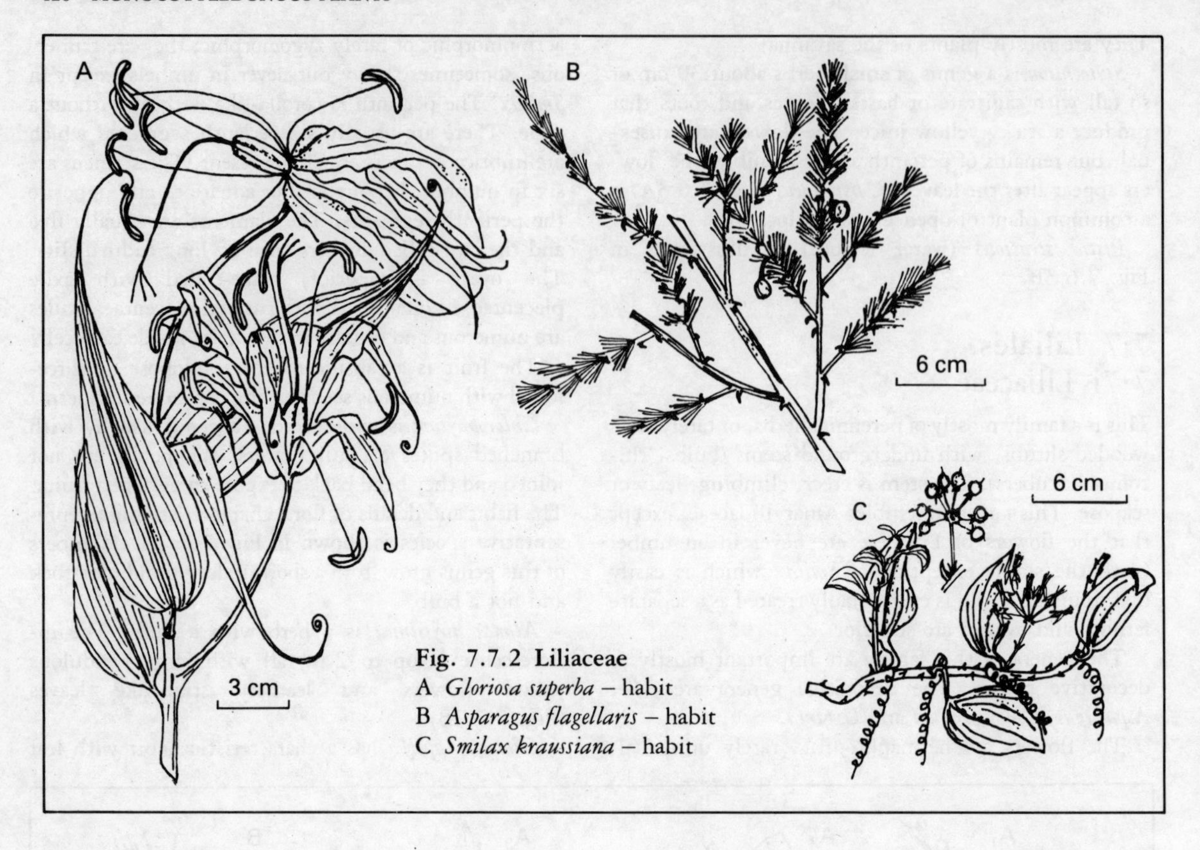

Fig. 7.7.2 Liliaceae
A *Gloriosa superba* – habit
B *Asparagus flagellaris* – habit
C *Smilax kraussiana* – habit

tendrils for climbing. The flower is very attractive (Fig. 7.7.2A).

Asparagus is a small genus of straggling and soft-wooded shrubs. All species are spiny, while the thin branches are green and often mistaken for the leaves. The flowers are tiny, white and unisexual. The female flowers produce green (ripening into orange) berries in axillary clusters. *A. flagellaris* is illustrated in Fig. 7.7.2B.

As we noted above, *Smilax kraussiana* is a yam-like, thorny regrowth species with typically liliaceous flowers in a distinct head carried in leaf axils; the woody tendrils on the petioles are also diagnostic (Fig. 7.7.2C). The genus is sometimes placed in a separate family (Smilacaceae).

Aloe schweinfurthii (like other West African species of *Aloe*) differs from all other members of this family because of its sisal-like leaves which are spiny at the margins. The plant is decorative with flowers on a branched scape.

7·7k Amaryllidaceae

This is a family of herbs with a bulbous rootstock of succulent scale leaves; it is rarely rhizomatous. The leaves are usually linear arising from ground level.

This is the family of onion (*Allium cepa*) which is widely cultivated throughout the drier parts of West Africa. Many species in this family are also decorative plants; these include *Hymenocallis littoralis* (spider lily), *Hippeastrum equestre* (harmattan lily) and *Zephyranthes tubispatha*. Many of the wild species are also very attractive.

The fruit is a capsule or a berry; when the fruit is a berry, it is usually one-seeded. A portion of the perianth tube is usually persistent on the fruit.

The flower is regular, bisexual and trimerous. The perianth lobes are six in number, petaloid, free or joined; the stamens are six in number; they are usually fused with the perianth tube or a staminal cup may be present. There are three joined carpels forming an inferior ovary except in *Allium*, where the ovary is superior. There are few to many ovules per cell in axile placentation. The inflorescence is umbelliform on a naked scape and the flowers are subtended by membranous spathes. (Fig. 7.7.3).

There are about seven species of *Crinum* in West Africa; they are plants of relatively damp soils with

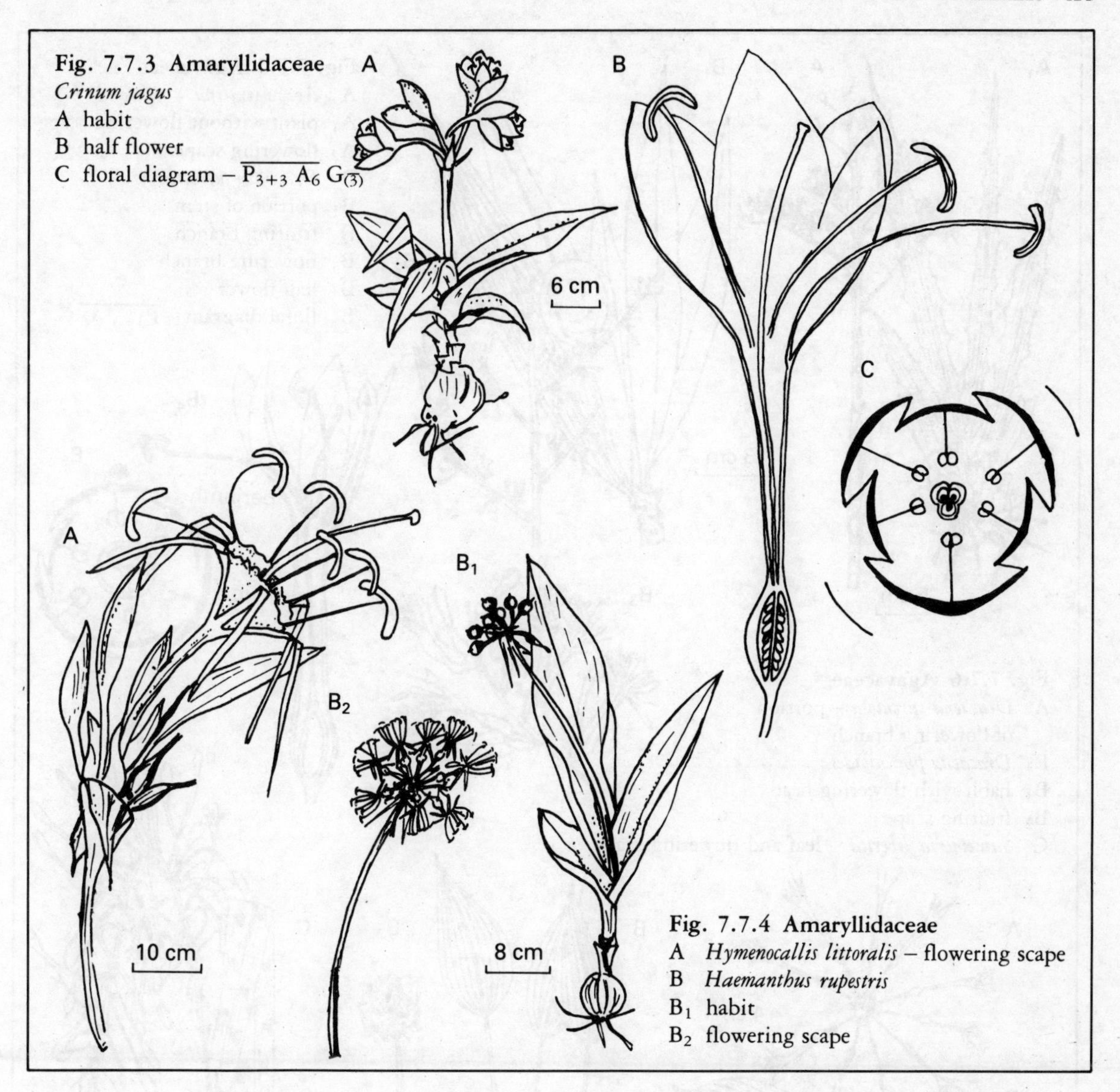

Fig. 7.7.3 Amaryllidaceae
Crinum jagus
A habit
B half flower
C floral diagram – $\overline{P_{3+3}\ A_6}\ G_{(3)}$

Fig. 7.7.4 Amaryllidaceae
A *Hymenocallis littoralis* – flowering scape
B *Haemanthus rupestris*
B_1 habit
B_2 flowering scape

showy flowers. *C. jagus* is a common plant of swampy locations with white flowers that appear in the dry season (Fig. 7.7.3). *C. ornatum* is a decorative plant in grassland swamp with pink stripes on the perianth lobes.

Hymenocallis littoralis is a common ornamental with stamens that unite basally to form a corona (Fig. 7.7.4A); it is introduced from America. *Pancratium* is a native genus very much like *Hymenocallis*, except that there are petaloid recesses from the corona and the stamens are not as long.

Haemanthus is the genus of the 'fire ball lilies' which flower after bush fires, especially in the grasslands and fringing forests. *H. rupestris* has lateral inflorescences about 30 cm tall with few flowers; it occurs in fringing forests (Fig. 7.7.4B). The bright red flowers vary in number from species to species. The fruits in *Haemanthus* are orange or yellow in conspicuous umbellate bunches.

7·71 Agavaceae

This is the sisal hemp family. The plants grow from rhizomatous underground stems, or the stem may be short or well-developed above ground level. The leaves are often crowded at the base of the stem or at the end

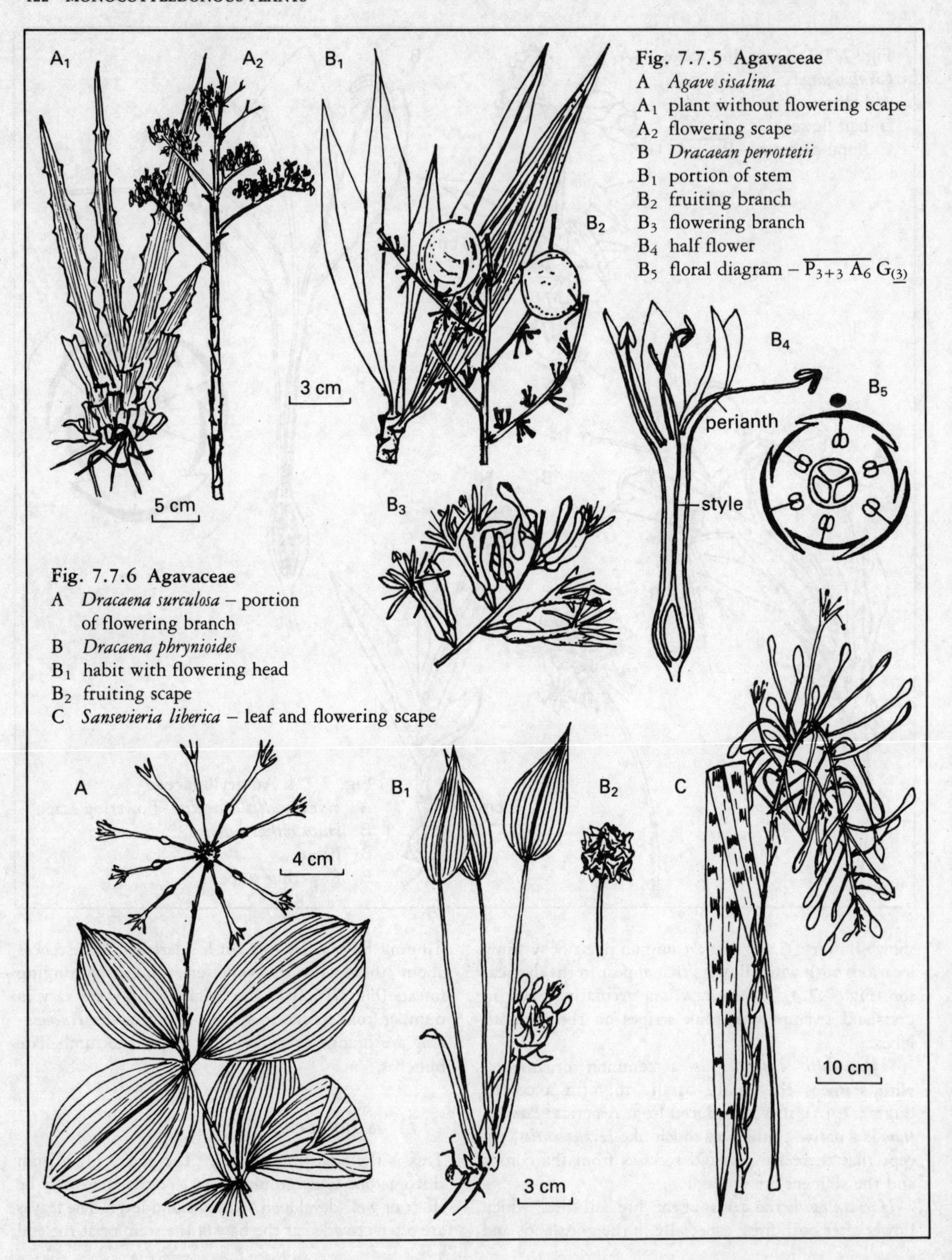

Fig. 7.7.5 Agavaceae
A *Agave sisalina*
A_1 plant without flowering scape
A_2 flowering scape
B *Dracaean perrottetii*
B_1 portion of stem
B_2 fruiting branch
B_3 flowering branch
B_4 half flower
B_5 floral diagram – $\overline{P_{3+3}\,A_6}\,G_{\underline{(3)}}$

Fig. 7.7.6 Agavaceae
A *Dracaena surculosa* – portion of flowering branch
B *Dracaena phrynioides*
B_1 habit with flowering head
B_2 fruiting scape
C *Sansevieria liberica* – leaf and flowering scape

of the branches. The leaves are often long, thick and leathery.

The economic species of importance is *Agave sisalina* cultivated for making rope. Ropes are also made locally from *Sansevieria liberica*. *Dracaena smithii* is planted for demarcating land boundaries and in shrines in southwestern Nigeria.

The flowers are bisexual, polygamous or dioecious in racemose, paniculate or umbelliform inflorescences. Perianth segments are joined into a tube; they usually occur in two whorls of three; corona is absent. There are six stamens which are epipetalous or attached to the base of the perianth segments. The filaments are free, filiform or stout. The anthers are introse, two-locular and open by longitudinal slits. The ovary is superior or inferior, three-celled and it may be beaked. The style is slender and the numerous ovules occur in axile placentation.

The fruit is a capsule or a berry and the seeds have copious endosperm.

Agave sisalina is a widely cultivated plant with very tall thorny leaf edges. The plant has a short creeping stem covered with old leaf bases. The leaves are about 50 cm long. The flowering stem is about 2 m tall with yellowish green flowers with six erect and narrow petaloid perianth segments (Fig. 7.7.5A). Plants with larger leaves, smooth leaf edges and much longer flowering scapes are also included in this species.

Dracaena is a very large genus of herbs, undershrubs and large trees, the latter with peculiar mechanisms of secondary thickening. *D. perrottetii* is a tree up to 20 m tall with white fragrant flowers in large panicles (Fig. 7.7.5B). *D. arborea* is a clean-boled, branched tree with deep green leaves crowded at the end of the branches. The fruits have a characteristic orange to red colour. *D. smithii* has crowns of long leaves (2 m long) at the ends of many stems arising from the ground level. *D. surculosa* (Fig. 7.7.6A) and *D. phrynioides* (Fig. 7.7.6B) are forest species, the latter found on forest floors with red angular fruits and characteristic foliage.

Sansevieria liberica is a robust herb with large leaves arising from underground rhizomes; the flowers are numerous in bracteate spikes (Fig. 7.7.6C).

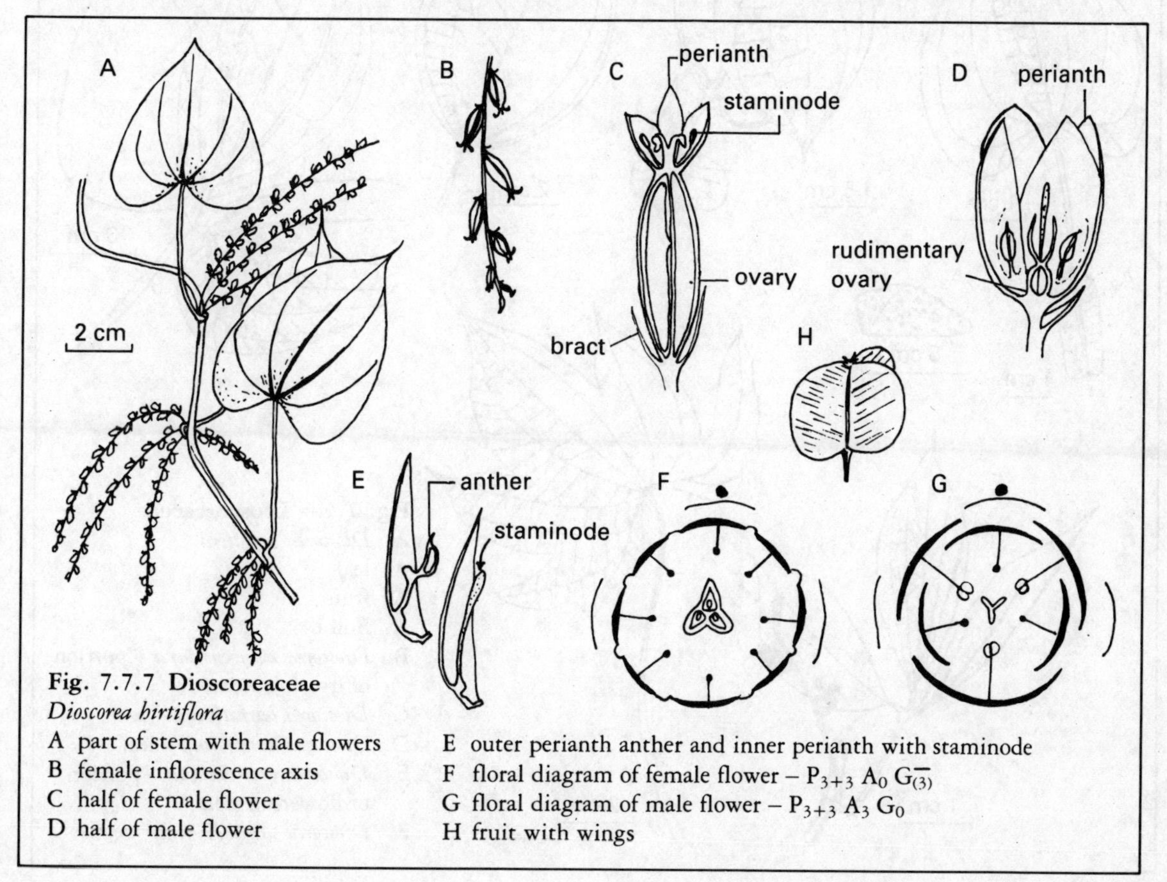

Fig. 7.7.7 Dioscoreaceae
Dioscorea hirtiflora

A part of stem with male flowers
B female inflorescence axis
C half of female flower
D half of male flower
E outer perianth anther and inner perianth with staminode
F floral diagram of female flower – $P_{3+3}\ A_0\ G_{\overline{(3)}}$
G floral diagram of male flower – $P_{3+3}\ A_3\ G_0$
H fruit with wings

7·7m Dioscoreaceae

This is a family of climbing annuals or perennials with perennial tubers of the type that is renewed annually. The tubers may be poisonous or edible and they are often protected by thorny roots. Aerial tubers are present in some species. The climbing stems are pilose, glabrous or protected by thorns. The stems may twine either clockwise or anticlockwise (i.e. 'dextrorse' or 'sinistrorse'): this is an important taxonomic character. The leaves are alternate, often cordate, entire, lobed, digitately divided or palmately compound. The leaves have prominent palmate main veins and regular transverse veins. The petiole is generally twisted. The family is represented in West Africa by only one genus and it occurs from forest to very dry northern locations.

This is the yam family and is of major economic importance as such. The species of major commercial importance are *Dioscorea dumetorum, D. alata, D. bulbifera* and *D. cayenensis* (the white Guinea yam).

The flowers are small, dull and actinomorphic in spicate, racemose or paniculate inflorescences. The perianth is cup-shaped or spreading, six-lobed in one or two series, free or joined. Male flowers occur on several spikes, panicles or racemes in leaf axiles; they are sessile or pedicellate, usually bracteolate (Fig. 7.7.7A). There are six or three stamens (with staminodes present or absent) which are free or partially connate (Fig. 7.7.7C, D and E). A rudimentary pistil is often present. The anthers are two-celled. The female flowers are borne on loose spikes which may be one to many in leaf axils (Fig. 7.7.7B). Staminodes are pre-

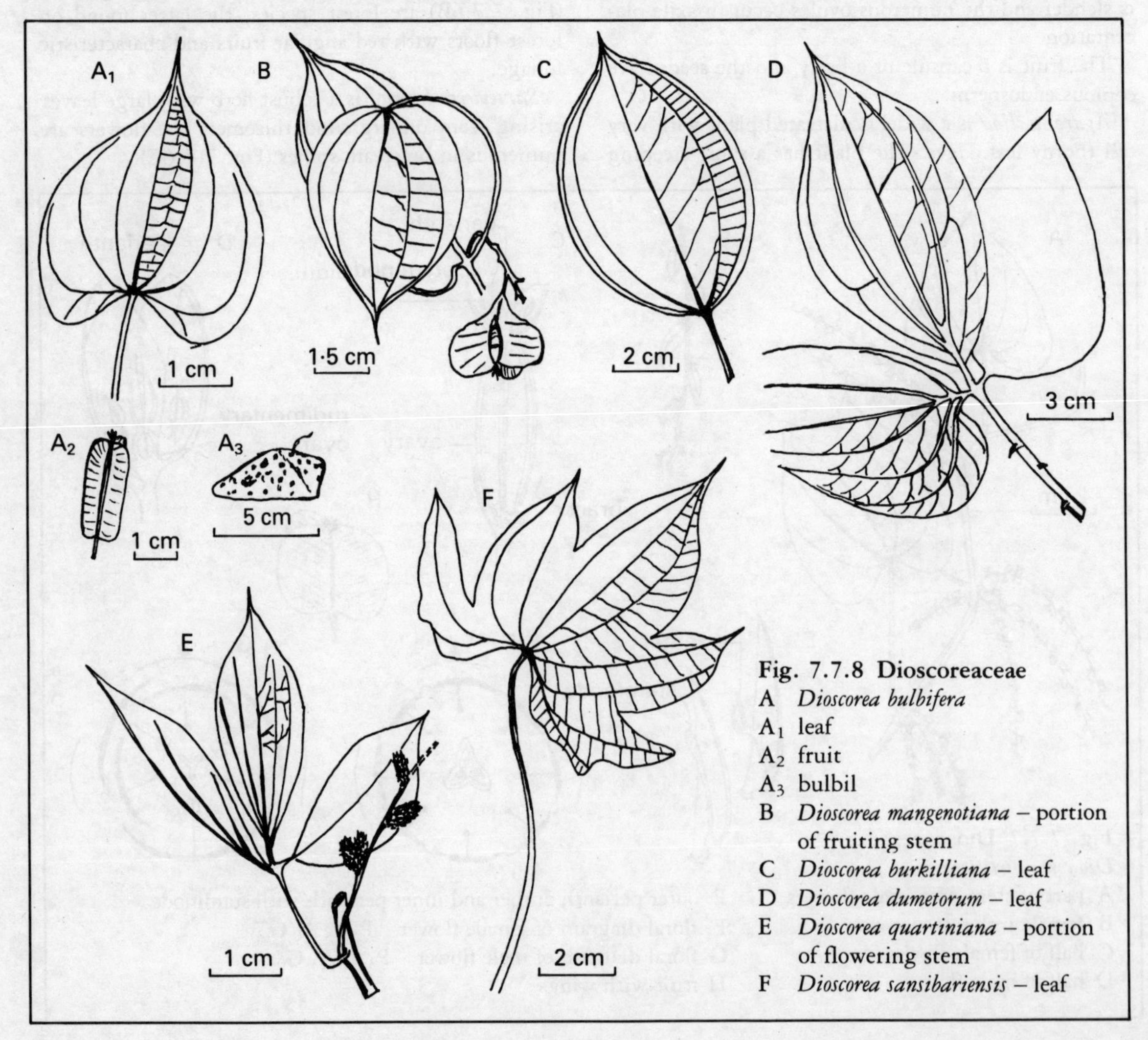

Fig. 7.7.8 Dioscoreaceae
A *Dioscorea bulbifera*
A_1 leaf
A_2 fruit
A_3 bulbil
B *Dioscorea mangenotiana* – portion of fruiting stem
C *Dioscorea burkilliana* – leaf
D *Dioscorea dumetorum* – leaf
E *Dioscorea quartiniana* – portion of flowering stem
F *Dioscorea sansibariensis* – leaf

sent or absent; when present, they are three or six in number. The ovary is inferior and three-celled; with axile placentae and two ovules per cell.

The fruit is a three-valved capsule with winged albuminous seeds. The shape and size of the fruits and seeds can be important taxonomic characters (Figs. 7.7.7 and 7.7.8).

Dioscorea is the only West African genus in this family and it is represented by about twenty species. *D. hirtiflora* is a pubescent climber about 7 m tall; bulbils are present or absent and the tubers are lobed and considered poisonous (Fig. 7.7.7). The leaf shape and fruits, as was noted above, are useful for taxonomic purposes; Fig. 7.7.8 shows one or both of these for each of *D. bulbifera*, *D. mangenotiana* with opposite leaves, *D. burkilliana* with unevenly-spaced main veins, *D. dumetorum*, *D. quartiniana*, and *D. sansibariensis*. Two forest species – *D. preussii* and *D. bulbifera* may be easily confused with each other and with *D. alata*. *D. preussii* is a larger plant with larger leaves and fruits than *D. bulbifera*; *D. alata* does not fruit; all the three have winged stems.

7·8 Orchidales

7·8n Orchidaceae

This is a cosmopolitan family of perennial herbs. The plants are terrestrial with tuberous or rhizomatous underground stems. They may be epiphytic with special aerial roots or they may be stragglers or climbers. Some species are totally leafless being an epiphytic tangle of wiry vegetation.

Many species of ground and epiphytic orchids produce remarkably attractive foliage and flowers such that they have considerable decorative potential. Many orchids are actually cultivated for their flowers or as pot plants.

The flowers of orchids are very specialised and very important in identification. The flowers are trimerous, hermaphroditic and zygomorphic with an inferior ovary. They are either solitary or they occur in racemose inflorescences. The details of floral structure vary considerably and we use the flower of *Aerangis biloba* to illustrate some of the basic botanical terms in orchid floral structure (Fig. 7.8.1). The perianth is in two

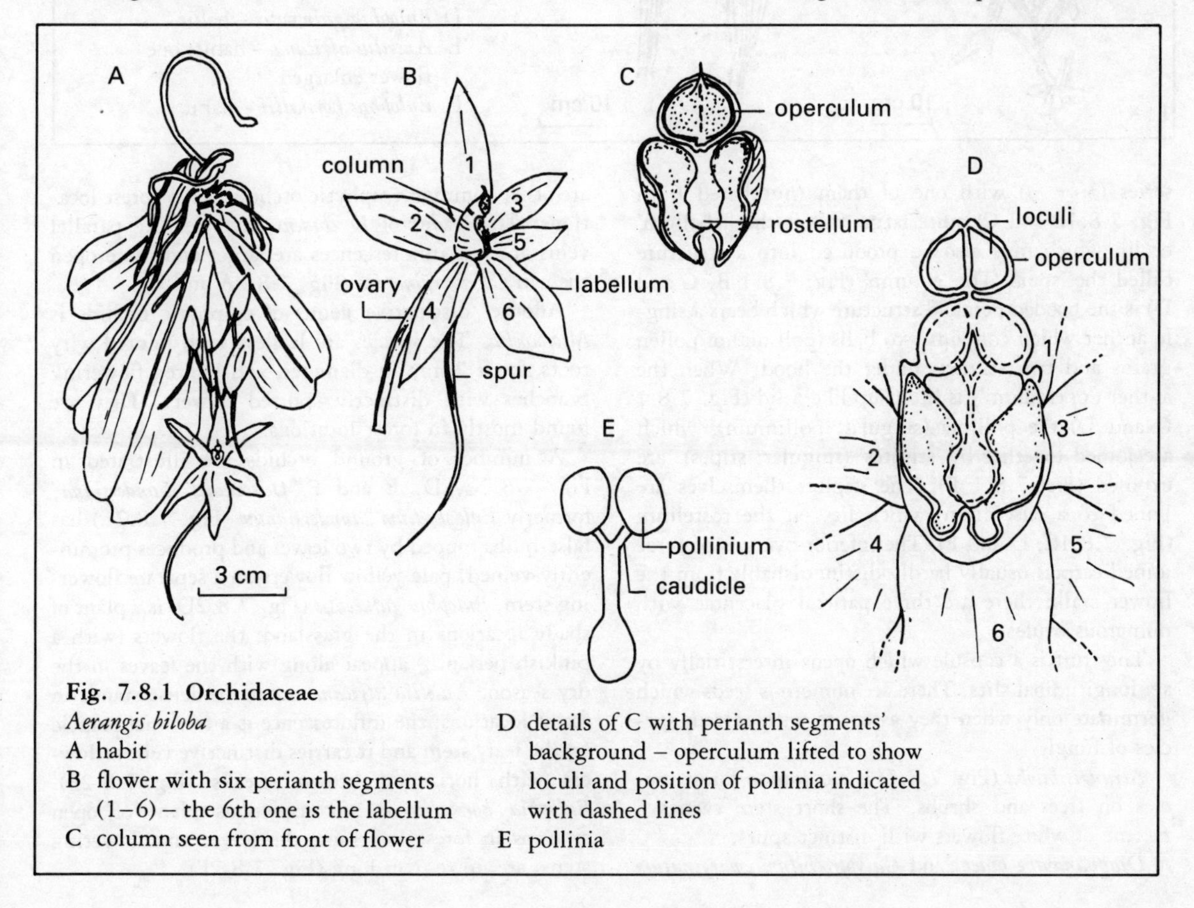

Fig. 7.8.1 Orchidaceae
Aerangis biloba
A habit
B flower with six perianth segments (1–6) – the 6th one is the labellum
C column seen from front of flower
D details of C with perianth segments' background – operculum lifted to show loculi and position of pollinia indicated with dashed lines
E pollinia

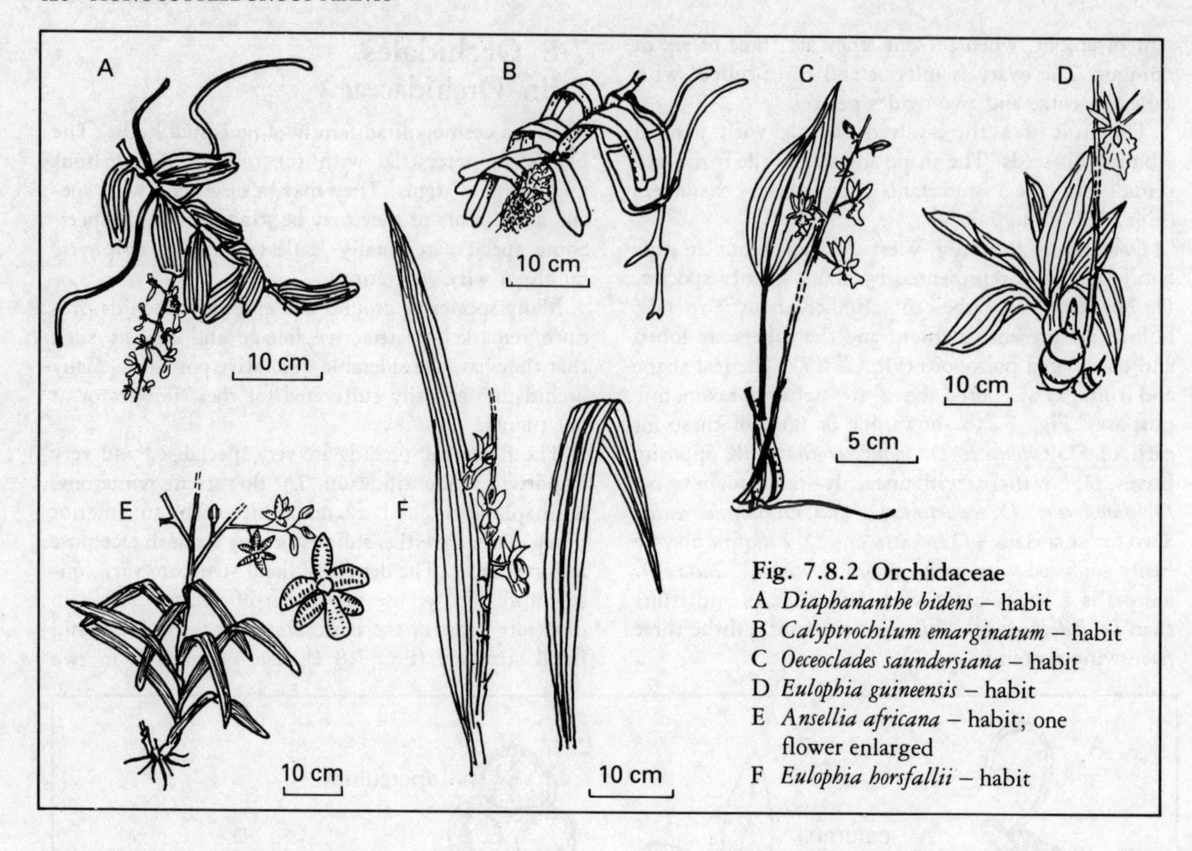

Fig. 7.8.2 Orchidaceae
A *Diaphananthe bidens* – habit
B *Calyptrochilum emarginatum* – habit
C *Oeceoclades saundersiana* – habit
D *Eulophia guineensis* – habit
E *Ansellia africana* – habit; one flower enlarged
F *Eulophia horsfallii* – habit

series (3 + 3) with one of them (numbered 6 in Fig. 7.8.1B and C) quite large; this is the 'labellum' or lip which may also be produced into a structure called the 'spur'. The 'column' (Fig. 7.8.1 B, C and D) is the hooded petaloid structure which bears a single anther which contains two balls (pollinia) of pollen grains and two stigmas under the hood. When the anther ('operculum') is lifted up like a lid (Fig. 7.8.1 C and D) the pollinia (singular: pollinium), which are joined together by 'stipites' (singular: stipes), are exposed (Fig. 7.8.1 E). The stipites themselves are joined to a 'viscidium' which lies on the rostellum (Fig. 7.8.1C, D and E). The inferior ovary is of three joined carpels usually hardly distinguishable from the flower stalk; there are three parietal placentae with numerous ovules.

The fruit is a capsule which opens interstitially by six longitudinal slits. There are numerous seeds which germinate only when they associate with certain species of fungi.

Aerangis biloba (Fig. 7.8.1) is a common forest species on trees and shrubs. The short stem carries a raceme of white flowers with distinct spurs.

Diaphananthe bidens and *Calyptrochilum emarginatum* are other common epiphytic orchids of dry forest locations. The leaves of *D. bidens* have definite parallel veins and the inflorescences are larger and more open than in *C. emarginatum* (Fig. 7.8.2A and B).

Another distinctive genus of epiphytic orchids is *Microcoelia*. The species are leafless with jointed wiry roots about 2 mm in diameter, and jointed flowering branches with distinctly spurred flowers. They are found mostly in forest locations.

A number of ground orchids are illustrated in Fig. 7.8.2C, D, E and F. *Oeceoclades saundersiana*, formerly *Eulophidium saundersianum* (Fig. 7.8.2C) has false-bulbs topped by two leaves and produces prominently-veined, pale yellow flowers on a separate flowering stem. *Eulophia guineensis* (Fig. 7.8.2D) is a plant of shady locations in the grassland; the flowers (with a pinkish perianth) appear along with the leaves in the dry season. *Ansellia africana* occurs in dense stands in forest locations; the inflorescence is a terminal panicle on the leafy stem and it carries distinctive yellow flowers with horizontal brown streaks (Fig. 7.8.2E). *Eulophia horsfallii* is a conspicuous plant of open swamps in forest locations; the leaves and flowering stems are up to 2 m long (Fig. 7.8.2F).

Selected references to Chapters 3–7

Cobley, L. S. 1956. *Botany of Tropical Crops*. Longman: London, UK.

Graham, V. E. 1963. *Tropical Wild Flowers*. Hulton Educational Publications: London, UK.

Hutchinson, J. and Dalziel, J. M. 1954–1972. *Flora of West Tropical Africa*. Vols I, II, III. Crown Agents for Overseas Governments and Administrations: London, UK.

Lowe, J. and Stanfield, D. P. 1974. *Flora of Nigeria: Sedges*. University of Ibadan Press: Ibadan, Nigeria.

Morton, J. K., 1961. *West African Lilies and Orchids*. Longman: London, U.K.

Porter, C. L., 1959. *Taxonomy of Flowering Plants*. W. H. Freeman & Company: San Francisco, USA.

Saunders, H. N., 1958. *A Handbook of West African Flowers*. Oxford University Press: London.

Stanfield, D. P., 1970. *Flora of Nigeria: Grasses*. University of Ibadan Press: Ibadan, Nigeria.

Stebbins, G. L. 1974. *Flowering Plants – Evolution above the Species Level*. Harvard University Press: Cambridge. Massachusetts, USA.

SECTION III

Appendices

APPENDIX I

An index to Nigerian names of some West African angiosperms

ABBREVIATIONS

Ful = Fulani
Hau = Hausa
Yor = Yoruba

Acacia senegal (Ful: debehi; Hau: dakwara; Kanuri: kolkol)
Acalypha ciliata (Igbo: abaleba; Yor: jinwinni)
Achyranthes aspera (Yor: eemo)
Adansonia digitata (Bini: usi; Hau: kuka; Nupe: muchi; Yor: oshe)
Adenopus breviflorus (Yor: tagiiri)
Aframomum melegueta (Bini: ehin edo; Hau: citta; Yor: ataare)
Afzelia africana (Bini: ariyan; Ful: gayohi; Hau: kawo; Igbo: akpalata; Nupe: bachi; Yor: apa)
Ageratum conyzoides (Bini: ebi-igbedore; Efik: otitida-hadaha; Igbo: akwukwu-nwaosi-naka; Yor: imi esu; ajihewu)
Albizia adianthifolia (Bini: uwowe nolaghbor; Igbo: evu; Nupe: sakanchi; Yor: ayunre)
Alchornea cordifolia (Bini: unwonwe; Hau: bambami; Igbo: ububo; Yor: ipa; esin)
Aloe schweinfurthii (Hau: zabo; zabuwa; kabar giwa)
Alstonia boonei (Bini: ukhu; Igbo: egbu; Ijaw: ndodo; Yor: ahun)
Alternanthera pungens (Yor: dagunro)
Amaranthus hybridus (Igbo: inine; Yor: tete)
Anchomanes difformis (Efik: ebaanangi; Hau: hantsaga-da; Igbo: oje; Yor: ogiri sako)
Andropogon gayanus (Yor: eruwa)
Andropogon tectorum (Yor: eruwa odo)
Aneilema spp. (Bini: ohiowu; Hau: balasa; Igbo: abala-ja; Yor: gbagodo odo)
Annona senegalensis (Hau: gwandar daji; Igbo: ubur-uocha; Yor: abo; ibobo)
Anogeissus leiocarpus (Ful: koloji; Hau: marike; Kanuri: annum; Yor: ayin)
Anthocleista nobilis (Yor: shapo)
Antiaris africana (Bini: ogi oyu; Igbo: oji anwu; Ijaw: alalawa; Yor: ooro)
Aphania senegalensis (Yor: eyindie)
Arachis hypogaea (Hau: jida; Yor: epa)
Asclepias curassavica (Hau: rizgar kureje)
Aspilia africana (Hau: kalankuwa; Igbo: iranjile; Yor: yunyun)

Baphia nitida (Bini: otua; Igbo: aboshi; Yor: irosun)
Berlinia grandiflora (Bini: ekpagoi; Hau: dokar rafi; Igbo: ububa; Nupe: baborochi bata; Yor: apado)
Blighia sapida (Bini: ukpe; Ful: feso; Hau: gwanja kusa; Nupe: ila; Yor: ishin)
Blighia unijugata (Bini: ukpe nehwi; Igbo: ojawala; Yor: ishin oko)
Bombax brevicuspe (Bini: ogiukpogha; Igbo: akpudele; Yor: awori)
Bombax buonopozense (Bini: obokha; Igbo: akpu; Yor: ponpola)

Bombax costatum (Ful: kuruhi; Hau: kurya; Nupe: kutukpachi)
Borassus aethiopum (Bini: urua; Efik: nsongo; Ful: dubbi; Nupe: gbachi; Tiv: kuvugh; Yor: agbon onidu)
Bosqueia angolensis (Bini: ukputu; Igbo: oze; Ijaw: binu; Yor: koko eran)
Bridelia spp. (Bini: ogangan; Hau: kirni; Igbo: aga; Yor: ira)
Butyrospermum paradoxum (Ful: kajere; Hau: ka'danya; Igbo: osisi; Yor: emi)

Caesalpinia bonduc (Hau: 'ya'yan dara; Yor: ayo; seyo)
Calotropis procera (Hau: tumfafiya; Yor: bomubomu)
Canna indica (Bini: osalebo; Hau: gwangwama; Igbo: manya ogolo; Yor: ido)
Capsicum spp. (Bini: isie; Efik: ntokon; Hau: barkono; tasshi; Igbo: ose; Yor: ata)
Cardiospermum halicacabum (Hau: garafunin fadama; Igbo: okulefo; Yor: saworo)
Cassia alata (Bini: akoria; Hau: filasko; Igbo: ogala; Yor: asunwon)
Cassia occidentalis (Hau: rai'dore; bajamfari; Igbo: akede agbara Yor: reere)
Ceiba pentandra (Bani: okha; Efik: ukem; Hau: rimi; Igbo: akpu; Tiv: vambe; Yor: araba)
Celosia spp. (Igbo: eri emi onu; Yor: sokoyokoto)
Celosia trigyna (Yor: ajefawo)
Celtis integrifolia (Hau: zuwo; Nupe: gimachi; Yor: ape)
Celtis zenkeri (Bini: ohia; Yor: ita)
Chlorophora excelsa (Bani: uloko; Hau: loko; Igbo: oji; Ijaw: olokpata; Nupe: roko; Yor: iroko)
Chrysophyllum albidum (Igbo: udala; Yor: osan; agbalumo)
Cola acuminata (Bini: evbe; Ibibio/Efik: ibong; Igbo: oji Igbo; Yor: obi gidi; abata)
Cola millenii (Yor: obi edun)
Cola nitida (Hau: goro; Igbo: oji Hausa; Nupe: chinga'bi; Yor: gbanja; goro)
Colocasia esculenta (Bini: iyekho; Efik: mkpong; Hau: gwaza, gwamba; Ibo: ede; Yor: isu koko)
Colocynthis vulgaris (Bini: ogi; Efik: ikon; Igbo: ogili; Hau: guna; agushi; Yor: egusi; bara)
Combretum glutinosum (Ful: buski; Hau: taramniya)
Combretum molle (Ful: damoruchi; Hau: muyan damo; Yor: aragba)
Combretum straggling species (Bini: oven; Hau: farargeza; Igbo: okolo; Yor: okan; ogan)
Commelina spp. (Bini: ohiowu; Hau: balasa; Igbo: mbong; Yor: gbagodo)
Corchorus spp: (Hau: kerkeshi; Igbo: ahiha; Yor: ooyo; ewedu)
Costus afer (Bini: ikweroha; Hau: kak'i-zuwa Hausa; Igbo: opete; Yor: tete egun)
Cucurbita pepo (Bini: eyen; Efik: nnangi; Hau: kabewa; Igbo: ugbugulu; Yor: elegede)
Cuscuta australis (Hau: soyaya; Yor: omonigelegele)
Crassocephalum biafrae (Yor: worowo)
Crassocephalum crepidioides (Yor: ebolo)
Crinum jagus (Hau: albasa kwa'di; Igbo: ede obase; Yor: isumeri)
Crossopteryx febrifuga (Ful: rima jogohi; Hau: kasfiya; Nupe: nembi sunsun)
Cyclodiscus gabunensis (Bini: okan; Efik: anyan; Igbo: uzi; Yor: olosan)
Cyperus esculentus (Efik: isip Accra; Hau: aya; chizp; Yor: ofio; imumu)

Daniellia ogea (Bini: ozia; Efik; ipayan; Igbo: abwa; Yor: ojia)
Daniellia oliveri (Bini: ozia; Ful: kaharlahi; Hau: maje; Igbo: ozabwa; Nupe: danchi; Yor: iya)
Dennettia tripetala (Bini: ako; Ibibio: nkarika; Igbo: nmimi; Yor: igberi)
Dialium guineense (Bini: ohiome; Igbo: icheku; Ijaw: akin; Hau: tsamiyan kurmi; Yor: awin)
Diaphananthe bidens and many other epiphytic orchids (Efik: nditek; Hau: murucin bisa; Igbo: ibisi; Yor: ela)
Dichrostachys cinerea (Ful: burli; Hau: dundu; Igbo: ami ogwu; Yor: kara)
Dioscorea alata (Bini: akenedo; Hau: sakata; Yor: ewura)
Dioscorea bulbifera (Hau: doyar bisa; Yor: isu alabahun; emina)
Dioscorea cayenensis (Bini: ikpen; Yor: alo; Igangan; aginnipa)
Dioscorea dumetorum (Bini: olimehi; Hau: rogwan biri; Efik: edidia iwa; Ibo: ono; Yor: esuru)
Dombeya buettneri (Yor: ewe ofo)
Dracaena arborea (Igbo: odo; Yor: ope kannakanna)
Dracaena smithii (Yor: peregun)

Elaeis guineensis (Bini: udin; Hau: kwakwa; Nupe: efu; Tiv: irile; Yor: ope)
Emilia spp; (Igbo: nte ene; Yor: odundun odo)
Entada africana (Ful: pade wanduhi; Hau: tawatsa; Yor: ogurobe)
Erythrina senegalensis (Efik: edeng; Hau: majiriya; Igbo: echichi; Tiv: showo; Yor: ologunsheshe)
Eulophia guineensis (Hau: gatarin kureje)
Euphorbia hirta (Bini: asin uloko; Hau: nonankurciya; Yor: emile)
Euphorbia hyssopifolia (Yor: emile)
Euphorbia kamerunica (Efik: akpa mbiet; Hau: kyanara; Tiv: agondo; Yor: oroadete)

Ficus capensis (Ful: rima bichehi; Hau: nwar yara; Nupe: glanchibokun; Yor: opoto)
Ficus mucuso (Yor: ogoro; obobo)
Ficus polita (Ful: durmihi; Hau: durumi; Tiv: kondam)
Ficus thonningii (Ful: biskehi; Hau: chediya; Yor: odan)
Funtumia elastica (Bini: onyan)

Gardenia spp. (Hau: gaude; Yor: oruwon; sanga)
Gliricidia sepium (Yor: agunmaniye)
Gloriosa superba (Bini: egwa rere; Hau: gundumar zomo; Yor: ewe aje)
Glyphaea brevis (Bini: uwenrhiontan; Igbo: anyansu; alo; Yor: atori)
Grewia mollis (Hau: daraji; Nupe: roronchi; Yor: ogbolo)
Grewia pubescens (Bini: evbare; Yor: ora igbo; afunforo igbo)

Haemanthus spp. (Hau: albasa kwa'di; Yor: toto odo)
Hibiscus cannabinus (Hau: rama; Yor: okun)
Hibiscus sabdariffa (Yor: isapa)
Hildegardia barteri (Igbo: ufuku; Yor: eso; shishi)
Holarrhena floribunda (Ful: niwahi; Hau: bakin mutum; Yor: irena)
Holoptelea grandis (Bini: olazo; Yor: inajoko)
Hymenocardia acidia (Ful: yawa sotoje; Hau: jan yaro; Igbo: ikalaga; Yor: orupa)
Hyphaene thebaica (Ful: gellohi; Hau: goriba; Kanuri: kerzum)

Imperata cylindrica (Hau: tofa; toha; Yor: eekan)
Ipomoea batatas (Bini: iyan ebo; Efik: bia-mbakara; Hau: dankali; Igbo: ji-oyibo; Yor: odunkun; anamo)
Ipomoea involucrata (Hau: duman kwa'di; Igbo: mgbanala; Yor: alukerese)
Isoberlinia doka (Hau: doka; Nupe: babarochi; Tiv: mkovol)

Jatropha gossypifolia (Bini: ora ebo; Hau: bi ni da guzu; Igbo: ake mbogho; Yor: lapalapa; botuje)

Khaya grandifoliola (Hau: male; Yor: oganwo)
Khaya ivorensis (Bini: ogangwo; Igbo: ono; Yor: oganwo)
Kigelia africana (Bini: ugbongbon; Ful: jirlghi; Hau: rawuya; Igbo: uturubein; Kanuri: bulungu; Nupe: bechi; Yor: pandoro)

Lactuca taraxacifolia (Hau: nonan'barya; Yor: yanrin)
Lagenaria spp. (Bini: okpan; Efik: iko; Hau: duma; Igbo: mbubu; Yor: igba; odo)
Lannea humilis (Hau: faru; Kanuri: kurubulul)
Lannea kerstingii (Hau: farun biri; farun doyan)
Lecaniodiscus cupanioides (Bini: utantan; Igbo: okpu; Yor: akika)
Lonchocarpus cyanescens (Bini: ebelu; Hau: talaki; Igbo: njasse; Yor: elu)
Lovoa trichilioides (Bini: apopo; Yor: akoko igbo)
Luffa aegyptiaca (Yor: kaninkanin arugbo)

Mallotus oppositifolius (Hau: kafar mutuwa; Igbo: okpokirinyan; Yor: pepe)
Manihot esculentus (Bini: igari; Hau: rogo; Igbo: akpu; Yor: ege; gbaguda)
Manilkara obovata (Igbo: ukpi; Ijaw: wono; Yor: emido)
Mansonia altissima (Yor: afon)
Markhamia tomentosa (Bini: ogie; ikhimi; Igbo: onyiri akikara)
Mezoneuron benthamianum (Yor: amureju)
Mimosa pudica (Efik: mbaba kudi; Igbo: anasieye; Yor: patanmo; aluro)
Mitragyna ciliata (Bini: eben; Igbo: uburu; Yor: abura)
Monodora tenuifolis (Bini: uyenghen; Igbo: ehuru ohia; Yor: lakesin)
Momordica charantia (Yor: ejinrin)
Morinda lucida (Igbo: nuke; eze ogu; Yor: oruwo)
Morus mesozygia (Yor: ewe aye)
Musa parasidiaca (Bini: ogeda; Hau: ayaba; Igbo: ogede jioko; Yor: ogede agbagba)
Musanga cecropioides (Bini: ogohen; Efik: uno; Igbo: oro; Yor: aga)
Myrianthus arboreus (Bini: ihi eghe; Igbo: ujuju; Nupe: ksapachi; Yor: ibishere)

Nauclea diderrichii (Bini: obiakhe; Igbo: uburu)
Nauclea latifolia (Hau: tafashiya; Yor: egbesi)
Nesogordonia papaverifera (Bini: urhuaro; Igbo: otalo; Yor: otutu)
Newbouldia laevis (Bini: ikimi; Efik: obot; Hau: adaruku; Igbo: ogirisi; Tiv: kontor; Yor: akoko)
Nicotiana tabacum (Bini: itaba; Hau: taba; Igbo: utaba; Yor: taba)

Ocimum spp. (Igbo: nchowun; Yor: efinrin)
Oryza sativa (Bihi: ize; Efik: edesi; Hau: shinkafa; Igbo: osikapa; Yor: iresi)

Parinari curatellaefolium (Ful: nawarre-badi; Hau: rura; Nupe: putu; Tiv: ibua; Yor: idofun)
Parinari polyandrum (Ful: chiboyi; Hau: kaikayi; Yor: idofun)
Parkia clappertoniana (Hau: dawadawa; Igbo: ogiri Hausa; Yor: igba; irugba)
Paullinia pinnata (Bini: aza; Hau: yatsa biyar; Yor: kakasenla)
Pennisetum americanum (Hau: gero)

Phoenix reclinata (Bini: ukukon; Efik: eyupinuen; Hau: kijinjiri; Igbo: ngala; Tiv: wure; Yor: ookun)
Phyllanthus discoideus (Bini: asiyin; Igbo: isi nkpi; Urhobo: ololo; Yor: ashansha)
Physalis angulata (Bini: ahapue-nebi; Hau: matsar mama; Igbo: putu; Yor: koropo)
Piliostigma thonningii (Ful: barkehi; Hau: kaego; Igbo: okpo otu; Nupe: bafin; Yor: abafe)
Piper guineense (Bini: ebe ahanhi; Efik: odusa; Hau: masoro; Igbo: ozeza; Yor: iyere)
Pistia stratiotes (Efik: ekpong; Hau: kainuwa; Yor: osibata)
Prosopis africana (Ful: kohi; Hau: kiriya; Igbo: ubwa; Tiv: kpaye; Yor: ayan)
Pseudocedrela kotschyii (Hau: tuna; Yor: emigbegi)
Pseudospondias microcarpa (Nupe: kekerakuchi; Yor: okika aja)
Pterocarpus erinaceus (Bini: ukpeka; Ful: banuhi; Hau: madobia; Igbo: aze agu; Yor: apepe; aara)
Pterocarpus mildbraedii (Bini: urube; Hau: madobia rafi; Ijaw: geneghar)
Pterocarpus santalinoides (Bini: akumeze; Hau: gunduru; Tiv: kereke; Yor: gbingbin)
Pupalia lappacea (Hau: marin kusu; Igbo: ose; Yor: eemo agbo)

Raphia hookeri (Yor: oguro; pako; eriko)
Raphia sudanica (Hau: tukuruwa)
Raphia vinifera (Bini: oha; Urhobo: ekian; Itsekiri: ebeje)
Rauvolfia vomitoria (Bini: akata; Efik: uto enyin; Igbo: akanta; Yor: dodo; asofeyeje)
Ricinodendron heudelotii (Bini: okhuen; Igbo: okue; Ijaw: ekengbo; Yor: erinmado)
Ricinus communis (Bini: era ogi; Hau: zurma; Ibo: ogilisi; Yor: laraa; laa)
Rothmannia spp. (Igbo: uli)

Saccharum officinarum (Bini: okwere; Igbo: okpete; Hau: areke; Yor: ireke)
Sansevieria liberica (Hau: moda; Igbo: ebube agu; Yor: oja ikooko)
Schrankia leptocarpa (Igbo: anasieye; Yor: aluro)
Scleria verruncosa (Efik: asai; Hau: kudundurin iya; badayi; Igbo: worowo; Yor: labelabe)
Sclerocarya birrea (Hau: danyan)
Sida spp. (Yor: osepotu)
Smilax spp. (Yor: kasan)
Solanum melongena (Bini: ekue; Efik: anyana; Hau: gauta; Igbo: anana; Yor: igba; ikan)
Spathodea campanulata (Bini: owewe; Efik: esenim; Igbo: imi ewu; Nupe: dimberi chamili; Yor: oruru)
Spondias mombin (Bini: okighan; Hau: tsadar masar)
Sterculia setigera (Hau: kukuki; Kanuri: sugubo; Nupe: kokongiga)
Sterculia tragacantha (Bini: oporipor: Hau: kukukin rafi; Ibibio: udot eto; Igbo: oloko; Nupe: nyichin kiso; Yor: ilaka)
Stereospermum acuminatissimum (Bini: oshuobon; Yor: eru iyeye)
Stereospermum kunthianum (Ful: golombi; Hau: sansami; Kanuri: golombi; Tiv: umana tumba; Yor: ajade)
Striga spp. (Hau: kuduji dawa)
Strophanthus hispidus (Hau: kwankwani tantsiya; Yor: isa; isa gere)
Strychnos spinosa (Ful: kumbija; Hau: kokiya; Nupe: manvovogi; Tiv: amaku; Yor: iguningunin)

Tabernaemontana pachysiphon (Bini: ibu; Igbo: petepete: Yor: dodo)
Terminalia glaucescens (Hau: baushe; Igbo: edo; Yor: idi odan)
Tetracarpidium conophorum (Yor: awusa; asala)
Tetrapleura tetraptera (Bini: ighimiaka; Ibo: oshosho; Ijaw: apapa; Nupe: ikoho; Yor: aridan)
Thaumatococcus daniellii (Yor: ketemfe; ewe eeran)
Tragia spp. (Igbo: abalagwo; Yor: esisi)
Treculia africana (Bini: ize; Efik: ediang; Igbo: ukwa; Yor: afon)
Trema guineensis (Bini: ehu ogo; Igbo: telemukwu; Yor: afunforo)
Trichilia heudelotii (Bini: oyallo; Yor: akika; ako rere)
Triplochiton scleroxylon (Bini: obeche; Igbo: okpobo; Urhobo: ewowo; Yor: arere)
Triumfetta rhomboidea (Hau: dakadafi; Igbo: odo; Yor: esura)

Uraria picta (Yor: alupayida)
Urena lobata (Bini: oronhon; Hau: rama-rama; Yor: ilasa agbonrin)
Urera spp. (Efik: ntan; Yor: ewe ina)
Uvaria chamae (Igbo: nmimi ohia; Yor: igberi akiti)

Vernonia amygdalina (Bini: oriwo; Hau: shiwaka; Igbo: onubu; Yor: ewuro)
Vernonia conferta (Bini: oriweni; Igbo: ubube; Yor: shapo)
Vigna multinervis (Yor: inabiri)
Vitex doniana (Ful: galbihi; Hau: dinyar; Igbo: ucha koro; Yor: oori)
Voacanga africana (Igbo: petepete; Yor: ako dodo)

Waltheria indica (Yor: imi omo)

Zingiber officinale (Efik: jinga; Hau: citta aho; Yor: atale)

APPENDIX II

Author citations

Abelmoschus esculentus (L.) Wight and Arn.
Abrus precatorius L.
Abutilon mauritianum (Jacq.) Medic.
Acacia Miller.
A. farnesiana (L.) Willd.
A. senegal (L.) Willd.
Acalypha L.
A. ciliata Forskal.
Acanthus L.
A. montanus (Nees) T. Anders.
Achyranthes aspera L.
Acioa Aublet.
Adansonia digitata L.
Adenopus Benth.
A. breviflorus Benth.
Aerangis biloba (Lindl.) Schltr.
Aframomum K. Schum.
A. daniellii K. Schum.
A. elliotii K. Schum.
A. melegueta (Rosc.) K. Schum.
A. sceptrum K. Schum.
Afzelia Sm.
A. africana Sm. ex Pers.
Agave rigida Mill.
Ageratum conyzoides L.
Alafia Thou.
A. barteri Oliv.
Albizia Durazz.
A. adianthifolia (Schum.) W. F. Wight.
A. zygia (DC) J. F. Macbr.
Albuca abyssinica Dryand.
Alchemilla L.
Alchornea Sw.
A. laxiflora (Benth.) Pax and Hoffm.
A. cordifolia (Schum. and Thonn.) Muell. Arg.
Allium cepa L.
Aloe L.
A. schweinfurthii Bak.
Alstonia boonei De Wild.
Alternanthera Forskal.
A. pungens H. B. et K.
A. sessilis (L.) R. Br.
Amaranthus L.
A. hybridus L.
A. spinosus L.
Ambrosia L.
Amorphophallus Blume ex Decne.
Amphiblemma Naudin.
A. mildbraedii Gilg. ex Engl.
Anacardium occidentale L.
Anchomanes hookeri Schott.
A. welwitschii Rendle.

Ancistrophyllum opacum (Mann and Wendl.) Drude.
Andropogon L.
A. gayanus Kunth.
A. tectorum Schum. and Thonn.
Aneilema R. Br.
A. umbrosum (Vahl) Kunth subsp. *umbrosum.*
Annona muricata L.
A. senegalensis Pers.
Anogeissus leiocarpus (DC) Guill. and Perr.
Ansellia africana Lindl.
Antherotoma naudini Hook.
Anthocleista Afzel. ex R. Br.
A. nobilis G. Don.
Antiaris africana Engl.
Arachis hypogaea L.
Asparagus L.
A. flagellaris (Kunth) Bak.
Aspilia Thours.
A. africana (Pers.) C. D. Adams.
Asystasia calycina Benth.
A. gangetica (L.) T. Anders.
Aubrevillaea platycarpa Pellegr.
Azadirachta indica A. Juss.

Baphia nitida Lodd.
Barleria flava Jacq.
Bauhinia L.
B. monandra Kruz.
B. tomentosa L.
Berlinia Solander ex Hook. f.
B. grandiflora (Vahl) Hutch. and Dalz.
Blighia sapida Koen.
B. unijugata Bak.
Boehmeria platyphylla Hamilton ex D. Don.
Bombax brevicuspe Sprague.
B. buonopozense P. Beauv.
B. costatum Pellegr. and Vuillet.
Borassus aethiopum Mart.
Borreria G. F. W. Meyer.
B. filifolia (Schum. and Thonn.) K. Schum.
B. filiformis (Hiern) Hutch. and Dalz.
B. octodon Hepper.
B. scabra (Schum. and Thonn.) K. Schum.
Bosqueia angolensis Ficalho.
Brachystegia eurycoma Harms.
Bridelia Willd.
Brillantaisia lamium (Nees) Benth.
Buchnera capitata Benth.
Butyrospermum paradoxum (Gaertn. f.) Hepper.

Caesalpinia bonduc (L.) Roxb.
C. pulcherrima (L.) Sur.
Cajanus cajan (L.) Millsp.
Caladium Vent.
Calliandra haematocephala Hassk.
Calopogonium mucunoides Desv.
Calotropis procera (Ait.) Ait. f.
Calyptrochilum emarginatum Schltr.
Canavalia ensiformis (L.) DC.
Canna indica L.
Capraria biflora L.
Capsicum L.
Cardiospermum halicacabum L.
Cassia L.
C. alata L.
C. hirsuta L.
C. mimosoides L.
C. occidentalis L.
C. rotundifolia Pers.
C. siamea Lam.
C. sieberiana DC
C. tora L.
Cassytha filiformis L.
Cayaponia Manso.
C. africana (Hook. f.) Exell.
Ceiba pentandra (L.) Gaertn.
Celosia L.
C. trigyna L.
Celtis integrifolia Lam.
C. zenkeri Engl.
Centrosema pubescens Benth.
Cephaelis peduncularis Salisb.
Chasalia kolly (Schumach.) Hepper.
Chlorophora excelsa (Welw.) Benth.
Chlorophytum Pohl ex DC.
Chrysophyllum albidum G. Don.
Cienfuegosia Cav.
C. digitata Car.
C. heteroclada Spraque.
Clappertonia ficifolia (Willd.) Decne.
Clematiopsis Bojer ex Hutch.
C. scabiosifolia (DC) Hutch.
Clematis grandiflora DC.
C. hirsuta Guill and Perr.
Clematis L.
Clerodendrum speciosissimum Van Geert.
C. speciosum D'Ombrain.
C. volubile P. Beauv.
Clitoria ternatea L.
Cocos nucifera L.
Codiaeum variegatum Blume.
Coffea L.
C. arabica L.
C. canephora Pierre.

C. liberica Bull. ex Hiern.
Cola Schott and Endl.
C. acuminata Schott and Endl.
C. gigantea A. Chev.
C. millenii K. Schum.
C. nitida (Vent.) Schott and Endl.
Colocasia esculenta (L.) Schott.
Combretum Loefl.
C. glutinosum Perr. ex DC.
C. molle R. Br. ex G. Don.
C. paniculatum Vent.
C. platypterum (Welw.) Hutch and Dalz.
C. racemosum P. Beauv.
C. smeathmannii G. Don.
Commelina L.
Conocarpus erectus L.
Corchorus aestuans L.
C. olitorius L.
C. tridens L.
Costus L.
C. afer Ker-Gawl.
C. englerianus K. Schum.
C. lucanusianus J. Braun and K. Schum.
C. schlechteri Winkler.
C. spectabilis (Fenzl) K. Schum.
Crassocephalum Moench.
C. biafrae (Oliv. and Hiern) S. Moore.
C. crepidioides (Benth.) S. Moore.
C. mannii (Hook. f.) Milne-Redhead.
Crinum L.
C. jagus (Thomps.) Dandy.
C. ornatum (Ait.) Bury.
Crossopteryx febrifuga (Afzel. ex G. Don) Benth.
Crotalaria L.
C. retusa L.
Croton L.
C. lobatus L.
Cucumeropsis edulis (Hook f.) Cogn.
Cucumis sativus L.
C. melo L. var. *agrestis* Naud.
Cucurbita pepo L.
Culcasia P. Beauv.
C. scandens P. Beauv.
Cuscuta L.
C. australis R. Br.
Cyanotis D. Don.
C. caespitosa Kotschy and Peyr.
Cyclodiscus gabunensis Harms.
Cyperus articulatus L.
C. dilatatus Schum. and Thonn.
C. esculentus L.
C. rotundus L.

Cyrtosperma senegalense (Schott) Engl.

Dactyloctenium aegyptium (L.) Beauv.
Daniellia Benn.
D. ogea (Harms) Rolfe ex Holl.
D. oliveri (Rolfe) Hutch. and Dalz.
Datura candida (Pers.) Safford.
Deinbollia pinnata Schum. and Thonn.
Delonix regia (Bojer ex Hook) Raf.
Delphinium L.
Dennettia tripetala Bak. f.
Desmodium Desv.
D. triflorum (L.) DC.
D. velutinum (Willd.) DC.
Detarium Juss.
Dialium guineense Willd.
Diaphananthe bidens (Sw.) Schltr.
Dichrostachys cinerea (L.) Wight and Arn.
Digitaria horizontalis Willd.
Dinophora spenneroides Benth.
Diodia Gronov ex L.
Dioscorea L.
D. alata L.
D. bulbifera L.
D. burkilliana J. Miege.
D. cayenensis Lam.
D. dumetorum (Kunth) Pax.
D. mangenotiana J. Miege.
D. preussii Pax.
D. quartiniana A. Rich.
D. sansibariensis Pax.
Dissotis Benth.
D. graminicola Hutch.
D. rotundifolia (Sm.) Triana.
Distemonanthus benthamianus Baill.
Dombeya buettneri K. Schum.
Dorstenia L.
Dracaena Vand. ex L.
D. arborea (Willd.) Link.
D. perrottetii Bak.
D. phrynioides Hook.
D. smithii Bak. ex Hook. f.
D. surculosa Lindl.
Duranta repens L.

Eclipta prosptrata (L.) L.
Elaeis guineensis Jacq.
Eleusine coracana (L.) Gaertn.
Emilia Cass.
Entada Adans.
E. africana Guill. and Perr.
E. gigas (L.) Fawcett and Rendle.

E. pursaetha DC.
Eragrostis Host.
E. ciliaris (L.) R. Br.
E. tenella (L.) P. de B. ex Roem. and Schult.
Eremospatha hookeri (Mann and Wendl.) Wendl.
Erythrina senegalensis DC.
Eulophia guineensis Lindl.
E. horsfallai (Batem.) Summerhayes.
Eulophidium saundersianum (Rchb. f.) Summerhayes.
Eupatorium odoratum L.
Euphorbia L.
E. heterophylla L.
E. hyssopifolia L.
E. kamerunica Pax.

Ficus L.
F. capensis Thunb.
F. carica L.
F. exasperata Vahl.
F. mucuso Welw. ex Ficalho.
F. polita Vahl.
F. pumila L.
F. thonningii Blume.
Fimbristylis Vahl.
F. dichotoma (Linn.) Vahl.
Forrestia tenuis (C.B.Cl.) Benth.
Funtumia elastica Stapf.

Gardenia Ellis.
Geophila D. Don.
Gliricidia sepium (Jacq.) Steudel.
Gloriosa superba L.
Glyphaea brevis (Spreng.) monachino.
Gmelina arborea Roxb.
Gomphrena celosioides Mart.
Gossweilerodendron balsamiferum (Verm.) Harms.
Graptophyllum pictum (L.) Griff.
Grewia carpinifolia Juss.
G. mollis Juss.
G. pubescens P. Beauv.
G. villosa Willd.
Guiera senegalensis J. F. Gmel.

Haemanthus L.
H. rupestris Bak.
Helianthus annuus L.
Hemizygia bracteosa (Benth.) Briq.
H. welwitschii (Rolfe) M. Ashby.
Hevea brasiliensis Muell. Arg.
Hewittia Wight and Arn.
H. sublobata (L. f.) Kuntze.
Hibiscus cannabinus L.
H. congestiflorus Hochr.
H. sabdariffa L.
H. surattensis L.
Hildegardia barteri (Mast.) Kasterm.
Hippeastrum equestre (Ait.) Herb.
Holarrhena floribunda (G. Don) Dur. and Schinz.
Holoptelea grandis (Hutch.) Mildbr.
Hygrophyla auriculata (Schum.) Heine.
Hymenocardia acidia Tul.
Hymemocallis littoralis Salisb.
Hyparrhenia Anderss.
Hyphaene thebaica Mart.

Imperata cylindrica (L.) P. Beauv.
Indigofera L.
I. spicata Forskal.
Ipomoea L.
I. aquatica Forskal
I. batatas (L.) Lam.
I. hederifolia L.
I. involucrata P. Beauv.
I. mauritiana Jacq.
I. pes-caprae (L.) Sweet.
I. quamoclit L.
Isoberlinia Craib and Stapf.
I. doka Craib and Stapf.
I. dalzielii Craib and Stapf.
Ixora coccinea L.

Jardinea congoensis Franch.
Jatropha L.
J. gossypifolia L.
Justicia L.
J. flava (Forskal) Vahl.
J. insularis T. Anders.

Kaempferia L.
K. nigerica Hutch.
Kerstingiella geocarpa Harms.
Khaya grandifoliola C. DC.
K. ivorensis A. Chev.
Kigelia DC.
K. africana Benth.
Kyllinga Rottb.
K. nemoralis (Forst.) Dandy ex Hutch.

Lactuca L.
L. sativa L.
Lagenaria Ser.
Lannea A. Rich.
L. fruticosa Hochst.
L. humilis (Oliv.) Engl.

L. kerstingii Engl. and Krause.
L. microcarpa Engl. and Krause.
L. schimperi (Hochst. ex A. Rich.) Engl.
Lantana camara L.
Laportea aestuans (L.) Gaudich.
Launaea taraxacifolia (Willd.) Amin ex C. Jeffrey.
Lecaniodiscus cupanioides Planch.
Leonotis nepetifolia (L.) Ait. f.
Lepistemon Blume.
L. owariense (P. Beauv.) Hallier f.
Leucaena leucocephala (Lam.) de Wit.
Leucas martinicensis R. Br.
Lindernia All.
L. diffusa (L.) Wettst.
L. numulariifolia (D. Don) Wettst.
Lonchocarpus cyanescens Benth.
Lovoa trichilioides Harms.
Luffa aegyptiaca Mill.
Lycopersicon lycopersicum (L.) Karst.

Mallotus Lour.
M. oppositifolius (Geisel) Muell. Arg.
Mangifera indica L.
Manihot Mill.
M. esculentus Crantz.
M. glaziovii Muell. Arg.
Manilkara obovata (Sabine and G. Don) J. H. Hemsley.
Mansonia J. R. Drumm.
Maranta L.
M. arundinacea L.
Marantochloa Brongn. ex Gris.
M. leucantha (K. Schum.) Milne-Redhead.
M. purpurea (Ridley) Milne-Redhead.
Mariscus Vahl.
M. alternifolius Vahl.
M. ligularis (L.) Urban.
Markhamia Seem. ex Baill.
M. tomentosa (Benth.) K. Schum.
Melochia Rottb.
Memecylon L.
M. fosteri Hutch. and Dalz.
Merremia Dennst.
M. aegyptiaca (L.) Urban.
M. dissecta (Jacq.) Hallier f.
M. tridentata subsp. *angustifolia* (Jacq.) Ooststr.
Mezoneuron benthamianum Baill.
Micrococlia Lindl.
Mikania Willd.
M. carteri Bak.
M. cordata (Burm. f.) B. L. Robinson.
Millettia Wight and Arn.
Mimosa L.
M. pigra L.
M. pudica L.
Mimusops L.
Mitracarpum scabrum Zucc.
Mitragyna ciliata Aubrev and Pellegr.
Momordica charantia L.
Monodora tenuifolia Benth.
Morinda L.
M. lucida Benth.
Morus mesozygia Stapf.
Musa L.
M. sapientum L.
M. paradisiaca L.
M. nana Lour.
Musanga C.Sm ex R. Br.
M. cecropioides C. Sm ex R. Br.
Mussaenda erythrophylla Schum. and Thonn.
Myrianthus P. Beauv.
M. arboreus P. Beauv.

Nauclea L.
N. diderrichii De Wild.
N. latifolia Sm.
Nelsonia canescens (Lam.) Spreng.
Neptunia oleracea Lour.
Nesogordonia Baill.
N. papaverifera A. Chev.
Neurada L.
Neuropeltis Wall.
N. acuminata (P. Beauv.) Benth.
Newbouldia Seem. ex Bur.
N. laevis Seem.
Nicotiana rustica L.
N. tabacum L.

Ocimum gratissimum L.
Oecoeclades saundersiana (Rchb. f.) Garay and Taylor.
Oldenlandia L.
O. corymbosa L.
Olyra latifolia L.
Omphalocarpum P. Beauv.
O. procerum P. Beauv.
Oreodoxa regia H.B.K.
Oryza L.
O. punctata Kotschy ex Steud.
O. sativa L.
Oxyanthus speciosus DC.

Pachycarpus E. Meyer.
Palisota Reichb.

P. hirsuta K. Schum.
Pancratium L.
Panicum L.
P. maximum Jacq.
Parinari Aubl.
P. curatellaefolium Planch. ex Benth.
P. polyandrum Benth.
Parkia clappertoniana R. Br.
P. bicolor A. Chev.
Paullinia pinnata L.
Pennisetum americanum (L.) K. Schum.
P. pedicellatum Trin.
P. purpureum Schum.
P. subangustum (Schumach.) Stapf. and Hubb.
Peperomia pellucida (L.) H.B.K.
Pergularia daemia (Forskal) Chiov.
Phaseolus lunatus L.
Phaulopsis falcisepala C. B. Cl.
Phoenix dactylifera L.
P. reclinata Jacq.
Phyllanthus L.
P. amarus Schum. and Thonn.
P. discoideus (Baill.) Muell. Arg.
P. muellerianus (Kuntze) Exell.
Physalis angulata L.
Picris humilis DC.
Pilea Lindl.
P. angolensis (Hiern) Rendle.
Piliostigma Hochst.
P. thonningii (Schum.) Milne-Redhead.
Piper L.
P. guineense Schum. and Thonn.
P. umbellatum L.
Piptadeniastrum africanum (Hook. f.) Brenan.
Pistia stratiotes L.
Pouzolzia guinensis Benth.
Prosopis africana (Guill. and Perr.) Taub.
Pseudocedrela kotschyi (Schroeinf) Harms.
Pseudospondias microcarpa (A. Rich.) Engl.
Pterocarpus Jacq.
P. erinaceus Lam.
P. osun Craib.
P. mildbraedii Harms.
P. santalinoides L'Her. ex DC.
Pterygota Schott and Endl.
Pueraria phaseoloides (Roxb.) Benth.
Pupalia lappacea (L.) Juss.

Quisqualis indica L.

Ranunculus L.
Raphia Beauv.
R. hookeri Mann and Wendl.
Rauvolfia vomitoria (Benth.) Swizzlestick.
Ravenala madagascariensis J. F. Gmel.
Rhoeo discolor Hance.
Rhynchospora corymbosa (L.) Britton.
Ricinodendron heudelotii (Baill.) Pierre ex Pax.
Ricinus communis L.
Rothmannia Thunb.
Rubus L.
Russelia equisetiformis Schlecht and Cham.

Sabicea Aubl.
S. calycina Benth.
Saccharum officinarum L.
Salvia L.
Samanea saman (Jacq.) Merr.
Sansevieria liberica Ger and Labr.
Sapindus abyssinicus Fresen.
Sarcophrynium brachystachyum (Benth.) K. Schum.
Sarcostemma viminale (L.) R. Br.
Schizachyrium Nees.
Schrankia leptocarpa DC.
Schwenckia americana L.
Scleria Bergius.
S. verrucosa Willd.
Sclerocarya Hochst.
S. birrea (A. Rich.) Hochst.
Scoparia dulcis L.
Securinega virosa (Roxb. ex Willd.) Baill.
Setaria Beauv.
S. anceps Stapf. ex Massey.
S. barbata (Lam.) Kunth.
S. chevalieri Stapf.
S. megaphylla (Steud.) Dur. and Schinz.
S. pallidefusca (Schumach.) Stapf. and Hubb.
Sherbournia G. Don
S. bignoniiflora (Welw.) Hua.
Sida L.
S. acuta Burm. f.
S. corymbosa R. E. Fries.
S. rhombifolia L.
S. veronicifolia Lam.
Smilax Linn.
S. kraussiana Meisn.
Solanum macrocarpon L.
S. melongena L.
S. nigrum L.
S. torvum Sw.
S. tuberosum L.
S. verbascifolium L.
S. wrightii Benth.
Solenostemon Thonn.

Sorghum arundinaceum (Desv.) Stapf.
S. bicolor (L.) Moench.
Spathodea P. Beauv.
S. campanulata P. Beauv.
Spigelia anthelmia L.
Spondias mombin L.
Stachytarpheta Vahl.
S. cayennensis (L. C. Rich.) Schau.
S. indica (L.) Vahl.
Sterculia setigera Del.
Stereospermum Cham.
S. acuminatissimum K. Schum.
S. kunthianum Cham.
Strelitzia Dryand.
Striga Lour.
S. hermonthica (Del.) Benth.
S. primuloides A. Chev.
Strophanthus DC.
S. hispidus DC.
Struchium sparganophora (L.) Kuntze.
Strychnos L.
S. spinosa Lam.
Stylochiton Lepr.
S. hypogaeus Lepr.
Synedrella nodiflora Gaertn.

Tabernaemontana pachysiphon Stapf.
Tectona grandis L.f.
Telfairia occidentalis Hook. f.
Tephrosia Pers.
T. bracteolata Guill. and Perr.
Terminalia L.
T. catappa L.
T. glaucescens Planch. ex Benth.
T. ivorensis A. Chev.
T. superba Engl. and Diels.
Tetracarpidium conophorum (Mull. Arg.) Hutch. and Dalz.
Tetrapleura tetraptera (Schum. and Thonn.) Taub.
Thalia welwitschii Ridl.
Thalictrum L.
Thaumatococcus daniellii (Benn.) Benth.
Theobroma cacao L.
Thunbergia erecta (Benth.) T. Anders.
T. grandiflora (Roxb. ex Rottl.) Roxb.
Tragia L.
Treculia africana Decne.
Trema orientalis (L.) Blume.
Trichilia heudelotii Planch ex Oliv.
Tridax L.
T. procumbens L.
Triplochiton K. Schum.
T. scleroxylon K. Schum.
Triumfetta L.
T. cordifolia A. Rich.
T. rhomboidea Jacq.
Trochomeria dalzielli Baker f. ex Hutch.
Toxocarpus wightianus

Uapaca Baill.
Uncaria Schreb.
U. africana G. Don
Uraria picta (Jacq.) DC.
Urena lobata L.
Urera Gaud.
U. rigida (Benth.) Keay.
Urginea Steinh.
Uvaria afzelii Sc. Elliot.
U. chamae P. Beauv.

Vernonia Schreb.
V. ambigua Kotshy and Peyr.
V. amygdalina Del.
V. biafrae Oliv. and Hiern.
V. cinerea (L.) Less.
V. colorata (Willd.) Drake.
V. conferta Benth.
V. nigritiana Oliv. and Hiern.
V. pauciflora (Willd.) Less.
V. tenoreana Oliv.
Vigna Savi.
V. multinervis Hutch. and Dalz.
V. unguiculata L.
Vinca rosea L.
Vitex L.
V. doniana Sweet.
Voacanga africana Stapf.
Voandzeia geocarpa (L.) DC.

Waltheria indica L.
Wissadula amplissima (L.) R. E. Fries.
Withania somnifera (L.) Dunal.

Xanthosoma sagittifolia Schott.
X. mafaffa Schott.

Zea mays L.
Zebrina pendula Schnizl.
Zephyranthes tubispatha Herb.

APPENDIX III

Artificial key to the groups of common dicotyledonous families of West Africa

Modified after M. S. Nielsen's *Introduction to the Flowering Plants of West Africa*. University of London Press: London, UK, 1965.

1. Gynoecium with 2 or more free carpels; carpels with separate styles and stigmas .... Group 1
1. Gynoecium with 1 carpel and 1 placenta in ovary; or with 2 or more joined carpels with free or joined styles and stigmas; or ovaries free with styles or stigmas joined with 2 or more placentae in ovary
  2. Ovules attached to wall or walls of ovary
    3. Ovary superior
      4. Petals present
        5. Petals free from each other .... Group 2
        5. Petals joined to each other .... Group 3
      4. Petals absent .... Group 4
    3. Ovary inferior .... Group 5
  2. Ovules attached to central axis or to base or apex of ovary cell
    6. Ovary superior
      7. Petals present
        8. Petals free from each other .... Group 6
        8. Petals joined to each other .... Group 7
      7. Petals absent .... Group 8
    6. Ovary inferior .... Group 9

GROUP 1 Plants with 2 or more free carpels; carpels with separate styles and stigmas; petals present or absent

1. Petals present
  2. Petals free from each other
    3. Leaves stipulate; stamens more or less joined into a column (androgynophore); calyx lobes valvate; trees, shrubs or rarely herbs .... Sterculiaceae
    3. Leaves exstipulate
      4. Flowers perfect or unisexual and monoecious
        5. Calyx and corolla 3-merous with petals in 2 series .... Annonaceae
        5. Calyx and corolla 5-merous
          6. Leaves simple; receptacle cupular .... Rosaceae
          6. Leaves compound; receptacle hypogynous
            7. Leaves gland-dotted .... Rutaceae
            7. Leaves not gland-dotted .... Connaraceae
      4. Flowers unisexual and dioecious .... Menispermaceae
  2. Petals more or less joined to each other
    8. Leaves pinnately-compound or 3-foliolate .... Connaraceae
    8. Leaves simple .... Annonaceae
1. Petals absent
  9. Leaves alternate
    10. Leaves stipulate; stamens joined in a column .... Sterculiaceae
    10. Leaves exstipulate; twiners .... Menispermaceae
  9. Leaves opposite; climbers with twining petioles or savanna herbs .... Ranunculaceae

GROUP 2 Plants with 1 carpel or 2 or more joined carpels; ovules attached to wall or walls of ovary; ovary superior; petals present and free from each other

1. Leaves opposite

  2. Stamens more or less joined into separate bundles opposite the petals; leaves often gland-dotted .... **Guttiferae**

  2. Stamens free or more or less joined into 1 or 2 groups; flowers papilionaceous .... **Papilionaceae**

1. Leaves alternate or all radical

  3. Stamens more than 10

    4. Filaments more or less joined into a tube; leaves bipinnate .... **Mimosaceae**

    4. Filaments free

      5. Ovary carried on a stalk; petals clawed

        6. Leaves simple or digitately compound .... **Capparidaceae**

        6. Leaves 1-foliolate or pinnately compound; flowers papilionaceous .... **Papilionaceae**

      5. Ovary sessile or nearly so; anthers straight

        7. Anthers opening by pores

          8. Seeds hairy .... **Cochlospermaceae**

          8. Seeds not hairy .... **Ochnaceae**

        7. Anthers opening by slits

          9. Flower actinomorphic

            10. Corona present .... **Passifloraceae**

            10. Corona absent

              11. Carpel single

                12. Receptacle a cup or tube .... **Rosaceae**

                12. Receptacle hypogynous, leaves bipinnate .... **Mimosaceae**

              11. Carpels single to several with the same number of parietal placentae, one-celled on a hypogynous receptacle .... **Flacourtiaceae**

          9. Flower zygomorphic; ovary of 1 carpel

            13. Odd petal adaxial; corolla papilionaceous .... **Papilionaceae**

            13. Odd petal abaxial; corolla not papilionaceous .... **Caesalpiniaceae**

  Stamens 10 or fewer

    14. Flowers zygomorphic

      15. Ovary with more than 1 carpel

        16. Fruit a berry or a drupe, often beaded .... **Capparidaceae**

        16. Fruit a 3-valved capsule .... **Violaceae**

      15. Ovary of 1 carpel

        17. Corolla papilionaceous .... **Papilionaceae**

        17. Corolla not papilionaceous .... **Caesalpiniaceae**

    14. Flowers actinomorphic

      18. Corona present .... **Passifloraceae**

      18. Corona absent

        19. Leaves compound

          20. Leaves bipinnate .... **Mimosaceae**

          20. Leaves pinnate .... **Anacardiaceae**

        19. Leaves simple or tri-foliolate

          21. Anthers opening by pores .... **Ochnaceae**

          21. Anthers opening by slits

            22. Plant a climber with tendril .... **Passifloraceae**

            22. Plant not a climber with tendril

              23. Anther connective produced .... **Violaceae**

              23. Anther connective not produced .... **Capparidaceae**

GROUP 3 Plants with 1 carpel or 2 or more joined carpels; ovules on wall of ovary; ovary superior; petals present, more or less joined

1. Stamens free from corolla tube
   2. Ovary of more than 1 carpel; stamens numerous, more than twice the number of petals; anthers appendaged at apex .... **Annonaceae**
   2. Ovary of 1 carpel; fruit usually a pod
      3. Plants dioecious twiners .... **Menispermaceae**
      3. Plants not dioecious twiners
         4. Flowers actinomorphic in heads or short spikes; corolla valvate .... **Mimosaceae**
         4. Flowers zygomorphic; corolla imbricate
            5. Adaxial petal internal .... **Caesalpiniaceae**
            5. Adaxial petal external, flower papilionaceous .... **Papilionaceae**
1. Stamens joined to the corolla tube
   6. Leaves alternate .... **Polygalaceae**
   6. Leaves opposite; carpels 2
      7. Stamens 5, ovaries 2 and free, stigmas joined
         8. Corona absent .... **Apocynaceae**
         8. Corona present .... **Asclepiadaceae**
      7. Stamens 4, ovary single with 2 parietal plancentae .... **Bignoniaceae**

GROUP 4 Plants with 1 carpel or with 2 or more joined carpels; ovules attached to wall or walls of ovaries; ovary superior; petals absent

1. Ovary of 1 carpel
   2. Flowers perfect; leaves usually compound .... **Caesalpiniaceae**
   2. Flowers unisexual on a swollen receptable .... **Moraceae**
1. Ovary of more than 1 carpel
   3. Ovary and fruit stalked or beaded .... **Capparidaceae**
   3. Ovary and fruit not stalked or beaded .... **Flacourtiaceae**

GROUP 5 Plants with 1 carpel or with 2 or more carpels; ovules on wall or walls of ovaries; ovary inferior; petals present or absent

1. Petals present
   2. Petals free from each other
      3. Flowers hermaphroditic; anthers straight and opening by pores .... **Malastomataceae**
      3. Flowers unisexual; anthers bent and opening by slits .... **Cucurbitaceae**
   2. Petals more or less joined
      4. Flowers hermaphroditic .... **Rubiaceae**
      4. Flowers unisexual .... **Cucurbitaceae**
1. Petals absent, calyx 1-sided, coloured .... **Aristolochiaceae**

GROUP 6 Plants with 1 carpel, or 2 or more joined carpels; ovules attached to central axis or to base or apex of ovary cell; ovary superior; petals present; free from each other

1. Perfect, stamens same number as petals and opposite them; leaves alternate (rarely opposite or all radical)
  2. Leaves not gland-dotted, simple, not bipinnate, lacking tendrils
    3. Disc absent; trees or shrubs; ovules 2 or more per cell .... **Sterculiaceae**
    3. Disc present
      4. Leaves stipulate
        5. Ovules axile; fruit a capsule .... **Sterculiaceae**
        5. Ovules basal; fruit a drupe .... **Rhamnaceae**
      4. Leaves exstipulate
        6. Calyx conspicuous; ovary 1 to 3-celled .... **Olacaceae**
        6. Calyx minute; ovary 1-celled .... **Opiliaceae**
  2. Leaves gland-dotted, usually compound and stipulate; inflorescence leaf opposed, often with a tendril .... **Ampelidaceae**
1. Perfect, stamens connate into bundles opposite the petals
  7. Leaves opposite exstipulate; calyx imbricate
    8. Styles free from the base or nearly so .... **Hypericaceae**
    8. Styles joined or stigma single and more or less sessile .... **Guttiferae**
  7. Leaves alternate stipulate; calyx valvate; fruit hooked .... **Tiliaceae**
1. Perfect, stamens same number as petals and alternating with them, or more numerous, rarely fewer
  9. Style basal; stamens numerous .... **Rosaceae**
  9. Style terminal
    10. Flowers zygomorphic; anthers opening by pores .... **Polygalaceae**
    10. Flowers actinomorphic
      11. Leaves opposite or verticillate, simple
        12. Stamens more than twice as many as petals, calyx valvate
          13. Flowers perfect, stipules interpetiolar .... **Rhizophoraceae**
          13. Flowers unisexual .... **Euphorbiaceae**
        12. Stamens not more than twice as many as petals
          14. Trees, shrubs and wood climbers
            15. Leaves stipulate, stipules not glandular
              16. Stipules interpetiolar
                17. Filaments joined .... **Rhizophoraceae**
                17. Filaments free .... **Malphighiaceae**
              16. Stipules not interpetiolar
                18. Flowers perfect, stamens 3 to 5 inserted on or below margin of disc .... **Celastraceae**
                18. Flowers unisexual .... **Euphorbiaceae**
            15. Leaves exstipulate or stipules gland-like
              19. Anthers opening at apex by pore .... **Melastomataceae**
              19. Anthers opening by slits; ovules few in each cell
                20. Sepals 2-glandular outside .... **Malpighiaceae**
                20. Sepals not so
                  21. Tendril climber .... **Icacinaceae**
                  21. Not so .... **Celastraceae**
          14. Herbs; leaves with 3 or more longitudinally parallel veins .... **Melastomataceae**
      11. Leaves alternate or all radical
        22. Stamens twice number of petals or fewer
          23. Leaves compound, or if 1-foliolate, then petiole swollen

24. Stamens joined into a tube; leaves exstipulate .... **Meliaceae**
24. Stamens free or joined only basally
  25. Leaves stipulate .... **Sapindaceae**
  25. Leaves exstipulate
    26. Leaves gland-dotted .... **Rutaceae**
    26. Leaves not so
      27. Ovary 1-celled
        28. Ovules axile or basal .... **Connaraceae**
        28. Ovules pendulous .... **Anacardiaceae**
      27. Ovary of 2 or more cells
        29. Wood resinous (never climbers) .... **Burseraceae**
        29. Not so .... **Sapindaceae**

23. Leaves simple
  30. Anthers opening by apical pores .... **Ochnaceae**
  30. Anthers opening by slits
    31. Shrubs and trees
      32. Leaves stipulate
        33. Flowers unisexual; disc present; petals entire
          34. Stipules persistent, conspicuous .... **Euphorbiaceae**
          34. Stipules deciduous, inconspicuous .... **Celastraceae**
        33. Flowers perfect
          35. Stipules intrapetiolar or encircling stem
            36. Ovary entire, 2-celled; fruit a drupe or samara .... **Irvingiaceae**
            36. Ovary 3 to 5-celled; fruit a capsule .... **Linaceae**
          35. Stipules not so; ovules erect .... **Celastraceae**
      32. Leaves exstipulate
        37. Stamens joined into a tube .... **Meliaceae**
        37. Stamens not so
          38. Stamens hypogynous and including sterile ones, twice the number of the petals
            39. Sepals with 2 large external glands .... **Malpighiaceae**
            39. Sepals not so
              40. Ovary 1-celled; leaves large; flowers in panicles .... **Anacardiaceae**
              40. Ovary 1-celled with a central placenta arising from the base, or 2- or more-celled .... **Olacaceae**
          38. Stamens same number as petals or fewer .... **Icacinaceae**
    31. Herbs
      41. Flowers perfect, disc absent
        42. Sepals valvate .... **Tiliaceae**
        42. Sepals imbricate; ovary not naked .... **Molluginaceae**
      41. Flowers unisexual .... **Euphorbiaceae**

22. Stamens more than twice number of petals
  43. Sepals valvate, or open in bud
    44. Anthers 2-celled; epicalyx absent
      45. Stamens free or joined only at base
        46. Calyx closed in bud
          47. Flowers perfect; leaves simple .... **Tiliaceae**
          47. Flowers unisexual; leaves simple, entire or 3-lobed; petiole with 2 glands at top .... **Euphorbiaceae**
          47. Flowers polygamous; leaves mostly compound; petiole not glandular .... **Anacardiaceae**
        46. Calyx open in bud; leaves simple alternate

48. Leaves stipulate .... **Dipterocarpaceae**
48. Leaves exstipulate.. **Olacaceae**
49. Fruit schizocarpic, forming samaras .... **Sterculiaceae**
49. Fruit with an enlarged calyx .... **Olacaceae**
45. Stamens joined into a sheath or separate bundles .... **Sterculiaceae**
44. Anthers 1-celled; epicalyx sometimes present
50. Trees; leaves compound digitate .... **Bombacaceae**
50. Herbs of undershrubs mainly; leaves simple; epicalyx sometimes present.. **Malvaceae**
43. Sepals neither valvate nor open in bud
51. Leaves compound, if 1-foliolate then with a swollen petiole
52. Ovule ascending
53. Leaves gland-dotted .... **Rutaceae**
53. Leaves not so
54. Wood resinous; leaves scented; style or stigma often to one side... **Anacardiaceae**
54. Wood not so; leaves not or rarely scented; style or stigma central .... **Sapindaceae**
52. Ovule(s) pendulous; bark bitter .... **Simaroubaceae**
51. Leaves simple
55. Flowers unisexual.. **Euphorbiaceae**
55. Flowers perfect.. **Ochnaceae**

GROUP 7 Plants with 1 carpel or 2 or more joined carpels; ovules attached to central axis or to base or apex of ovary cell; ovary superior; petals present, more or less joined

1. Stamens same number as, and opposite to, corolla lobes
  2. Ovules solitary in ovary or in each cell of ovary
    3. Petals imbricate .... **Sapotaceae**
    3. Petals valvate
      4. Inflorescence leaf-opposed; leaves mostly compound; tendrils often present .... **Vitaceae**
      4. Inflorescence not so; leaves simple; tendrils absent .... **Olacaceae**
  2. Ovules 2 or more in each ovary cell; inflorescence leaf-opposed; leaves mostly compound; tendrils often present .... **Vitaceae**
1. Stamens more than twice as many as corolla lobes; leaves stipulate.......... **Euphorbiaceae**
1. Stamens up to twice as many as corolla lobes
  5. Stamens as many as corolla lobes or more numerous
    6. Flowers zygomorphic
      7. Petals joined only at base.......... **Polygalaceae**
      7. Petals joined into a definite tube.......... **Verbenaceae**
    6. Flowers actinomorphic
      8. Leaves opposite or verticillate; anthers opening by slits
        9. Corona present .... **Asclepiadaceae**
        9. Corona absent
          10. Leaves exstipulate
            11. Corolla contorted .... **Apocynaceae**
            11. Corolla not so (stipules minute) .... **Salvadoraceae**
          10. Leaves stipulate or sheathing .... **Loganiaceae**
      8. Leaves alternate or all radical or reduced to scales
        12. Twining parasite without chlorophyll .... **Convolvulaceae**
        12. Not so
          13. Leaves stipulate; flowers unisexual .... **Euphorbiaceae**
          13. Leaves exstipulate
            14. Stamens separate from corolla or joined only to its base; anthers opening by slits
              15. Leaves compound .... **Connaraceae**
              15. Leaves simple .... **Ebenaceae**
            14. Stamen joined to corolla tube
              16. Styles 2, gynobasic .... **Convolvulaceae**
              16. Style 1, terminal; corolla imbricate; ovules numerous in each cell.......... **Solanaceae**
  5. Stamens fewer than corolla lobes
    17. Flowers actinomorphic
      18. Stamen 1; climbing shrub .... **Loganiaceae**
      18. Stamens more than 1
        19. Stamens 6 to 8, opposite inner lobes of corolla.......... **Sapotaceae**
        19. Stamens 2
          20. Inflorescence a whip-like spike .... **Verbenaceae**
          20. Inflorescence a branched cyme.......... **Solanaceae**
    17. Flowers zygomorphic
      21. Ovary deeply 4-lobed; stem usually 4-sided .... **Labiatae**
      21. Ovary not so
        22. Ovules numerous in ovary or in each ovary cell; or if only 2, then 1 placed above the other
          23. Leaves pinnately compound .... **Bignoniaceae**
          23. Leaves simple .... **Acanthaceae**
        22. Ovules solitary in each ovary cell; or if 2, then side by side.......... **Verbenaceae**

GROUP 8 Plants with 1 carpel or 2 or more joined carpels; ovules attached to central axis or to base or apex of ovary cell; ovary superior; petals absent

1. Calyx absent from male flower and often from female flower as well
  2. Inflorescence a cyathium .... **Euphorbiaceae**
  2. Inflorescence not so
    3. Flowers in dense spikes .... **Piperaceae**
    3. Flowers on a common open receptacle, which forms part of the fruit .... **Moraceae**
1. Calyx present, sometimes petaloid
  4. Leaves opposite or verticillate, never all radical
    5. Leaves stipulate
      6. Flowers perfect; trees or shrubs with tiny flowers; leaves simple........ **Ulmaceae**
      6. Flowers unisexual
        7. Ovary 2- or more-celled .... **Euphorbiaceae**
        7. Ovary 1-celled
          8. Ovule erect
            9. Filaments incurved........ **Urticaceae**
            9. Filaments not so........ **Moraceae**
          8. Ovules pendulous; filaments incurved; fruit a small achene........ **Moraceae**
    5. Leaves exstipulate; flowers perfect; ovary 1-celled; stamens erect in bud .... **Amaranthaceae**
  4. Leaves alternate, or all radical or reduced to scales
    10. Leaves stipulate
      11. Stamens joined in 1 group
        12. Flowers perfect........ **Sterculiaceae**
        12. Flowers always unisexual
          13. Ovary 3-celled .... **Euphorbiaceae**
          13. Ovary 1-celled .... **Myristicaceae**
      11. Stamens free or joined only basally
        14. Stamens same number as sepals and alternate with them .... **Rhamnaceae**
        14. Stamens same number as sepals and opposite them, or more numerous, or fewer
          15. Stipules sheathing round stem .... **Polygonaceae**
          15. Stipules not so
            16. Flowers polygamous, style basal .... **Rosaceae**
            16. Flowers unisexual
              17. Ovary 3-celled........ **Euphorbiaceae**
              17. Ovary 1-celled
                18. Trees and shrubs
                  19. Female inflorescence forming the fruit .... **Moraceae**
                  19. Not so, fruit a drupe or samara........ **Ulmaceae**
                18. Herbs .... **Urticaceae**
    10. Leaves exstipulate
      20. Leaves compound; trees and shrubs
        21. Trees with milky latex .... **Euphorbiaceae**
        21. Trees or shrubs; latex not milky........ **Sapindaceae**
      20. Leaves simple
        22. Flowers in large bracteate heads........ **Proteaceae**
        22. Flowers in a cyathium........ **Euphorbiaceae**
        22. Flowers not as above
          23. Ovules solitary
            24. Ovules pendulous........ **Euphorbiaceae**
            24. Ovules basal........ **Amaranthaceae**
          23. Ovules 2 in each cell; fruit 3-winged .... **Sapindaceae**

GROUP 9 Plants with 1 carpel or 2 or more joined carpels; ovules attached to central axis or to base or apex of ovary cell; ovary inferior; petals present or absent

1. Petals absent
  2. Leaves stipulate
    3. Flowers unisexual .... **Moraceae**
    3. Flowers perfect .... **Rhamnaceae**
  2. Leaves exstipulate; stamens incurred in bud; fruit winged .... **Combretaceae**
1. Petals present
  4. Petals more or less joined
    5. Leaves opposite
      6. Leaves stipulate .... **Rubiaceae**
      6. Leaves exstipulate
        7. Anthers free from each other .... **Myrtaceae**
        7. Anthers joined around style .... **Compositae**
    5. Leaves alternate
      8. Corona present .... **Lecythidaceae**
      8. Corona absent
        9. Anthers joined around style; inflorescence a capitulum .... **Compositae**
        9. Anthers free from each other; inflorescence not a capitulum .... **Cucurbitaceae**
  4. Petals free from each other
    10. Leaves opposite or verticillate, never all radical
      11. Leaves stipulate
        12. Stamens same number as, and opposite to, petals .... **Rhamnaceae**
        12. Stamens alternate with petals or more numerous; ovary of 2–6 carpels; ovules pendulous .... **Rhizophoraceae**
      11. Leaves exstipulate
        13. Stamens numerous .... **Myrtaceae**
        13. Stamens up to twice as many as petals
          14. Anthers opening by terminal pore or anthers bent .... **Melastomataceae**
          14. Anthers opening by slits, anthers straight .... **Combretaceae**
    10. Leaves alternate or all radical
      15. Flowers unisexual; actinomorphic .... **Cucurbitaceae**
      15. Flowers perfect; trees and shrubs
        16. Leaves stipulate
          17. Leaves simple .... **Rhamnaceae**
          17. Leaves compound .... **Araliaceae**
        16. Leaves exstipulate
          18. Leaves simple
            19. Ovary 1- or 2-celled
              20. Ovary 1-celled .... **Combretaceae**
              20. Ovary 2-celled .... **Lecythidaceae**
            19. Ovary 3- to 4-celled .... **Rhizophoraceae**
          18. Leaves compound .... **Umbelliferae**

APPENDIX IV

Artificial key to the groups of common monocotyledonous families of West Africa

Modified after M. S. Nielsen's *Introduction to the Flowering Plants of West Africa*. University of London Press: London, UK, 1965

1. Ovary superior
   2. Carpels 2 or more per flower, free with separate styles; or 1 carpel per flower; perianth present, if inconspicuous then flowers not in axils of dry scale-like bracts .... Group 1
   2. Carpels 2 or more per flower and joined; perianth present or absent or represented by bristles or scales
      3. Perianth lobes all similar, petaloid or inconspicuous .... Group 2
      3. Perianth lobes differentiated into calyx and corolla, or perianth absent or represented by bristles or scales .... Group 3
1. Ovary inferior or half-inferior .... Group 4

GROUP 1 Plants with superior ovary; carpels 2 or more per flower, carpels free with separate styles, or 1 carpel per flower; perianth present, if inconspicuous then flowers not in axils of dry scale-like bracts

1. Flowers in axils of a bract
   2. Plants tree palms .... Palmae
   2. Plants not so
      3. Plants herbs of watery places with green leaves .... Alismataceae
      3. Plants saprophytic .... Triuridaceae
1. Flowers without bracts
   4. Plants of freshwater habitats
      5. Flowers in racemes or spikes
         6. Inflorescence one-sided .... Aponogetonaceae
         6. Inflorescence not so .... Potamogetonaceae
      5. Flowers axillary, solitary or in small cymes
         7. Carpels 2 or more .... Zanichelliaceae
         7. Carpels solitary .... Najadaceae
   4. Plants of marine or brackish habitats
      8. Flowers in terminal spikes .... Ruppiaceae
      8. Flowers axillary .... Zanichelliaceae

GROUP 2 Plants with superior ovary; carpels 2 or more per flower, carpels joined; perianth present, perianth lobes all similar and petaloid or inconspicuous

1. Perianth lobes similar and petaloid
  2. Flowers in an umbel, with dry, conspicuous, spathaceous bracts .... **Amaryllidaceae**
  2. Flowers not in umbels, or if subumbellate, then bracts not spathaceous
    3. Plants aquatic; inflorescence subtended by a spathe-like leaf sheath .... **Pontederiaceae**
    3. Plants not aquatic
      4. Plants straggling or climbing with prickly stems, stipular tendrils and reticulately-veined leaves .... **Smilacaceae**
      4. Plants not so
        5. Anthers opening by pores; perennial with a corm .... **Tecophilaeaceae**
        5. Anthers opening by slits
          6. Leaf tip tendril-like
            7. Ovules numerous .... **Liliaceae**
            7. Ovules 1 per cell .... **Flagellariaceae**
          6. Leaf tip not so
            8. Plants pineapple-like; fruit a capsule in which the thin fruit wall falls away from a berry-like group of seeds .... **Agavaceae**
            8. Plants not so
              9. Perennial herbs with a bulb, corm, rhizome or tuber; fruit a capsule or berry, many-seeded .... **Liliaceae**
              9. Perennial herbs, shrubs or trees with 1-seeded berries .... **Agavaceae**
1. Perianth lobes all similar, very inconspicuous
  10. Leaves folded in bud .... **Palmae**
  10. Leaves not so
    11. Plants shrubs or trees, leaves with prickly margin; often with stilt roots .... **Pandanaceae**
    11. Plants herbs or herbaceous climbers; inflorescence a spadix .... **Araceae**

GROUP 3 Plants with superior ovary; carpels 2 or more per flower, carpels joined; perianth differentiated into calyx and corolla, or absent, or represented by bristles or scales

1. Perianth present and differentiated into calyx and corolla
  2. Plants tall palms with leaves folded in bud, flowers often in panicles with large spathaceous bracts .... **Palmae**
  2. Plants not so
    3. Herbs, with leaf tips tendril-like .... **Flagellariaceae**
    3. Herbs, leaf tips not so .... **Commelinaceae**
1. Perianth absent or represented by bristles or scales, flowers in axil of scaly bracts, carpels 2 or 3
  4. Inflorescence a spadix .... **Araceae**
  4. Inflorescence not so
    5. Stem 3-sided, leaf sheath closed, ligule absent .... **Cyperaceae**
    5. Stem cylindrical, leaf sheath open, ligule present .... **Gramineae**

GROUP 4 Plants with inferior or half-inferior ovary; perianth with separate calyx and corolla or perianth lobes all similar and petaloid; corolla actinomorphic or zygomorphic

1. Perianth of separate calyx and corolla
  2. Corolla actinomorphic
    3. Petaloid staminodes absent, flower perfect .... **Iridaceae**
    3. Petaloid staminodes present, often more conspicuous than corolla; stamen simple
      4. Anther 2-celled; sepals joined .... **Zingiberaceae**
      4. Anther 1-celled; sepals free or joined only at the base .... **Marantaceae**
  2. Corolla zygomorphic
    5. Flowers orchidaceous .... **Orchidaceae**
    5. Flower not so
      6. Stamen 1; petaloid staminodes present .... **Cannaceae**
      6. Stamens 3 to 5; petaloid staminodes absent
        7. Stamens 5 plus 1 staminode; petals not fused into a tube .... **Musaceae**
        7. Stamens 3 .... **Iridaceae**
1. Perianth lobes all similar, usually petaloid
  8. Inflorescence umbellate
    9. Flowers orchidaceous .... **Orchidaceae**
    9. Flowers not so; perianth actinomorphic .... **Amaryllidaceae**
  8. Inflorescence not so
    10. Plants saprophytic
      11. Flowers orchidaceous .... **Orchidaceae**
      11. Flowers not so
        12. Ovary and fruit winged .... **Burmanniaceae**
        12. Ovary and fruit not winged .... **Thismiaceae**
    10. Plants green
      13. Plants climbing with leafy stems; flowers inconspicuous, small; seeds winged .... **Dioscoreaceae**
      13. Plants not so
        14. Flowers orchidaceous .... **Orchidaceae**
        14. Flowers not so
          15. Stamens 3 .... **Iridaceae**
          15. Stamens 6, or stamens 5 plus 1 staminode
            16. Stamens 6 .... **Hypoxidaceae**
            16. Stamens 5 plus 1 staminode .... **Musaceae**

General index

Index of scientific names

References to illustrations are shown in bold type